HUMANS *and* OTHER ANIMALS

HUMANS *and* OTHER ANIMALS

Edited by **Arien Mack**

Foreword by **Marc Bekoff**

OHIO STATE UNIVERSITY PRESS
Columbus

Copyright © 1995 by The New School for Social Research.
Foreword and index © 1999 by The Ohio State University.
All rights reserved.

Originally published 1995 as a special issue of *Social Research* entitled *In the Company of Animals*.
Ohio State University Press edition first published 1999.

Library of Congress Cataloging-in-Publication Data

Humans and Other Animals / edited by Arien Mack.
    p. cm.
   Includes bibliographical references and index.
   ISBN 0-8142-0817-7 (cloth : alk. paper).—ISBN 0-8142-5017-3 (pbk. : alk. paper)
   1. Human-animal relationships.   I. Mack, Arien.
QL85.I5 1999
304.2'7—dc21   98-37449
                                                    CIP

Printed in the United States.

9 8 7 6 5 4 3 2 1

Contents

SAMENESS AND DIFFERENCE

Foreword

What a pleasure it is to write a foreword for this much needed book. *Humans and Other Animals* (a collection of essays that originally appeared as a special issue of the journal *Social Research* entitled *In the Company of Animals*) is a timely volume that is a gold mine of diverse and important information on humans and their relationships with nonhuman animals (hereafter animals). The study of human-animal interactions is a hot interdisciplinary area. Many researchers are interested in this topic (see, e.g., Beck and Katcher, 1996, and the journals *Anthrozoös* and *Society and Animals*), as are numerous nonresearchers, all of whose lives will be affected by what we learn about human-animal partnerships. As I write this foreword, the American Academy of Religion is organizing a caucus to discuss interactions between humans and animals in the context of theological and religious studies, and a new volume concerned with cruelty to animals and its relationship to human violence has just been published (Lockwood and Ascione, 1998).

In her introduction, Arien Mack describes the volume's purpose as to examine "our relationship with other animals over time and in different cultures" and to study "the current, and often inflamed, debate about what our proper relationship to other animals ought to be." Almost all the essays in this collection inform readers not only about how humans have struggled with their relationships with animals but also about how humans have struggled with their own views of themselves. Although there are numerous differences between humans and other animals, it is clear that in many important ways, "we" are one of "them" and "they" are one of "us."

One aspect of this book that makes it so appealing is that most of the essays are easy to read for nonspecialists. Much debate about human-animal relationships takes place in academic settings, but we can no longer sit back and luxuriate in merely academic discussions. We must accept that all people can be the voices for other animals and that all animals' lives must be taken seriously.

Although there is a lot of interest in the nature of human-animal relationships, change has been slow in coming in one important area, specifically, the speciesist line drawing by which people place themselves and other animals on a hierarchical scale using such criteria as cognitive capacities or the ability to experience pain, anxiety, or suffering. There is little consensus about where such lines should be drawn, except among those who preach humanocentrism, according to which humans are placed above ("higher") and apart from other ("lower") animals. Because there is a lot at stake for individuals of species who fall below some rigid boundary, great care must be taken in drawing lines because of the consequences for how animals are treated when they are welcomed into our company.

The context in which animals are used or encountered can also inform attitudes that people have even to individuals of the same species. For example, those who train animals may view them differently from the way scientists who use them do. Scientists also show different attitudes toward animals depending on whether they are encountered in the laboratory or at home. Many scientists who name and praise the cognitive abilities of the companion animals with whom they live are likely to leave this sort of "baggage" at home when they enter their laboratories to do research with other members of the same species. Based on a series of interviews with practicing scientists, M. T. Phillips reported that many of them construct a "distinct category of animal, the 'laboratory animal,' that contrasts with namable animals (e.g., pets) across every salient dimension . . . the cat or dog in the laboratory is perceived by researchers as ontologically different from the pet dog or cat at home" (1994, p. 119). Recognizing and honoring differences among those who are concerned with human-animal relations and supporting a pluralistic approach should, in the future, help us learn more about the animals with whom we share this planet.

The way people refer to animals—recognizing them as individual subjects by naming them or referring to them as nameless objects—also is closely related to how they view and treat them. In much science, naming animals is taboo: it's nonscientific, for it might color otherwise objective endeavors. Historically, it is interesting to note that Jane Goodall's first scientific paper dealing with her research on the behavior of chimpanzees was initially returned by the *Annals of the New York Academy of Sciences* because she named, rather than numbered, the chimpanzees she observed. The journal also wanted her to refer to the chimpanzees using "it" or "which" rather than "he" or "she" (Montgomery, 1991, pp. 104–5). Goodall refused to make the requested changes, but the journal published her paper anyway. Given that the goal of many studies of nonhuman animal cognition is to come to terms with nonhumans' subjective experiences—their points of view—making nonhuman animals subjects rather than objects seems a move in the right direction.

Another issue that is of great importance in the study of human-animal relations is the attractiveness of animals to humans. Certain animals—the cuddly and fuzzy ones, those with whom we are most familiar, or those who resemble humans behaviorally or phenotypically—seem to be welcomed into humans' company more than others who are not as cuddly or who are unfamiliar. Certainly, unfamiliarity can breed contempt. We need to expand the circle of animals who are welcomed into humans' company as friends. Overt characteristics alone should not guide these sorts of decisions.

Not only do overt features influence how humans relate to other animals, but often people allow legal precedent to sway their views. It remains an open question whether animals really are protected by existing laws, but as long as they are viewed as property or as mere resources for human consumption, it seems unwise that animals should feel comfortable in the courtroom.

Clearly, much interest in human-animal relationships centers on how humans treat and use their kin, and whether animals have moral standing. Some of the thorniest issues that affect how humans view other animals involve conceptions of animal consciousness, pain, and suffering. Are some animals conscious, do some animals feel pain, can some animals experience anxiety, do some animals suffer? If answers to these difficult questions are "yes" or "no" or "maybe," so what? Putting common sense aside, which is probably ill-advised, even experts do not agree on the correct answers. Much more work on a wide variety of organisms has to be done concerning the nature of animal minds and on the existence of animal pain and suffering if cognitive capacities and sentience are to be used to dictate the treatment to which animals can be exposed. We need to expand our taxonomic interests. Of course, questions about cognition and sentience are moot if animals are granted rights. For it is the compromising of individual lives that needs to be taken seriously, not whether individuals are "smart" or able to experience pain, anxiety, or suffering.

There seems to be little doubt that how scientists and other academics portray themselves and their feelings about animals to nonscientists and to nonacademics influences how animals are viewed outside the academic towers in which much discussion about human-animal relations occurs. Do (some) scientists who often come across as removed and heartless really care about the nonhuman animals who they use and kill? What do scientists (really) think about science? What do nonscientists think about science and scientists? There has been increasing discontent among nonscientists concerning science and scientists. In fact, some scientists who kill animals and use them in morally questionable ways really do seem to care about the animals who they maintain in captivity, use, or kill, and some scientists, when they let their hair down, confess to being deeply concerned about the ethics of their activities (Kummer, 1995; Bekoff, 1998a) and are as fed up with the arrogance of science and scientists as are nonscientists.

Another area in the arena of human-animal relationships that needs much more research concerns the treatment of individuals in relation to the treatment of whole species of animals. Speciesists often use taxonomic or behavioral closeness to humans, similar appearance, or the possession of various cognitive capacities displayed by normal adult humans to draw the line that separates humans from other animals. Cognitive abilities include the capacities to exhibit self-consciousness, to engage in purposive behavior, to communicate using a language, to make moral judgments, and to reason. The British government recently declared a ban on the use of great apes in research, in part because it has been argued that these primates deserve special treatment because of their cognitive capacities. This decision is speciesist.

Under the above criteria, most animals cannot qualify for protection. But there also are some humans (young infants and senile adults) who cannot qualify either, and this can be a problem for speciesists who rely on cognitive capacities. If people make speciesist decisions that, for example, all and only humans or all and only members of certain other species are to be protected from certain types of treatment, then these claims need to be supported by informed arguments. These decisions should not be made because it is convenient to do so. Furthermore, there is marked within-species variation in behavioral and other phenotypes, and it seems too fast a move to conclude that any sort of speciesistic account should be accepted and implemented without serious and informed debate. The view from the top, a human-centered "them" versus "us" perspective, can be difficult to apply consistently.

Taken together, the topics considered in this collection are important ones that demand careful consideration. The problems with which we are faced concerning human-animal relationships are numerous, diverse, complex, and contentious (Bekoff, 1998b), and keeping open minds and, more important, open hearts is essential.

Let me end this brief foreword with a thought-provoking quotation that is related to many of the topics in this book. Paul Shepherd has noted: "There is a profound, inescapable need for animals that is in all people everywhere, an urgent requirement for which no substitute exists. This need is no vague, romantic, or intangible yearning, no simple sop to our loneliness or nostalgia for Paradise. . . . Animals have a critical role in the shaping of personal identity and social consciousness. . . . Because of their participation in each stage of the growth of consciousness, they are indispensable to our becoming human in the fullest sense" (1996, p. 3). We and the animals with whom we share the planet and whom we use ought to be viewed as partners and friends in a joint venture. What a privilege it is to share our company, our time, and our space with such wonderful individuals.

There is no doubt that this book will go a long way in facilitating the development of strong and intimate bonds between humans and other animals. Expanding our circle of respect and understanding will help bring us all together. The animals "out there" will become the animals "in here"—in our hearts. And as we break bread in the company of many diverse and interesting animals, and as we appreciate them for who they are, our feelings will be incorporated into action for the betterment of all animals.

Marc Bekoff

References

Beck, Alan, and Katcher, Aaron, *Between Pets and People: The Importance of Animal Companionship,* rev. ed. (West Lafayette, IN: Purdue University Press, 1996).

Bekoff, Marc, "Deep Ethology." *AV Magazine* (published by the American Anti-Vivisection Society) 106 (1998a): 10–18.

Bekoff, Marc, ed., *Encyclopedia of Animal Rights and Animal Welfare* (Westport, CT: Greenwood, 1998b).

Kummer, Hans, *In Quest of the Sacred Baboon: A Scientist's Journey* (Princeton, NJ: Princeton University Press, 1995).

Lockwood, Randall, and Ascione, Frank R., eds., *Cruelty to Animals and Interpersonal Violence* (West Lafayette, IN: Purdue University Press, 1998).

Montgomery, Sy, *Walking with the Great Apes: Jane Goodall, Dian Fossey, and Biruté Galdikas* (Albany: SUNY Press, 1991).

Phillips, M. T., "Proper Names and the Social Construction of Biography: The Negative Case of Laboratory Animals." *Qualitative Sociology* 17 (1994): 119–42.

Shepard, Paul, *Traces of an Omnivore* (Washington, D.C.: Island Press, 1996).

Editor's Introduction

The original versions of the papers collected in this issue were given at a conference, *In the Company of Animals*, held at the New School for Social Research in April 1995 which was part of a multi-institutional collaboration organized under the auspices of *Social Research*. It was the third in a series of such collaborative projects, all of which have certain common features. Each has entailed a public conference at the New School and collaborations with other New York City cultural institutions around a shared theme. Each of them has been motivated by a serious and contested contemporary social issue, but none of these projects focuses on the policy issues. Rather, they attempt to examine the cultural and historical roots of these issues, which are frequently forgotten in the heat of the on-going debates, by approaching them from many different perspectives.

In the Company of Animals examined our relationship with other animals over time and in different cultures through a public conference at the New School, a poetry reading organized by the American Academy of Poets, and exhibits and other public programs at the Pierpont Morgan Library, The Asia Society, The Museum of African Art, and the Jewish Museum.

The motivating issue behind *In the Company of Animals* is the current, and often inflamed, debate about what our proper relationship to other animals ought to be. The papers given at the conference, and revised for this issue, explore how our relationships with animals have evolved over time and place, and how they reflect different understandings of what it means to be human. What becomes clear in reading the papers in this issue, if it is not clear already, is that the delineation of human/animal relationships occurs in all cultures, and in all cultures this boundary is a matter of great significance.

The question of our proper relationship to other animals is a question with a long history—as long as the history of our species. Throughout history and in all places, animals have been an important part of human culture. They have been hunted and domesticated, befriended and eaten, worshiped and feared, romanticized and demonized, studied and mythologized. Reflections upon our relationships with them have been continuous and are expressed in our traditions, arts, literature, religions, and sciences.

How we live with animals has changed dramatically in the course of human history. In agrarian societies, non-human animals were an integral part of daily life. In contemporary, post-industrialized society, we know animals primarily as pets and objects of occasional observation in moments of leisure. These changes have had profound effects on how we conceptualize and understand our relationships with them. Because our contemporary attitudes have deep roots in the past, an important aspect of this project is the illumination of the close relationship between how we live and the ways in which we have understood our relationships to other animals.

The last session of the conference directly confronted some of the currently contested issues surrounding the question of our proper relationship with other animals. This session of the conference had a different format than all the others which are reflected in this issue. In the concluding session, panelists responded to a series of questions posed by Andrew Rowan, the moderator of this session. For this issue, the panelists were invited to respond in writing to the questions posed by Rowan rather than presenting formal papers. This discussion was deliberately placed at the end of the conference, and is the last section of the issue, because we hoped that embedding the policy discussion about our rights and responsibilities to other animals in its cultural context would increase the likelihood of a more dispassionate and informed discussion. I think you will see that it did.

The conference would not have been possible without the

generous support of the Howard Gilman Foundation, The National Endowment for the Humanities, The Esther and Joseph Klingenstein Foundation, and, finally, Ms. Caroline Williams. We are deeply indebted to them.

Arien Mack

CATEGORIES

Introduction

It is appropriate to begin a wide-ranging exploration of the relationship between people and other animals with a section on classification. Whether acknowledged or not, this powerful intellectual act necessarily precedes any further dealings with animals, however utilitarian. Without ordered categories, there is no way to make sense of information, which is almost the same thing as saying there is no information. Both the title and the subject of this collection, and of the conference in which it originated, include significant taxonomic discriminations: the identification of animals as a significant, coherent group, and the tacit exclusion of human beings from that group.

Most bodies of material neither define their own boundaries nor provide their own indices, although this taxonomic neutrality may not be obvious to those who use them. Different people identify and structure such bodies of material in different ways, reflecting their various interests, needs, social contexts, and historical experiences. Alternative solutions to any classificatory problem are likely to be revealing. Why is one person's sacred cow another's *Bos taurus*? The interest and significance are heightened in proportion to the cultural centrality of the material to be classified. As the essays in this section demonstrate, animals have been among the core concerns of Western culture for at least as long as it has left records, and there is no reason to assume that they have had less importance in other cultures or at earlier periods.

Animal taxonomies reveal much more than what animals are like, or what they do, or how they are related to each other. For example, in the heroic founding period of scientific taxonomy—that is, the eighteenth century—the ability to deploy this much ballyhooed, ostensibly new intellectual technique became a marker for the classification and ranking of groups of naturalists, as well as groups of animals. In the age of Newton, structured schemes of classification gave natural history some claim to disciplinary dignity, and distinctively separated its practitioners from the unscientific chaos that they attributed to medieval and renaissance bestiaries. The extent to which they valued this distinction is indicated by the claim of one English interpreter of Linnaeus, that whoever could not give an animal "its true name according to some system . . . does not deserve the name of naturalist" (Linnaeus, 1759, pp. xx-xxi). Without systematic classification, it was feared, naturalists might be considered "mere collectors of curiosities and superficial trifles . . . objects of ridicule rather than respect" (Pulteney, 1805, p. 11).

In part because of its enthusiastic appropriation by enlightenment natural history, taxonomy is often considered to be the preserve of botanists and zoologists. But scientists are not the only ones keenly interested in the animal world; indeed the nature of their interest can be seen as partial and idiosyncratic, and their taxonomy, therefore, as an analysis produced by a particular intellectual elite. Farmers and poets, hunters and trainers, among others, have evolved classifications that express and expedite their own relationships to animals. These classifications exist in parallel to scientific taxonomy and do not yield often to its authority, even in a direct confrontation. It is, after all, thousands of years since Aristotle opined that dolphins were not fish, but many people remain unpersuaded. And non-scientific taxonomies can be as illuminating as scientific ones with regard to both classifiers and classified.

Harriet Ritvo

References

Linnaeus, Carolus, *Miscellaneous Tracts . . .*, Benjamin Stillingfleet, trans. and ed. (London: R.J. Dodsley, 1759).

Pulteney, Richard, *A General View of the Writings of Linnaeus* (London: J. Mawman, 1805).

Aristotle, the Scale of Nature, and Modern Attitudes to Animals

Juliet Clutton-Brock

GEORGE GAYLORD SIMPSON, one of the great biologists of this century, began his book on *The Principles of Taxonomy* (1961) with the statement that, "Any discussion should start with a clear understanding as to what is to be discussed." What will be discussed here is how the animal kingdom has been classified in the European-speaking world since the time of Aristotle, and how these classifications have affected our attitudes to animals today. I will start with the dominating role that Aristotle played in European civilization for an incredible length of time. Then I will go on to discuss how this dominance began to crack in the eighteenth century, and how it was finally broken apart in the nineteenth century, nearly 2000 years after Aristotle's death. Broken apart it may have been, yet Aristotle's philosophy is still with us today; it is the backcloth to our attitudes to animals, and, in fact, to the whole way we live and think, even if we do not agree with the *Encyclopaedia Brittanica*, which as late as 1875 claimed that many of Aristotle's works, "make an excellent curriculum for training young men and fitting them for the superior business of life."

Aristotle's Life and His Books on Natural History

Aristotle was born in 384 BC and died, aged 63, in 322 BC. Like Darwin, whose breadth of knowledge on natural history may have been almost comparable, Aristotle was described as always having "weak health."

After the death of his father in 367 BC, Aristotle, who was then 17 years old, went to live in Athens. There he spent the next twenty years studying under Plato. Plato died in 348 BC, aged 81, and then Aristotle went to live in Lesbos, where at the request of King Phillip of Macedon, he became tutor to Phillip's son, the young Alexander, later to become Alexander the Great. In 336 BC, Aristotle returned to Athens where he established his school in the garden Lyceum, teaching as he walked about, from which his school of philosophy came to be known as *Peripatetic*. Aristotle's books would probably have been written first on papyrus or parchment, and it has been claimed that they were the basis of the famous library in Alexandria which was later to be destroyed. It is not known if any of his books survive in their entirety; it is more likely that what remain are mostly notes and materials for lectures that Aristotle gave to his students. However, the knowledge about zoology that is contained in the works we have will never cease to amaze the reader who can learn an enormous amount, even today, from his translated books. I will quote just one example from the *History of Animals* on the incubation habits of birds to give a taste, which I hope will inspire more people to read the works of Aristotle for information as well as pleasure:

> I mentioned when speaking of pigeons that the male and female take it in turns to sit on the eggs. Most other birds do the same, but the males of some kinds sit only long enough to give the female time to get herself some food. The nests of all marsh birds are built near swampy and grassy places, and as a result of this the birds can remain sitting quietly on the eggs and get some food for themselves and so not go without eating altogether. Among crows, too, the females only sit on the eggs, and remain on them from start to finish: the males fetch food for them and feed them. The female wild pigeon begins to sit in the afternoon and stays on the eggs all night until breakfast time; the male sits for the remainder of the day (Peck, 1970, VI. viii 564a 10–15).[1]

When discussing the truly remarkable influence of Aristotle

it is first of all necessary to remember that he lived more than 300 years before the birth of Christ. He had no microscope, no knowledge of the circulation of the blood, or of gravity, or of the solar system, or even that the world was round. He believed that the Earth was stationary and the center of the world; he thought the seven planets (including the sun and moon) moved around the earth in oblique courses to the left, while the outer heaven or sphere of the stars, composed not of perishable matter but of divine ether, moved from left to right with perfect and regular motion returning on itself, deriving its motion from an encompassing essence which itself was not moved.

Aristotle's Belief in the Four Causes

Aristotle, and probably most of his contemporaries, believed that the universe was a continuous chain; at one end was the purely potential, matter without form or qualities; at the other end was the actual, which was ever existent. The actual always had to precede the potential. Thus, the seed was the potentiality of the plant, and the plant must always have preceded the seed, the fowl the egg, and so on. This was the system of cause and effect, which made up what Aristotle called "Nature" and which he believed was of eternal duration, although it could be modified and altered by two unpredictable elements of Causation, chance and the will of Man. Aristotle's beliefs about the natural world thus were based on the philosophy of Causes, or what we might call today "reasons." To know, said Aristotle, was to know by means of Causes. A thing was explained when you knew its Causes, and a Cause was that which was responsible, in any of four senses, for a thing's existence (Peck, 1970, p. xxxviii)."If we take a pigeon as an example, then the four Causes that explain its existence would be: (1) The Motive Cause: the parent pigeon which produced an egg; (2) The Material Cause: the pigeon's egg and its nourishment; (3) The Formal Cause: the egg as it developed and hatched into a chick that had the characteristics proper for a pigeon; (4) The Final Cause: the end towards which the process advanced, the perfected pigeon.

The Final Cause was the one of paramount importance to Aristotle and the one which dominated every process. His approach to the natural world was, therefore, teleological, that is, he believed that everything in Nature had a purpose, and this purpose was for the benefit of Mankind. He wrote, "plants are evidently for the sake of animals, and animals for the sake of Man; thus Nature, which does nothing in vain, has made all things for the sake of Man" (Peck, 1970, p. xli). And please note that whenever I say Man, I mean men and not women.

So far, I have given a very brief summary of how Aristotle and probably all his contemporaries looked at the natural world, and there does not seem to be anything very spectacular and unalterable about this outlook. However, I want to come now to the great importance of Aristotle's zoological works, which is that they were the first attempt in Europe to observe and describe individual living animals in a scientific way. He wished to deal with what can be known for certain and to express this in exact language, and his method of obtaining information was by observation and by the dissection of dead animals in the same way that zoologists learn today.

Aristotle's investigations into zoology are compiled into a series of books, known as the *History of Animals* (Peck, 1965, 1970; Balme, 1991), the *Generation of Animals* (Peck, 1990), and the *Parts of Animals*, the *Movement of Animals*, and the *Progression of Animals* (Peck and Forster, 1983). He wrote about more than 500 species, including shellfish, insects, birds, reptiles, and quadrupeds, with humans being treated in the same way as all other animals. The breadth of knowledge covered by these books is so great that it is hard to believe that one man could have learned so many facts if it were not that we know some naturalists in later periods, including Charles Darwin, were equally erudite and prolific. All the same, it is often claimed that Aristotle's works are the compilations of many authors. Certainly, his descriptions of animals were much quoted in the later classics, such as Pliny's *Natural History* and Aeolian's *On Animals*.

According to Aristotle, the purpose of his History of Animals was to obtain information through investigation, that is, to ascertain facts about each kind of animal, and then, as a second stage to find out the Causes of these observed and recorded differences (Peck, 1970, I.vi 491a.10). In the *History of Animals*, the parts themselves are described, for although this work is to some extent physiological, its main object was to deal with the anatomy of organisms. However, a great deal is also written about the habitats of animals and their behavior, for example:

> Again, some are mischievous and wicked, e.g., the fox; others are spirited and affectionate and fawning, e.g., the dog; some are gentle and easily tamed, e.g., the elephant; others are bashful and cautious, e.g., the goose; some are jealous and ostentatious, like the peacock.[2] The only animal which is deliberative is Man. Many animals have the power of memory and can be trained; but the only one which can recall past events at will is Man Peck, 1965, I.i 488b.20–25).

In the *Parts of Animals* (Peck and Forster, 1983), Aristotle takes the view of life known as internal finality, that is that each individual, or, at any rate, each species, is made for itself, that all its parts conspire for the greatest good of the whole and are organized in view of that end but without regard for other organisms or kinds of organisms.

The Scale of Nature

I come now to the views of the ancient Greeks on taxonomy and the classification of animals. Division of the animal kingdom is older than Aristotle; in Plato's philosophy the highest genus was divided by means of differentiae into subsidiary genera, and each of these was then divided and subdivided by dichotomy until the ultimate species was reached (which is the opposite system of classification to that of Linnaeus, as I will show later). Aristotle clearly disagreed with Plato's system of dichotomous subdivisions (Lloyd, 1968, p. 86). At the upper end of Aristotle's scale he had main groups, such as bird and fish, which were his genera, and at the lower end the commonly named animals such as dog, cat, eagle, and so on, which were his species, but normally the intermediate stages are missing (Peck, 1965, I.vi. 490b.10).

Aristotle did recognize a Scale of Nature, but the rungs of his ladder were not the stages of a taxonomic scheme, and there is no evidence that he felt they should be. His purpose was not to construct a taxonomic system but to collect data for ascertaining the Causes of the observed phenomena; this was to be done by looking to see whether certain characteristics are regularly found in combination: this was how the clues to the Causes would be brought to light. Aristotle believed that human beings were animals, but, at the same time, he was certain that all other animals existed for the sake of Man. He asserted that it was impossible to produce a neat hierarchical order on the basis of obvious physical differences because these cut across each other. This view runs right through all Aristotle's writings on zoology and is closely argued in the *Generation of Animals*, for example:

> Actually there is a good deal of overlapping between the various classes. Bipeds are not all viviparous (birds are oviparous) nor all oviparous (Man is viviparous); quadrupeds are not all oviparous (the horse and the ox and heaps of others are viviparous) nor all viviparous (lizards and crocodiles and many others are oviparous). Nor does the difference lie even in having or not having feet (Peck, 1990, II.i. 732b.15–20).

And so Aristotle goes on and on describing the similarities and differences between different kinds of animals, and for 2000 years he had no notable successor. In the eighteenth century, animals were still described in Aristotle's terms, as, for example, in a treatise on domestic pigeons which states: "All animals are distinguished into three sorts; oviparous, or such as are formed from an egg; viviparous, or such as are produced from the uterus alive and in perfection; and vermiparous, or such as are formed from a worm" (Moore, 1735).

This mention of the *perfection* of viviparous animals brings me to the last point that I shall discuss in Aristotle's philosophy. Although he made no clearly defined statement about a taxonomic Scale of Nature, it could be said that, in Aristotle's words, his most important tenet was: Nature's rule is that the perfect offspring shall be produced by the more perfect parent (Peck, 1990, II.i. 733a.5). This idea of perfection was tied in with Aristotle's idea of the "elements," earth, air, fire, and water, and he believed that the "hotter" beings were more perfect than others. Thus, warm blooded animals were more perfect than cold blooded, but also men were "hotter" than women and were the "natural rulers" because they were the most perfect of all animals (Lloyd, 1968, p. 252). However, not all men were equal, and slaves were clearly less perfect than their owners.

Since long before the time of Aristotle, the civilizations of Mesopotamia, Egypt, and Greece had developed into stratified societies ruled by powerful hierarchies and in which all manual work was carried out by slaves. It is, therefore, only to be expected that the Greek philosophers would view the natural world as a gradation from the lowest to the highest, or as a Scale of perfection, which was to become known as the Scale of Nature or the Great Chain of Being.

The Principle of Plenitude

The legacy of Plato to European thinking about the natural world has been expressed as the Principle of Plenitude

(Lovejoy, 1936; Rolfe, 1985). This was the belief that all possible kinds of things exist in the world already and nothing more can be created. Aristotle's legacy was more complicated but has been summarized by Rolfe as the concept of continuity and gradation between adjacent kinds of being when hierarchically arranged (1985, p. 300). Together, the Principle of Plenitude and the Great Chain of Being led to the belief, from medieval times, that a continuous chain extended from the inanimate world of non-living matter, such as earth and stones, through the animate world of plants, zoophytes, and the lowest forms of animal life, upwards to the quadrupeds and eventually through Man to the realms of angels and finally to the Christian God. This belief also entailed the view that just as nothing new could be created, neither could anything be exterminated, since this would counteract the will of God (Rolfe, 1985, p. 10).

Before the eighteenth century, writers about animals such as the Swiss naturalist, Conrad Gesner (1516–1565), and Edward Topsell, who published his *Historie of Foure-Footed Beasts* in 1607, viewed the world from an essentially human point of view. They had three categories of animals: edible and inedible; wild and tame; useful and useless (Thomas, 1983, p. 20). This belief in Man's supremacy over everything else in the world was to continue despite the very great increase in the writings about Nature by philosophers and naturalists in the seventeenth and eighteenth centuries. There was also a great increase in the efforts to classify plants and animals, notably by the botanist, John Ray (1627–1705)

The Five Predicables

In the European-speaking world, until well into the eighteenth century, the method of classification of all organisms was based on the Five Predicables. This was a hierarchical system that had been adapted from Aristotle's classification of logic, as written in his work known as the *Topics*. The five predicables are genus, species, differentia, property, and accident. They have been clearly defined by Simpson:

> Genus = that part of the essence shared by distinct species, that is it was a group of species with some attributes in common. Species = a group of things similar in essence. Differentia = that part of the essence peculiar to a given species and therefore distinguishing it from other species. Property = an attribute shared by all members of a species but not part of its essence and not necessary to differentiate it. Accident = an attribute present in some members of a species but not shared by all and not part of its essence (1961, p. 24).

Under this system humans can be classified thus: genus = animal, species = human being, differentia = rational, property = capacity for laughter, accident = say black or white skin.

Incidentally, it is interesting to look at the definitions of a species and a genus given in Dr Johnson's dictionary (1755) as this provides the scholastic view of classification at the time that the fundamental changes in biological classification were proposed by Linnaeus, who would have been familiar with these definitions:

> *Species*: Class of nature; single order of beings. A sort; a subdivision of beings. A special idea is called by the Schools a Species ; it is one common nature that agrees to several singular individual beings; so horse is a special idea or species, as it agrees to Bucephalus, Trot, and Snowball [the first being Alexander the Great's horse, the last two being famous horses of Johnson's time].
>
> *Genus*: A class of being comprehending under it many species; as quadruped is a genus comprehending under it almost all terrestrial beasts. A general idea is called by the Schools genus and it is one common Nature; so animal is a genus because it agrees to horse, lion, whale, and butterfly.

Linnaeus and Binomial Classification

I come now to the work of Linnaeus, who, it can be argued, has had as profound an influence on biological thinking as Aristotle had 2000 years before him and Charles Darwin had 100 years after him.

Carl Linnaeus was a Swedish botanist and explorer who lived from 1707–1778. He was an organizer who classified not only the plant and animal kingdoms but also the minerals and the kinds of diseases known in his day. Since the time of Aristotle animals and plants had been named in Latin by using the genus and the differentia from the five predicables of classification. The two together made up the definition which could be used as the name. However, with the classification and naming of more and more species over time, the differentia often became very long. The great innovation of Linnaeus was in creating the binomial or binary system by taking the old name for the genus and adding a single name from the many that had been used in the differentia, as the species, for example, *Felis catus* for the domestic cat.

The first edition of Linnaeus's classification of the animal kingdom, the *Systema Naturae* was published in 1735 and his definitive tenth edition in 1758, 101 years before Darwin's Origin of Species (1859). The book was written in Latin and the long introduction has been seldom translated, although it is full of fascinating comments on eighteenth-century attitudes to animals, as well as the first use of the term Mammalia,[3] which initiated the separation of the mammals from the rest of the Quadrupeds. The translation of Robert Kerr (1792) has the title *The Animal Kingdom or Zoological System of the Celebrated Sir Charles Linnaeus*. After the short introduction there is a chapter translated as "The Empire of Nature," which begins with quotations from Aristotle on the Causes and from the Roman writers Seneca (4 BC-AD 65) and Pliny the Elder (AD 23–79). Linnaeus followed Aristotle in believing that the three kingdoms of nature (minerals, vegetables, and animals) met together in the Order of Zoophytes, and also in the belief that everything in the world was created for Man. Linnaeus clearly believed that every person since the beginning of time had, like himself, a passion for classifying, for he wrote: "Hence one great employment of man, at the beginning of the world, must have been to examine created objects, and to impose on all the species names according to their kinds."

Before Linnaeus, most naturalists started their classifications by dividing all the known organisms into large groups and then subdividing these into progressively smaller groups. Unlike his predecessors, Linnaeus saw that the unit of classification had to be the species, that is, a population or group of organisms that are systematically related to each other. He then organized these species into larger groups or genera, arranged analogous genera to form Families, and related Families to form Orders and Classes. Thus, he produced a strict hierarchical classification which ended at its summit with the Kingdom. Linnaeus summarized his ideas as follows:

The science of Nature is founded on an exact knowledge of the nomenclature of natural bodies, and of their systematic arrangement; this enables a philosopher to travel alone, and in safety, through the devious meanderings of Nature's labyrinth. In this methodical arrangement the Classes and Orders are the creatures of human invention, while the division of these into Genera and Species is the work of Nature. All true knowledge refers finally to the species of things, while at the same time, what regards the generic divisions is substantial in its Nature. . . . God, beginning from the most simple terrestrial elements, advances through Minerals, Vegetables, and Animals, and finishes with Man. Man on the contrary, reversing this order, begins with himself, and proceeds downwards to the materials of the earth. The framer of a systematic arrangement begins his study by the investigation of particulars, from which he ascends to more universal proportions; while the teacher of this method, taking a contrary course, first explains the general propositions, and then gradually descends to particulars (Kerr, 1792, pp. 22–3).

Eighteenth-Century Naturalists

Linnaeus and his followers were, however, only one section of a large number of naturalists and philosophers who were attempting to understand the living world in the eighteenth century, and who were writing essays, tracts, and books about their ideas. One of the most influential of these was the French naturalist Georges-Louis Leclerc Buffon (1707–1788) was born in the same year as Linnaeus and, if anything, he worked even more prodigiously than his Swedish counterpart. Buffon's aim was to describe the whole of natural history in 50 volumes, of which, at the time of his death, he had completed 36. It was not this great production of work that was to have a lasting influence, but his construction of geological stages and his radical acceptance that species could become extinct. Until the mid-eighteenth century, firm believers in the Principle of Plenitude and the Great Chain of Being, such as Ray, had forcefully argued against the possibility that species could become extinct and they produced all sorts of explanations for the existence of fossils. This was because inherent in the Principle of Plenitude was the idea that all the possible kinds of things exist, and extinction would not be possible because it would counteract the will of the all-powerful Creator.

Buffon was an aristocrat, and, as might well happen today, his total absorption with the animal world was considered to be very eccentric by his peers. Madame du Deffand, for example, is recorded as having remarked: "He concerns himself with animals; he must be something of one himself to be so devoted to such an occupation" (Times Literary Supplement, 27 January 1995, p. 29).

The Missing Link

There was a paradox in the thinking of many of these naturalists because although they believed in the separate creation of every species, their overall anthropocentric view led them to search for links in the chain between the newly discovered great apes and humans. For example, the Swiss naturalist Charles Bonnet (1720–1793) wrote a detailed account of the *orang-outang* and did not hesitate to claim its close relationship with Man, albeit with the "lowest races" of the human species.

Nineteenth-Century Biology and the Origin of Species

By the nineteenth century, naturalists were vigorously looking for an alternative theory to that of special creation and the rigid ranking of every known species of plant and animal. They had finally realized that Aristotle's legacy and the traditional ways of defining and categorizing living organisms in a Scale of Nature no longer fitted their modern world with its ever-increasing discoveries of fossil and living forms. Although the Principle of Plenitude and the Scale of Nature had provided for centuries a framework for the definition of living organisms by careful analysis, no account was taken of individual variation. Within this framework, the so-called doctrine of internal finality also decreed that every organism was complete in itself, and every organ or part of the body had its own peculiar function. This meant that there could be no relationship between different species, one result of this concept being that the existence of rudimentary or vestigial organs was never recognized.

It was to his great advantage that Charles Darwin was not a classical scholar and had not been influenced by these age-old precepts, although he was much impressed by the works of Aristotle which he read in translation, long after the publication of the *Origin of Species*. The following excerpt from a letter that Darwin wrote to William Ogle in 1882 after he had read *The Parts of Animals*, which Ogle had translated, confirms this:

I have rarely read anything which has interested me more,
though I have not read as yet more than a quarter of the book
proper. From quotations which I had seen, I had a high notion
of Aristotle's merits, but I had not the most remote notion what
a wonderful man he was. Linnaeus and Cuvier have been my
two gods, though in very different ways, but they were mere
schoolboys to old Aristotle (F. Darwin, 1887, III:251).

Like Darwin, the few people who understood his theory of
the origin of species in the mid-nineteenth century were
empiricists who did not believe in the fixed essence of each
form. One of the most influential of these free-thinking people
was Thomas Henry Huxley, although even he had difficulty at
first in making the jump from belief in the immutability of
species to the concept of evolution. He wrote in a letter to
Darwin after first reading the *Origin of Species* and finally
understanding the full extent of its implications, "How stupid
not to have thought of that."

The following is a quick summary of the discovery that has
made Darwin the god of many biologists today: innumerable
natural variations occur in the progeny of nearly all living
organisms. In the natural world, the struggle for existence
ensures that only the fittest survive and reproduce, so that
variations are discarded or retained by natural selection. Over
time, these and other forces, such as reproductive isolation and
sexual selection, lead to the origin of new species.

The great dilemma that faced every thinking person of the
time was that if they accepted the theory of evolution, they had
to face the fact that human beings were not a special creation
made in the image of God but the result of a slow process of
transmutation from so-called lower forms of life. That is, the
Scale of Nature was not a ladder in which each step was the
result of a discrete and separate creation but was indeed more
like a Great Chain of Being in which the links were connected
to each other.

It is hard for us now to imagine the bitterness of the battle
that raged over the *Origin of Species*. It was exemplified in the

famous Oxford meeting of 1860, when the Bishop of Oxford assured his audience that there was nothing in the idea of evolution. He pronounced that, "rock-pigeons were what rock-pigeons had always been," and then turning to his antagonist, T. H. Huxley, with a smiling insolence, he begged to know was it through his grandfather or his grandmother that he claimed his descent from a monkey? To which Huxley made the famous retort the essence of which was that he was not ashamed to have a monkey for his ancestor, but he would be ashamed to be connected with a man who used great gifts to obscure the truth (Huxley, 1903, I: 267).

Education in nineteenth-century England was still strongly based on the classics and the teaching of science in schools was in its infancy. Among those who abhorred the turning of the tide towards mechanical power and scientific progress was William Morris (1834–96) who wrote prophetically in 1894:

Apart from the desire to produce beautiful things, the leading passion of my life has been and is hatred of modern civilization . . . What shall I say concerning its mastery of and its waste of mechanical power, its commonwealth so poor, its enemies of the commonwealth so rich, its stupendous organisation . . . its eyeless vulgarity . . . the place of Homer [is] to be taken by Huxley (MacCarthy, 1994, p. 261).

The Legacy of Aristotle and Modern Attitudes to Animals

Despite all the changes that have taken place in biological science over the last 100 years, most people still believe that the world is ordered according to a hierarchical Scale of Nature with unicellular organisms at its base and Man at the top. This is not surprising since, from the moment of birth, people in the Western world are ruled by hierarchies, first in the family, then in education, and on through adulthood. It can be argued whether social ranking is the result of the natural evolution of complex societies or whether it is a legacy from the slave states of the ancient world. But, whatever its origins, this hierarchical view of life has affected the way humans have lived in the European-speaking world since before the time of Aristotle. Today, social ranking still affects everything we do and it dominates our attitudes to other animals which are classified and named in the hierarchical system created by Linnaeus in his "Empire of Nature." It is very unlikely that taxonomists will ever escape from classifying the natural world in hierarchies, for as Keith Thomas has written:

> . . . all observation of the natural world involves the use of mental categories with which we, the observers, classify and order the otherwise incomprehensible mass of phenomena around us; and it is notorious that, once these categories have been learned, it is very difficult for us to see the world in any other way (1983, p. 52).

Aristotle believed that every living organism was made for the sake of Man, and little has changed in this attitude over the last 2000 years. In one way or another, in most parts of the world, people believe that animals are there for the benefit of the human species. Individual domestic animals are treated as substitute children, being loved, cared for, talked to, and chastised when they "do wrong," while livestock animals are farmed for food and other resources. Wild animals are still killed for sport, and until very recently it was accepted that the natural world also existed for the benefit of humans, as somewhat unbelievably summed up in 1969 by the International Union for the Conservation of Nature (IUCN), which defined conservation as the rational use of the environment to achieve the highest quality of living for mankind (Thomas, 1983, p. 302). Fortunately, over the last 30 years this view has been moderated, although there can be few wild places that are subject to no human intervention. By 1991, IUCN had changed its mission statement to include the premise that: " . . . Nature must be cared for as a life-renewing system to which we belong, and not just as a means of satisfying human needs" (Munro and Holdgate, 1991).

The attitudes of people to animals throughout the Western world is still a mass of contradictions. Ethologists tend to believe that "only man is vile" and concern themselves with the problem of whether animals have consciousness and whether animal societies can have a culture. Meanwhile, newspaper reports of ill-doings by and to humans ring with cries of denigration such as: "He was as empty of morals as any animal," or "He behaved like an animal," or "We were treated like animals." This is a direct result of the inherent belief in the Scale of Nature with its ranking of animals as far below humans in moral status.

Another great paradox can be seen in the different attitudes people, and this often includes biologists, have to the wild and to the domestic. This attitude is not inherited from Aristotle, who made no distinction in his descriptions between the wild and the tame. The ethologist will gain status by studying the behaviour of lions but none at all from observations on the family cat. The pitiful and often painful deaths of millions of domestic animals make an impact on only a tiny minority of people, but furious arguments rage about the rights and wrongs of hunting.

In the ancient world, people lived very close to their livestock, often sharing the same house with them, and however cruel they may have been at times, they treated their animals as individuals who could suffer like themselves. But it is inevitable that once the numbers of animals owned become large, say in the thousands, their individual identities are lost. This has been the inevitable result of industrial farming in the modern world, where the vast numbers of domestic animals have become like animate vegetables, all bred to look alike and reared in confined spaces for maximum yield at the lowest cost. Yet each of these animals does remain an individual with feelings and with a temperament of its own.

The question of whether animals have rights has been discussed by philosophers for hundreds of years, but the arguments for and against have remained mainly of academic interest. The time has now come, however, when we must accept that the animals for whom we are responsible do have rights This is an ethical issue of our time, and we must take it seriously, just as the question of the rights of human slaves had to be faced 150 years ago. Slavery throughout the world was not officially abolished until 1833, and even today there are huge numbers of people who are living as virtual slaves. It has therefore taken more than 2000 years to accept that, at least in theory, every human being has equal rights. We have to hope and expect that it will not be another 2000 years before we acknowledge that every animal too has rights. Maybe we can be optimistic about this for, at least in Britain, there are such widespread, ongoing protests against the cruelty of exporting thousands of live calves for the veal trade and live sheep to be slaughtered on the Continent of Europe that the time may be hastened when animals are indeed recognized as sentient beings.

Following these protests from a wide spectrum of people in Britain, the Labour Party published the pledge that they would curb the export of live animals for slaughter, and in the long term they would want to have farm animals redefined under the Treaty of Rome as "sentient beings" rather than as "agricultural produce." It is indeed a great step forward when a government party, even if in opposition and covering itself with "in the long term," appears to bear in mind what the radical philosopher Jeremy Bentham (1748–1832) wrote nearly 200 years ago: "The important question is not *can* they reason, nor *can* they talk, but *can* they suffer" (Harrison, 1948, cited in Thomas, 1983, p. 365, n14)?

Notes

[1] It is noteworthy that white doves (pigeons) in a dovecote in an English garden today still follow this night and day schedule for incubation, with the female sitting during the night and the male during the day.

[2] Note that neither the elephant nor the peacock were native to Greece in classical times, but they must have been brought from India from time to time.

[3] Kerr (1792, p. 33) has the following footnote after Linnaeus's "Class I. Mammalia":

The term Mammalia, here used, signifies such animals as feed their young by milk derived from proper glands situated on the mother, and furnished with teats or paps. There is no single English word by which this can be translated; Quadrupeds would exclude the Cetaceous order, which, from giving milk, are arranged by Linnaeus in this class.

References

Balme, D.M., trans., Aristotle, *Historia Animalium*, Books VII-X, Loeb Classical Library, (Cambridge, MA: Harvard University Press, 1991).

Bonnet, C., *Oeuvres d'Histoire Naturelle* (Neuchatel: S.Fauche, 1779).

Buffon, G. L. L., *Histoire Naturelle* (Paris: General et Particuliere, 1804).

Darwin, C., *On the Origin of Species by Natural Selection* (London: John Murray, 1887).

Darwin, F., *The Life and Letters of Charles Darwin* (London: John Murray, 1887).

Harrison, W., ed., *Jeremy Bentham, An Introduction to the Principles of Morals and Legislation* (Oxford: Oxford University Press, 1948).

Huxley, L., *Life and Letters of Thomas Henry Huxley* (London: Macmillan, 1903).

Kerr, R., *The Animal Kingdom or Zoological System of the Celebrated Sir Charles Linnaeus; Class I Mammalia* (London: J. Murray, 1792).

Linnaei, C., *Systema Naturae. A photographic facsimile of the first volume of the tenth edition (1758) Regnum Animale* (London: British Museum [Natural History], 1956).

Lloyd, G.E.R., *Aristotle: The Growth and Structure of His Thought* (Cambridge: Cambridge University Press, 1968).

Lovejoy, A.O., *The Great Chain of Being a Study of the History of an Idea* (Cambridge, MA: Harvard University Press, 1936).

MacCarthy, F., *William Morris a Life for our Time* (London: Faber & Faber, 1994).

Moore, T., *A Treatise on Domestic Pigeons*, Facsimile printed by Paul P.B. Minet (Chicheley, Buckinghamshire, United Kingdom, [1765]1972).

Munro, D.A. and Holdgate, M., eds., *Caring for the Earth a Strategy for Sustainable Living* (Gland, Switzerland: IUCN, 1991).

Peck, A. L., trans., Aristotle, *Historia Animalium*, Books I-III, Loeb Classical Library (Cambridge, MA.: Harvard University Press, 1965).

Peck, A. L., trans., Aristotle, *Historia Animalium*. Books IV-VI, Loeb Classical Library. Cambridge, MA: Harvard University Press, 1970).

Peck, A. L., trans., Aristotle, *Generation of Animals*, Loeb Classical Library (Cambridge, MA: Harvard University Press, 1990).

Peck, A. L. and Forster, E. S., trans., Aristotle, *Parts of Animals, Movement of Animals, Progression of Animals*, Loeb Classical Library (Cambridge, MA.: Harvard University Press, 1983).

Rolfe, W. D. I., "William and John Hunter: Breaking the Great Chain of Being," in W. F. Bynum and R. Porter, eds., *William Hunter and the Eighteenth-Century Medical World* (Cambridge: Cambridge University Press, 1985).

Simpson, G. G., *Principles of Animal Taxonomy* (New York: Columbia University Press, 1961).

Thomas, K., *Man and the Natural World: Changing Attitudes in England 1500–1800* (London: Allen Lane, 1983).

A Taxonomy of Knowing: Animals Captive, Free-Ranging, and at Liberty

Vicki Hearne

My title is deceptive, because it implies a balanced interest in the three ways of knowing animals it lists. There already exist rigorous and elaborate descriptions and accounts of captive animals and free-ranging animals, but there is very little formal knowledge or, for that matter, acknowledgment of the kinds of knowledge people who work with animals at liberty have. The term "at liberty" here does not mean "free." Indeed, an animal at liberty, whose condition frees her to make the fullest use of some or all of her powers—in, say, search and rescue or in a "clever disobedience act"—may seem to be the most restrained of animals, just as the person whose submission to discipline may, paradoxically, free him to otherwise unattainable achievements.

Unlike most taxonomical terms, mine do not describe permanent or inalienable conditions. An Airedale may be captive, free ranging, or at liberty. Indeed, the *same* Airedale may at different times and in relationship to different people be in any of these conditions. There lived in Holland a police dog, a German Shepherd, named Albert. In his work with his handler, he was "at liberty," but when he was taken to the laboratory of the Dutch psychologist Buytendijk and put into a highly controlled situation in order that his scenting powers could be dispassionately scrutinized, he was captive in relationship to Buytendijk, though still at liberty in relationship to his policeman handler. If an anthropologist were to follow Albert and his handler about, making notes, both dog and handler would be "free-ranging" *in relationship to the observer*.

That is, my terms describe not so much various conditions in which animals in themselves might be as conditions *we* are in with the animals, social and grammatical conditions and circumstances. A scientist's rats may be in a maze or a Skinner box while the scientist is free to leave the building, even neglect the animals, but the kind of knowledge the scientist can have is in part a function of, and limited by, the kinds of circumstances with which the animals are dealing. (The scientist, more and more these days, is recognized as part of the animal's environment, including the animal's social environment.) The walls and grids that restrain your animals restrain also your own knowledge. Of course, restraints and constraints can be fruitful.

When animals are kept or observed for the purposes of ethology or psychology, free ranging or captive, the knowledge that emerges from the scientist's relationship to them, ideally, is received and expressed in the third person, which is one of the reasons mechanism, which takes third person description as far as it can, has had so long a run. This is true even, say, of hand-reared wolves; it is the scientist's sporting goal to remain ignorant of anything that cannot be known in the third person. I think, for example, of a woman in Boulder, Colorado named Susan Townsend who studies Mexican gray wolves, many of them hand-reared. It came about that her mated pair had a litter of two male cubs. One was bold, outgoing, dominant. The other was timid, submissive, shy. The shy one seemed to her to have a lesser chance of survival since he would not be so determined in his pursuit of food. The bold one was so adamant about his rights that he growled once at the alpha male, Pancho, when Pancho forbade him access to a carcass.

One morning Townsend found one of the wolf cubs dead, killed by one or more of the adult wolves. The remaining cub was shy, timid, and so on. She is morally certain that the cub who was killed was the bold one, a fact that could have consequences for her evolutionary speculations, but since the cubs had not been marked, she is scientifically ignorant of which cub survived. She knew the cubs as individuals but not as subjects of dispassionate scrutiny, and it is, as we say, in "theory" possible that the bold cub became shy as a result of seeing his brother killed, being impressed by the lengths to which the adults were willing to go. It is gallant of science so to handicap herself, and much of great value has come from this gallantry, but it is not the only possible gallantry of the intellect.

The trainer, whose goal is animals working at liberty, is a different case. I take the expression "at liberty" from circus tradition, where it refers to horses who work without physical restraints—without tack, that is. They, rather, are restrained by the perimeters and terms and grammars of cooperation, by understanding. The understanding is a constraint in the way understanding of music is a constraint on the violinist who is not at war with herself. Because circus language and tradition are so lovely, I do not mean to take this term with me permanently—my usual term for the condition is "off lead." But I am borrowing it just for the moment, since it points to some characteristics of liberty among humans and also, perhaps, to what the conditions of true peace are; willing cooperation is a restraint of a sort, certainly, but is not coercion. It may take the form of an artistic perimeter. A condition for such work is that the dog and the human, or any other animals involved, be motivated and organized in their work by a social gravity which keeps them turning toward each other and/or mutually turning toward an object of work—a function the stock serve, for example, in the case of a handler and a herding dog. The ideas, the understanding, are social [Levinas here?]. The absence of physical restraints is, therefore, only a clue about, and not a criterion of, at liberty work. Grand Prix dressage, eventing and stadium jumping horses, guide dogs, sled dogs, and weight pull horses in harness, thus, are all liberty animals, when things are going as they should according to the notions that frame the work.

This condition is no more a guarantee of moral congruence than anything else. But its forms are the forms of a moral congruence between people and animals not otherwise possible, in the way the forms of language are forms of a moral congruence, or, for that matter, incongruence, not otherwise possible. And with unique possibilities of congruence come also unique possibilities of incongruence and even tragedy, which is why the best dog and horse stories are about working relationships of one sort or another. For both horse and human, dreaminess of an afternoon in a meadow is one thing but quite another at Badminton.

The understanding I have in mind is a particular grasp on and of the nature and boundaries of the world. It is something that happens between the animal and the handler, an overlapping awareness that grows as the work grows. An example of such grasp is the search and rescue dog who in training is taught that "Search!" means find anyone who is out there, but who sometimes, once she has made a real find, seems to understand that she is after someone in distress. Some search dogs can work through suburban backyards, ignoring people going about their business. Here, a leap of understanding on the dog's part sparks a leap on the handler's part, if the handler is on the ball.

Another example. One day I was teaching a lesson and had my Airedale, Texas, with me for backup. The handler was having trouble with her dog, a Mastiff, so I had to take over. The Mastiff went for me. Texas left his "down" and put himself between me and the Mastiff. Once things were under control again, he, without prompting from me, returned to his "down" position at the edge of the training area. Here the simple "down" exercise became the lively and thoughtful posture police dog handlers sometimes call the "strategic down." It was the dog's grasping and acting on his understanding that here expanded meaning. Before the incident, "Down" for this young dog had simply meant, "Lie down and hold still until you hear from me again." In the course of the incident with the Mastiff, Texas both created and learned a strategic down. A police officer, or a soldier, or a robber might leave her dog on a down with a view of one entrance while she went to cover the other; with a little experience, dogs come to understand the strategy in question. Once they grasp the point of the arrangement, they are controlled by their own understanding of the strategy, just as their handlers are.

The details and forms of this understanding come to the handler largely in the second person, a form of knowledge Buber uses the "I-Thou" to name. I do not mean that handlers never make inferences, never deduce anything about their animals or make observations, to themselves or to each other or to the world at large, and I do not suggest they live in a preternaturally lively world and are not subject to the deadliness of what Stevens called "the inert savior." I only mean that the liveliness of this knowledge is something particular. Training does not *give* the understanding that delineates the thinking and action of both animal and handler; it *is* that understanding which, when achieved, compels the dog and the handler in somewhat the way your understanding of color compels you to see grass as green rather than bluish-yellow.

An animal's capacity for liberty work is not, I should say, a function of docility; it is disastrously weak or absent in many wild animals, also in some wolves and wolf hybrids, who are extremely timid and submissive. (This is not an idea of home, either, though concepts of homes of several sorts are involved; it is an understanding of *work* [as opposed, say, to drudgery]. Having never been, say, behind a plow horse, I cannot report on what the terms of that would be, or whether the situations that look like drudgery always were or are. Though on occasion the work is heroic, at least if you are willing to follow me in doubting that fear is the measure of heroism.) A logging or circus elephant can be at liberty, but from what I can tell, research animals who tend to stay near an area because there is a feeding station, such as Jane Goodall's chimps, are not at liberty, because food, rather than a particular social understanding, is the gravity that keeps the animals near the human. Some would argue, against me, that the food is a token of a social understanding, but there is recent work, including Sarah Boysen's with chimps and Irene Pepperberg's with the parrot named Alex, that suggests food, especially food rewards, actively interferes with social understanding—in these two cases with language conceived of as social understanding.

In the case of dogs, a clue, a pretty strong one, is whether the dog, in the open, with squirrels, kitty cats, delivery personnel, and other enticements, comes when called. In the case of some trailing hounds, however, this is fairly irrelevant, even though it is important that the *capacity* for coming when called be there. My husband's Plott Hound, Lucy Belle, came when called; few Plott Hounds do.

Consider how it goes when I am working with the Bernese Mountain Dog named Sampson and his handler, Martha Goodman. When I gave this material as a talk, I could save a great deal of time simply by asking Sampson and Martha to demonstrate some of what I have in mind, but the fact of dogs working at liberty is a kind of secret in this culture, or is, in any event, arcane, and recent political interest in animals has done nothing to alter the concepts and policies that forbid a civilized dog such as Sampson entry into lecture halls. If there really were an animal rights movement, someone would be doing something about all the public buildings and the trains and taxis and airplane cabins where Sampson is forbidden to go.

Sampson is a large fellow, though quite agile. As I have said, he is civilized, meaning, among other things, civil. But I do not mean he is just anybody's dog, despite the usual public assumption in this culture by which a negative answer to the question, "Does he bite?" is construed as an invitation to pat, fondle, chuzzle, and coo.

He seems to be very aware of his own dignity and particular about his friendships—he did not accept me as a pal until over a year into my weekly work with him and his handler when I handled him myself, the day we introduced him to a couple of basic search and rescue exercises. For a dog with the capacity for it, search work is thrilling, transcending, and Sampson's realization that I knew about such wonders finally prompted him to decide that I was worth the effort of further acquaintance.

Sampson works off lead, at liberty. The gravity, or, as it were, specific density, of the grammar between an animal and a human that makes such work possible, so far as I am aware, has not been examined systematically or scientifically, unless obedience trials in the west constitute such an examination, where the trial ring boundary is marked only by a rope about waist high, more often than not they occur out of doors, and the events are characterized by the close presence of a great many quite active dogs and people as well as, sometimes, other animals and bands. Stanley Coren has, in *The Intelligence of Dogs*, treated such trials as a source of data, but he does not distinguish between indoor and outdoor trials, or between trials held behind gates and western trials in which there is only the frailest of psychological barriers to discourage the dog from bolting. Herding dogs and search and rescue dogs, as well as hunting dogs, can be working a mile or more from their handlers without losing track of what I am here calling a grammar.

It is possible, as I have ill-temperedly suggested, that the existence of *work* "at liberty" or "off lead" is one of the best kept secrets in the world, despite the fact that anyone who has seen a sheepdog trial or movies in which dogs appear as characters has evidence of the phenomenon. One of the reasons the fact of off lead work is not often brought home to the intellect or to the culture is that some dog trainers in effect conspire to keep it a secret by not teaching people how to achieve it, sometimes because they are unaware that it exists, sometimes out of superstitious disapproval. Also, I have seen a number of sheets of advice from veterinary behavior consultants and have never read, "Seek a trainer who teaches off lead control." I have called the fact that there are such dogs as Sampson a kind of secret, in that what he does is unintelligible to the culture. That does not mean there is something magical going on, that the abilities of trainers work on animals the way penicillin or lasers do. Not all animals, and indeed not even all dogs, can learn work at liberty; certainly not all animal-human teams can learn this work (it takes two to heel or do a recall). But *some* animal-human teams, and even groups, *can* learn it, in some cases virtually as one learns a native tongue, and this is the fact I report on when I say that Sampson works off lead.

The trainer's world is different from the scientist's in that, as I have said, here in the trainer's world trainableness and docility or submissiveness are not at all the same thing as they are in scientific accounts of formal work, or attempts at formal work, with chimps, wolves, and, especially, octopuses I have read. The resistance of the latter to any sort of adjustment to human terms is such as to prevent a great deal of work. It is different with many, if not most, dogs, horses, kitty cats, donkeys, and elephants. With these animals, the capacity for cooperation that makes work at liberty possible is often enough expressed as robustness, courage, a refusal of incoherent terms of work and relation. A friend of mine in her sixties, whose Pit Bull, Monkey, is currently APBT of the month for being the first Pit to earn her Agility Trial Championship, reports that nothing bothers her new puppy, Ziggy, nothing is a punishment for him, including, in recent California storms, having had the porch collapse next to him. He is busy, fearless, uncowable—I have a snapshot of him at 9 weeks dragging a one gallon water can over a four foot

A-frame. He is driving my friend nuts, but she says, "Of course, they have to be like that if you're going to do anything with them." "Doing anything" here means obedience and agility; that courage and, as I have heard it called, "overflowing temperament" are needed for the higher accomplishments in obedience seems like a paradox only in a culture in which work at liberty is, as I have said, largely invisible, read either as magic or else as coercion.

But back to Sampson. He is daily becoming a more and more able scent discrimination dog. In this exercise in competition, the judge places on the ground or floor eight articles. In this country, half are made of leather, half of metal; in Canada, wooden articles are also required. The group of articles is called a "pile" even though they are not touching each other and, perhaps, might be better called a "scatter." The dog and handler are allowed to watch the judge place the articles in the pile.

Then the dog and the handler turn their backs to the pile, and the handler scents another article, leather or metal, and gives it to the judge, who handles it carefully, often with tongs, to prevent her scent from fouling the article, as she places it in the pile. At the direction, "Send your dog," the dog and handler pivot in place; then the dog is given a command, such as "Find It!" and the dog walks or trots to the pile, or occasionally canters, checks the articles with his nose, and brings back the article that has been scented by the handler (or so the handler prays).

Notice that there are a total of nine articles for the dog to examine, in an untidy array. The articles have numbers stenciled on their ends, to enable the people involved to tell them apart, but I have never seen evidence that a dog is doing "number discrimination," though I have seen, early in training, "string" discrimination when the method of tying down the wrong article is used to begin to give the dog the idea of what is wanted, and also "praise discrimination," when the handler is too quick to praise, with the result that the dog simply sniffs around until she hears praise and then grabs that article. In order to respond meaningfully to the dog, the handler must fluidly "read" the animal from moment to moment, and the animal must as fluidly "read" the handler. (One of the phenomena, besides ill temper, stomach ache, fatigue, and hunger, that can interfere with this as with other kinds of reading is unconscious cuing. For reasons that I do not have time to go into here, unconscious cueing on the part of both dog and handler *interfere* with the work—with even the appearance of work—and a great deal of advice, as well as competitive rules, about handling are designed to eliminate this incoherence. Why and how this is so is beyond my current topic. Suffice it to say that in a trial the "right" article is frequently placed in such a way that the handler cannot read its number, and so does not know which one it is.)

One day recently I noticed, or rather recalled, that Sampson seemed to be doing something that I cannot do, and that I doubt many humans can do or do handily, and that is in some way mapping the scent pile. What he seemed to me to do was to check each article once and only once. Like many dogs I have seen, if he has not finished his scrutiny of the pile and comes to the right article, he continues until he has checked the whole pile. Once he has done this, he returns to the "right" article and brings it back to Martha.

Since I noticed this, I have not had the opportunity to observe Sampson, much less enough other dogs, to make a general observation about this ability. I have not even made enough observations to apply for funds, if I were a scientist, in order to collect the nuts and bolts data, much less manipulate or interpret it. The important fact here is that it was not while I was thinking about or working with Sampson that this phenomenon impinged itself on consciousness as possibly something worth formal study, but rather while I was reading a great deal of ethology and ecology and philosophy—that is to say, in terms I take from Martin Buber and Emmanuel Levinas, while I was in an "I-it" rather than an "I-Thou" relationship to Sampson. I had in effect stepped back, out of range of Sampson's Thou, as if behind the lens of a video camera or behind a field notebook or tape recorder. While I was in this mood, subject to this muse, Sampson was no longer at liberty in relationship to me but rather—what? Not captive in the sense I have in mind, because his environment was not particularly controlled. He was free ranging, an "It" to my "I," and it was while I was in *this* mood that Sampson's impressive ability to map the scent pile came to the forefront of consciousness. I was thinking, say, analytically, objectively, as, say, a Connoisseur rather than a Beholder of Sampson's work, to take a crucial pair of terms from the philosopher Stanley Cavell.

The scientist may love and be in awe of the objects of her scrutiny, and that relation may even be a source of power for scientific thought, as people say it was in the case of Barbara McClintock. But the scientist who fails to step out of the grammar of the "I-Thou" and into the "I-it" is not a scientist, cannot have *that* knowledge. This claim is only *suggested* by my example.

Another point: for training purposes it does not matter whether or not Martha ever learns how Sampson learns to solve the various problems of the scent pile, including the one I am here, by a visual metaphor, calling "mapping." It only matters that Martha knows how to stir, urge, prompt him to solve the problems and to acknowledge him when he does. It does not matter in the same ways Wittgenstein has pointed out it does not and does matter how a pupil learns to continue a number sequence. It only seems to matter if things go wrong; and Wittgenstein makes us eerily aware, if we read him as Cavell has, that when things go wholly wrong there looms the dismissive diagnosis: this person is incapable, or insane, is not fully human. This is not what one says about failures in scent work or, say, sheep herding—that is, one does not, or does not necessarily, use a diagnostic language that erases the person-hood of the pupil who is showing no signs of learning, but rather that this particular dog, or this particular student, does not have this particular talent. It is more about what one might say about my failure to learn to sculpt well rather than about a child's failure to learn to speak or count. It is in novice work, as it is called, in the more fundamental grammars of exchange in which advanced commands such as "Find it!" are embedded, that occasions arise to suspect that a particular dog has a screw loose.

The teaching of scent work, like other dog work, goes in many cases somewhat the way things go in the *Meno*, when Socrates concludes from the slave boy's ability to solve the mathematical problem that there are forms. That is, what you teach, or think you teach, the dog never wholly accounts for what the dog learns, as though teaching were, as some say it is, a matter of creating forms in the world within which the dog's powers can become actual.

Sampson and Martha pivot toward the pile, Martha says "Find it!" and Sampson sets out. Once in a while Sampson is still a touch dishonest here—that is, he deliberately picks up the wrong article or just goes out and grabs at an article at random, making no attempt to sort things out. By and large nowadays, however, he is pretty steadily honest, and he is honest on purpose at the task of mapping the articles more tidily than, say, I can.

There are various forces and lines of discourse that prevent Sampson's work from being visible most of the time, despite the fact that you can go to an obedience trial every weekend of the year if you like and watch such work. But you can count on your thumbs the number and kinds and conditions of humans who have *that* kind of interest in obedience work.

What is this work? Well, to use a term of Levinas', even though I am not willing to follow Levinas all the way home, knowing is all in the greeting of the other, the stance toward the other. If I pursue my observations about dogs' patterns of scent discrimination, I may make some discoveries about dogs' capacities that I cannot make while Greeting or that I cannot go about systematically trying to make, but *this* will not be the Greeting, the "I-Thou," the receptive Beholding. As Cavell points out, Beholding is not always a morally, aesthetically, or intellectually privileged (even in cases where it is prior) mode of knowledge—he notes that we want our doctors to be Connoisseurs rather than Beholders of pain. But sometimes Beholding, or Greeting, is a privileged form of knowledge. It is in Beholding and Greeting, for example, that it is easiest to tell whether or not a dog is being honest—it takes an enormous amount of expertise to know this about an animal with whom you are not actually working, and the greatest experts can be wrong, which is why the horse trainer sometimes must get in the saddle in order to find out what is going on with a particular horse and rider.

I have said in various ways elsewhere that this kind of work cannot occur unless deliberately or unwittingly mechanomorphic ideas about the dog are kept at bay. They are in a bottle at the back of the cupboard; we take the bottle out when we are in trouble, as Auden, I believe, said in an earlier era about socialism.

For science until fairly recently, and still for many scientists, it is the "I-Thou" relation that is the bottle kept in the back of the cupboard and rarely looked at, in part because psychology and ethology have tasks that are hard enough without that complication. And even among scientists who build their work around an "I-Thou," the "I-it" in science is, as Santayana had it skepticism was, a chastity of the intellect and a necessary condition of science. But it is profligacy and self-indulgence to suppose that knowledge stops at the perimeters of the study. Many scientists have begun to look at the "I-Thou" and devise their studies and thinking around it, but even the signing chimps and Alex the talking (and not just parroting) parrot are not reported as having the ability to go for a walk with the scientist off lead—at liberty. I do not mean that there is no such capacity developed in or even built on in psychology or ethology, but rather that that capacity is by and large not now accommodated by the discourses in which science is reported.

What Socrates called "forms" in accounting for the slave boy's ability I am tempted to call *intelligence*, not by way of making any particular claims about a dog's, a shrimp's, a parrot's I.Q., but rather by way of reaching for terms and notions that unbind thought from the linear concepts of high and low intelligence, the concept of I.Q.—not to erase those linearities, but to say that they miss by discounting some things. Sampson is not in most ways as smart as I am, and is not, again in most ways, as smart as, say, the chimpanzees in Ohio who, with Sarah Boysen in charge, are demonstrating capacities for abstract thought traditionally thought to be peculiar to *Homo sapiens*. My casual report from the field is that it looks as though Sampson can map the articles in the scent pile better than I can, despite the fact that dogs take a while to learn purely visual tasks. It is as though the exercise created routes of thought by means of which the dog's relatively huge scent brain is enabled to augment his visual powers, even to the point of being able to do something most humans would find demanding—map the scent pile.

But I keep remembering that all of this is invisible in the culture, or cultures, in which most educated as well as illiterate people have their being. So I must again bring up the issue of cruelty. There are those who perceive any interaction between humans and animals as at least "exploitation," and who perceive all forms of training as *ipso facto* and *a priori* cruel, because they are seen as interventions in nature, and so violent, in ways not entirely unlike the violence against Final Causes that Aristotle will not put up with, at least in some cases. In other cases, what is presented as concern for animals seems to be more like a sadistic fantasy, or at least a fantasy of the possibility of a human being having Christ-like powers to be kind to the entire planet.

Sampson is, among other things, a scent-discriminating dog who takes his work to heart. You cannot be fully kind to

Sampson without yourself becoming a handler knowledgeable about scent discrimination.

The idea that training, of people, cheetahs, or elephants, is *a priori* cruel, like the idea that language is *a priori* cruel, has its source in a mistake about freedom and, in particular in Western thought, a mistake about the nature of the freedom that is knowledge. We cannot go about knowing things and creatures at random anymore than we can go about marrying at random, which is to say, at will. Formality of one sort or another is necessary for knowledge to exist at all, just as grammar is required for there to be language, or even just "wordness," a term I take from Levinas and am here using to mark the places where we are uncertain whether or not to call something language proper. Wordness brings with it form, whether in a relationship with a leopard or in a poem—two cases where the status of what is happening *as* language is variously brought into question—and form repels the slave as well as the person whose imagination can accommodate only the possibility of slavery, the one who knows work only as chains, rather than as a world-making energy. We think of form as a constraint on the will, and some forms certainly are this, mere or even dreadful restraints, deadenings, but the will finds its freedom in the world—its capacity to know the world as particularity, in meeting the forms of the world with its own forms. When the forms of two creatures meet in the right ways, the result can be friendship, or sometimes art, and, in the case of animal training, it can be both at once on occasion. (When I use the word "form" I am speaking as a poet. I request that the philosophers in the audience forget their upbringings for the moment and remember music or dance or at least the forms of greeting.)

The effect of the "I-Thou" is not to guarantee the liberty of the Thou; Martin Buber was perhaps over enthusiastic about

that or, anyhow, misleading. Also, the animal and the handler
may be, often are, confused, not fully individuals, not yet
capable of being the kind of promising animal that can fully
inhabit the kind of wordness of which Sampson's scent
discrimination is an instance. There is an issue I have not
touched on—the issue of belief. In order for this work to come
into being, Martha must say "Find it!" with belief in her dog's
power and honesty before there is any particular evidence of
that power and honesty. And Sampson must make his move
toward the articles, believing in the congruence of Martha's
participation (believing, for example, that there will be one and
only one article out there) without much or any evidence that
Martha can competently refer to a scent problem. They send
and receive each other's wordness in the darkness, as though
by the word the darkness could be made light. Here, the word,
when achieved, is the faith of the deeds of both human and
animal. Here, truth and knowledge both depend on a witting
refusal not only not to tell a lie but not to live one.

"I Named Them
As They Passed":
Kinds of Animals
and Humankind

John Hollander

At a significant moment in what we are told is the one book Darwin carried always with him aboard, and on treks ashore from, *The Beagle*,[1] Adam recounts to the angel Raphael what he remembers of his life. In Book VIII of *Paradise Lost* (345–8) he recalls how God commanded him, as Lord and possessor of whatever lived in every element, to behold each bird and beast "after their kinds": God also explained that the fish were also to be understood and named, but that they could not be present for this event, being unable to "change / Their element to draw the thinner air." (The text of Genesis 2.20, "And Adam gave names to all cattle, and to the fowl of the air and to every beast of the field," omits the aquatic creatures, and Milton hastens to provide a midrashic gloss.) Adam continues:

As thus he spake each bird and beast behold
Approaching two and two, these cowering low
With blandishment, each bird stooped on his wing.
I named them as they passed, and understood
Their nature, with such knowledge God endued
My sudden apprehension . . . (349ff).

Naming, and thereby "understanding their nature," was at once the first poetry and the first zoology: they occur, in fact, before he would view or name or know, in either epistemological or sexual senses, his wife. Francis Bacon notes this emphatically, observing that "the first acts which man

41

performed in Paradise consisted of the two summary parts of knowledge; the view of the creatures and the imposition of the names" (*Advancement of Learning*, I, vi 6). For Milton, this leads to a different matter entirely. For in the following lines Adam expresses the acute want of a name to give God. This is a subtle poetic move, and it is tempting to see in this the paradigm of a transition which would occur "later," in natural human history, with animistic or totemic idolatry becoming implicitly superseded, and nature itself giving rise to a realm that transcends it.

But Adam's Paradise was not yet a state of nature: the natural world only came into being with the fall. The first human signs of this new condition are Adam and Eve's feelings of sexual guilt and of being shamefully naked. Among the animals, the first predatory actions are seen: eagle and lion assert their new figurative reign over a kingdom of blood and pursue their prey. This implies that in the unfallen world, neither humans nor beasts were carnivores (XI, 185ff). It also seems clear that Milton feels that all the enmity among creatures—signs of the new state of nature in which Hobbes famously described the life of man as being "solitary, poor, nasty brutish and short"—also serve as complex emblems of how all natural life is constructed by death:

> . . . we are dust,
> And thither must return and be no more.
> Why else this double object in our sight
> Of flight pursu'd in the air and on the ground
> One way the self-same hour . . . (XI, 200ff).

But in the as yet unfallen world, neither animals nor humans need to be hunter-gatherers; there is no labor, of the hand or of childbearing, and the relation of beings to ambience is seamless.

The animals are observed in another mode of activity, giving pleasure to themselves and to the humans who have named and understood them:

. . . About them [Adam and Eve] frisking played
All beasts of th' earth, since wild, and of all chase
In wood or wilderness, forest or den;
Sporting the lion ramped, and in his paw
Dandl'd the kid; bears, tigers, ounces, pards
Gambol'd before them, th'unwieldy elephant
To make them mirth us'd all his might, and wreath'd
His lithe proboscis . . .

And indeed, this calls to mind a passage on the beauty of animals from Edmund Burke's *A Philosophical Enquiry into Our Ideas of the Sublime and Beautiful* (Part I Section X): " . .when other animals give us a sense of joy and pleasure in beholding them (and there are many that do so), they inspire us with sentiments of tenderness and affection toward their persons; we like to have them near us, and we enter into a kind of relation with them, unless we have strong reasons for the contrary." This formulation seems haunted by Milton's vision of the peaceable kingdom. And yet in the next sentence, Burke speaks for the fallen world, from a sense of almost Darwinian functionalism:

But to what end, in so many cases, this was designed, I am unable to discover; for I see no greater reasons for a connection between man and several animals who are attired in so engaging a manner, than between him and others who entirely want this attraction, or possess it in a far weaker degree. But it is probable, that providence did not make even this distinction, but with a view to some great end, though we cannot perceive distinctly what it is, as his wisdom is not our wisdom, and our ways not his ways.

In Paradise, man rules over the animals almost as *primus inter pares*: there is no agriculture, no livestock, no use to which living or dead animals are put, whether material, religious, or even imaginative. And significantly, there is no economic or culturally established distinction between wild and domestic creatures: the very need to draw one only comes into being with the Fall. What then, might we ask, of Adam's beholding the animals "after their kinds" and naming them as a result of his beholding? What would unfallen naming, and unfallen kinds, be like?

Let us consider the matter of naming first. Philo of Alexandria, the first-century Platonist scriptural commentator, remarked that when Adam called out the name of an animal—say, the cow—for the first time, the cow responded to it as if she had known it all her life (Philo, 1929, I, pp. 116–19). As if, in other words, the name were part of her nature and its form a necessary one. By this Philo meant, among other things, to suggest that the first act of naming was not the arbitrary sort we do in the natural world, in which a name, whether conventional or ad hoc, is *applied* to something rather than having been somehow elicited by its very character. (Indeed, a rabbinic account of the same passage in Genesis maintains that the angels were ordained to name the newly created animals, "but they were not equal to the task. Adam, however, spoke without hesitation: 'O Lord of the World! The proper name for this animal is ox, for this one horse, for this one lion, for this one camel.' And he called all in turn by name, suiting the name to the peculiarity of the animal" (Ginzberg, 1909, I, p. 61–2).

Today we might rather want to remark on another peculiarity of the naming of animals in Eden: the act is both a personal and an implicitly taxonomic one, in that the fauna in Paradise are unique specimens, the exemplars and templates for the species into which they are fruitfully to multiply. Adam's names may have been personal ones—"Here, ox!" "Nice horse!" and so on—but they were necessarily taxonomic in that there was at first only one pair of everything, and there was to be no subsequent renaming until, presumably, after the second fall, that of the Tower of Babel and the fracturing of a monolithic original tongue into human languages. Once in the natural world, we see proper names given to domestic animals or indeed to gods, whether animal-like, anthropomorphic— but often with emblematic or familiar animals accompanying them—or, as in Genesis, abstract. But species names are given, as will be later observed, as proper names to humans in many cultures.

Names can individuate, define, label or tag, interpret, classify, and so forth. Humans know their proper names; dogs and some horses recognize theirs; cats may appear to but not in the way dogs do—but this is part of the complex uncanniness of feline nature which is too complex to go into here. And certainly the role of assigning proper names to domestic or otherwise familiar animals is an act of crucial importance in constructing a unique kind of working relationship, even if the animal's work is simply to be there and to be itself with a particularly gratifying beauty. (I leave it to Vicki Hearne to deal with the importance of the relation between a human and a working dog as a paradigmatic one.) But it is the ways in which general names can invoke or construct systems of kind to which we might now turn.

Many readers of literature, and—because of how much is made of it by Michel Foucault at the very opening of *Les Mots et les choses* —a good number of younger professional scholars, are acquainted with Jorge Luis Borges' citation of a grotesque taxonomy of what we would call the animal kingdom, purportedly from "a certain Chinese encyclopedia" (Borges, 1965, pp. 106–10; Foucault, 1973, p. xv). It gambols about among classes of species, genera of phylum, families of order, and so forth, but perhaps a crucial matter of what makes it so funny is the very rhythm of its narrative unfolding, and the way each accruing category seems to mock the very mode of category of the preceding ones. You may remember that the catalogue of classifications of animals goes as follows: "(a) those that belong to the Emperor, (b) embalmed ones, (c) those that are trained, (d) suckling pigs, (e) mermaids, (f) fabulous ones, (g) stray dogs, (h) those that are included in this classification, (I) those that tremble as if they were mad, (j) innumerable ones, (k) those drawn with a very fine camel's hair brush (l) others, (m) those that have just broken a flower vase (n) those that resemble flies from a distance."

The beauty of this play of classifying aside, another interesting question about kinds of kinds emerges, namely that of what kinds are for, anyway? Where we have a choice of assigning creatures to kind, we find taxonomies arising ad hoc to some pragmatic occasion. A revealing parallel to the Borgesian citation is provided by the old anecdote of the British railway baggage-clerk puzzling over the freight rates for a creature unlisted in his schedule: "Dogs is dogs and cats is dogs, and squirrels in cages is parrots, but this here turkle is a hinsect. No charge."[3] And we are reminded that the hilarious and beautiful uncanniness of the list in the "certain Chinese encyclopedia" is the secondary puzzle that unfolds just beneath the surface of the primary one: not just "what sorts of kinds" govern the construction of this list, but also "for what possible purpose could these sorts of kind be conjured up and deployed?" (Is this pragmatic consideration characteristically absent from Foucault's discussion?) Or I recall E.O. Wilson remarking casually—and here it would be interesting to try to reconstruct the context, which I cannot remember—"I guess you could say that an ant is a small social wasp without wings." (With respect to *what* did he want to observe that you could say it?)

But one can find fantastic taxonomy almost everywhere. A rabbinic commentary on Genesis 1.20 has it that one Quintus, a Roman general, asked Rabbi Johannan ben Zakkai: "One verse speaks of fowl as having been created out of the waters of the sea [let the waters bring forth abundantly . . .fowl that may fly above the earth]. But another verse speaks of them as having been created out of the earth [out of the ground the Lord formed every fowl of the air—2.19]" Johannan ben Zakkai answered: "Both are true; fowl were created out of alluvial mud." On which Rabbi Samuel of Cappadocia commented: "Just the same, fowl are related to fish, for the skin of chickens' feet resemble the scale-covered skin of fish [and so fowl are considered fishlike."[4] While arguments like this may have left their traces in Sephardic Jewish dietary laws that ruled fowl not to be meat, as it was considered elsewhere, their remarkable improvisatory character, and the free play of analogy, can often seem like unintentional parody of biological taxonomies. Intentional parodies and anomalies, of course, begin to appear in poetry early on.

From the point of view of our own conceptual and linguistic tradition, the very fabric of other languages seems like a kind of poetry. One language will romance another's idioms and modes (for example, Pound on Chinese pictograms and so on, even though we could apply this estrangement to ourselves diachronically). Even such nuances of usage as different prepositional idioms in somewhat related languages might be seen as performing as much of that contingent segmentation of nature as taxonomic vocabularies of any sort (not having a generic word for what we would call "ice" and other such sensational examples. R. Bulmer's fascinating essay, "Why Is the Cassowary Not a Bird"(1973) (at least, for the taxonomi-cally-marked language of the Karam of New Guinea), is only one of a host of explorations of the ways in which the animal realm is so variously and so remarkably mapped.

Closer to our own linguistic home is the kindergarten story of language and culture that my generation learned from reading Scott's *Ivanhoe.* Consider the following pairs of English words: pig or swine/pork; sheep/mutton; calf/veal; ox or cow/beef—there is a total alignment of Norman French with butchered meat as opposed to Anglo-Saxon with animal on the hoof, and a record of historical cultural hegemony (who got to eat meat, who did not but provided it) was presumably inscribed in that pairing. (The ways in which these linguistic usages tempt us to construct models of cultural consciousness, one the one hand, and to model their basis in material practice, on the other, remains, of course, a central concern of various social sciences.)

The issue of names of and for animals, and the ways in which they may or may not embody analyses of kind, might show up in a consideration of two versions of the topic, naively put, of "animals in the bible." The first would involve that fascinating relation between taxonomy and linguistics, scholarship which allows us to observe that in the entire Hebrew bible there seems to be only one general term—*dagah* —for "fish," but eleven different species of owl identified by different names.[5] And, incidentally, no cats at all (it is thought that the mongoose was used against snakes and mice in its place.) This might be like a literary language in English that, in various texts composed over many centuries, had only the general term "bug" for all insects.

The second question of "animals in the bible" engages a secondary taxonomy, a purely textual one. It would concern the literary uses to which animals were put, and the imaginative, poetic, or narrative constructions of creatures we find in all subsequent literary fiction. Most literally, domestic animals are naturalistically chronicled for their economic status; predatory and passive characteristics of other creatures are given ad hoc allegorizations in their nonce symbolic uses—such as in the associations of animals with the tribes of Israel in Genesis 49 or the use of animal metaphors in prophetic poetry or in the parables and proverbs of wisdom-literature. Probably unique in the Hebrew Bible is the wonderful fable of Balaam's ass (Numbers 22), who with characteristic asinine stubbornness shied away three times from an angel—invisible to her rider—that she saw blocking her path. Balaam beats her until, suddenly being given the gift of speech, she protests reasonably against the injustice of this; when he acknowledges it, he is able to see the messenger of the Lord who gives him a crucial mandate. The ass did all she could without the power of speech, and that was given her when her prudence was violently misconstrued.

I am not a zoologist, nor a philosopher or historian of science, and I cannot speak in this case either for theories of ordering or for the many-structured profusion of living creatures. But as a poet I would like to talk about imaginative constructions of the animal kingdom (to call it that, of course, implies either the rule of man, or, if the lion is the king of the beasts, he is viceregent). These constructions could be ranged along a spectrum of fictionality, as it were, from the greatest to the least acknowledgment of natural fact. (Would some reductive, post-Linnaean description framed as a genetic code be a contemporary extreme? After all, Linnaean names, for ages so central to the development of descriptive biology and more, may now seem to be quaint relics of an older heroic age—the inscription of a discoverer's name to mark a taxonomic difference—as opposed to an easily digitalized code.)

Homer's enchantress, Circe, turned men into the beasts who served as conventional emblems of their behavior (Odysseus' men were turned into pigs, and he narrowly escaped becoming the fox that he—Homer calls him *polytropon*, "of many turns"—was so like). I have remarked elsewhere how in this regard she seems to be the first satirist, and that in any case she lays the groundwork for the more homely beast-fables in prose traditionally assigned to Aesop, of which more in a moment. The animals in Ovid's great poem about changes of shape, *The Metamorphoses*, are frequently forms taken by the Olympian Gods for their rapes of humans, a strange form of bestiality. (Satan, in *Paradise Lost*, incarnating himself in the serpent is made by Milton to be a prototype of all these, as well as a daemonic parody of the Christian incarnation.) On the other hand, Ovidian victims frequently become transformed into plants in his stories, as do the escaped objects of direct erotic pursuit (Daphne and Syrinx, for example). There are also animal-human transformations in creatures of popular folk-lore, like werewolves, cat-people, vampires—the result of internal enchantment—or fairly-tale frog-princes, and so on—the result of external enchantment. But Ovid has a poetic and satiric agenda based on marvelous particularity: Arachne was a superbly talented weaver whose skill was so great that the goddess Minerva, angered, destroyed her web; Arachne hanged herself by a strong thread and became metamorphosed into a spider.[6] His stories partake of aetiological fable of the sort that is always revisionary, first of prior fable and, later in history, of aetiologies squaring with some scientific model.

Hegel in *The Phenomenology of Mind* talks about a stage in the history of consciousness represented by deities, such as Egyptian ones, with animal heads on human bodies, a sort of allegory of the blending of animal form with the higher symbolizing faculties of human thought (Hegel, 1967, p. 706). But this is also an excellent metaphor for the constructions of much of the animal kingdom by human knowledge and writing and visual representation before the paradigm of scientific zoology distinguished between lore (that word originally means "learning," by the way) and fact. The "lewd" (in the older sense of those without "lore") probably had access to a great deal of fact—in regard to domestic animals, for example, livestock, predators, all the creatures they lived and worked among—and pre-eighteenth-century illiteracy might frequently guarantee a higher level of empirical knowledge of animals and their ways. The world of writing and organized knowledge, so often mediated and indirect, consisting of so much hearsay before the dawn of the criterion of evidence, continually mythologized living creatures. The medieval bestiaries in the tradition of the allegorizing *Physiologus*, for example, interpreted the natures of actual and purely mythical creatures in the same way and with the same authority.[7] A web of textual hearsay conflated beasts that you had never seen but knew of, and those that nobody ever could see. And, of course, literary fictions about animals do not have to compete with canonical biological knowledge about them until the end of the seventeenth century.

In the history of many religious fictions, animals move from being Gods to being chimerical half-personed gods to residing as ancillary to particular gods and then to attributes or emblems. Then there are mythological composite animals, and the point in mythographic history when composites like the Chimera and Pegasus form a fable of truth and falsehood (the triune monster, lies; the winged horse, poetry—itself far from literal, from zoologically normal—but you need to fly above the ground with Pegasus in order to kill falsehood!). There are so many different pre- or non-biological kinds of zoological taxonomy: those assigned, as in the passage in Genesis we started with, to the elements of air, earth, and water (thus, birds, beasts, and sea-creatures); creatures grouped by scale with respect to the human module; wild, tame, domestic. Among the wild, predator and prey, with human persons balanced between these; malignant and benign. Among the domestic, livestock—cattle, sheep, goats, horses, swine, fowl, bees, and so forth—working and companionable creatures; or—in various social conditions in various cultures—outdoor versus indoor. Or milch cows versus steers.

But like Hegel's Egyptian statues, even the most natural of creatures can form a spooky composite with its mythical identity. Consider three groupings of creatures about which much has been said and written (for convenience sake, I shall keep to avian creatures here). They comprise: Class I: Phoenix, Roc, Tsits; Class II: Penguin, Oriole, Osprey—and we have no problem in differentiating the zoological realm here from the purely fictional one. But then consider the third grouping (whose membership I extend beyond a token three: Class III: Pelican, Eagle, Sparrow, Owl, Nightingale, Skylark (and, in the United States as a deliberately contrived analogue of these last two, Mockingbird). These are purely natural creatures which have, in the history of our texts and discourse over nearly three millennia, doubled as allegories. They inherit conventional readings derived from pure fiction, from natural fact, and from a blend of both. For example, the Pelican was considered in the bestiaries to be a type of Christ because it was falsely believed that she would stab her own breast to feed her young, thus becoming a type of self-sacrifice. On the other hand, the nightingale does indeed sing most beautifully out of the darkness. The skylark sings uncannily, almost invisibly, from such a height that it is hard to discern so small a speck as the source of so much song. The first comes to our literature trailing clouds of Ovid's remarkable story of the raped and silenced Philomela, who regains her voice only after metamorphosis into a bird. The second, in a different context, is thought of as singing "at heaven's gate." Both become surrogates, in the poetry of the renaissance and after, for poetry itself.

But this poetic taxonomy might extend further. Class I—the purely mythical beasts—might be broken down into orders, families, and genera. An ordering of fabulous creatures might start with a classification into simple and composite. Basilisks— Chimera (bad), Pegasus (good) (is it because a winged horse is not a riddling composite but simply an adapted horse, like a unicorn?) The Hippocampus, a mermaid-like conjunction of horse and fishtail (simply one of Neptune's steeds and unallegorized at all). On the other hand, there are what we might call recombinant composites, like the Hippogriff of Ariosto's *Orlando Furioso* —sired by a Griffin—itself a mythical composite, part lion part eagle—upon an actual filly, but in another sense a revisionary descendent of Pegasus, but associated with love instead of with poetry. Yet again, the occasion of this very discussion might call for a crucial distinction, in the order of the mythical composites, between the animal composites and those that are part animal, part human.

There is the amusing side-issue here of the occasions on which mythology eerily seems to anticipate scientific zoology, for example, in the case of the musician Arion, cast overboard like Jonah from a ship but rescued by a dolphin who was charmed by his playing; this is influenced by a mythological tradition by which dolphins were held to have been men originally: as the third-century Hellenistic poet Oppian put it (as here translated in the eighteenth century by William Diaper):

> Dolphins were men (Tradition hands the Tale)
> Laborious Swains bred on the *Tuscan* Vale:
> Transform'd by Bacchus, and by Neptune lov'd
> They all the Pleasures of the Deep improv'd
> (Diaper, 1951, p. 162).

Modern oceanography has itself become our new Arion.

For Western poetry, there co-exist widely differing modes of signification in identifying people with animals. These can range from tribal totemic animals or characterological designations to traditions of animal-names totally devoid of significance. The most obvious instance is the case of modern European surnames, for which common usage implicitly contracts to pretend that (and I shall only use English ones here) proper names like Trout, Pike, Roach, Tench, Salmon, and Bass are only homonyms of generic names of fish, and that to pun on them is crude. This has resulted from a wearing away, over time, of the signified animal (and, of course, a foreign name like Hecht will not even be recognized as that of a pike in the first place). More intriguing is the apparent practice in Gothic of naming children with a pair of animal names, devoid of any emblematic significance and seemingly random in their conjunction: an amusing instance is that of a girl's name, hroslind (hros + lind, *horse + snake*), which got charmingly botanical when in Spain a romance language construed it as Rosalinda, or pretty rose.[8] But animal names can also conjoin humans and beasts in a metaphorically potent way, and fictions can manipulate these in almost Aesopian fashion. The actual characters of animal fables themselves are caricatures not of beasts, but of the way characteristic vices lead humans to behave; nevertheless, they make pointed use of physical and ethological particularities of the animal species involved and are compounded equally of fictive wit and the results of careful observation. Nicholas Howe, in *Fabling Beasts:*

Traces in Memory (1995), has many enlightening things to say in the matter of beast-fable. I shall only observe here that what is crucial about the Aesopian creatures is that they talk. Which brings up the question of what enabled Adam to do the naming.

We name animals, but if any of them name us—dolphins or gorillas, perhaps—the system has yet to be represented. The linguistic silence of the so-called dumb beasts has also been explained as post-lapsarian, which is to say, natural. The apocryphal Book of Jubilees (111, 8) proclaims that on the day of Adam's expulsion from Paradise, "the mouths of all beasts were closed . . . so that they could no longer speak. For they had all spoken one with another with one lip and with one tongue." It is as if, for all the animals, the human fall constituted their acutely undeserved Tower of Babel. The only two animals in the Hebrew Bible who speak are the serpent in Paradise and Balaam's ass, mentioned earlier.

When human fictions cause animals to speak, they can question our own moral sensibilities in human society, and they can question our sensibilities with respect to animals themselves—and perhaps also our previous constructions of and attitudes toward their natures: romantic, naturalistic-cum-visionary, modern, and so forth. Ovidian metamorphs may speak in Renaissance poetry as people, or as unique creatures (whether the first of their kind or merely exemplary of it). But only in the later nineteenth and twentieth centuries do poems start to get non-mythological first-person animal speakers, in which the persona of the lyric will be the bird or beast speaking of and for itself. And only then do encounters with animals in poems start to become moral showdowns, such as are seen in some of D.H. Lawrence's great poems from the volume called *Birds, Beasts and Flowers*, for example.

Medieval fable has its heroes learning the language of animals and birds. Conversely, Virgilian pastoral poetry has, in his first eclogue, the poet inscribing meaningful sound in natural noise, playing on his pipe and teaching the woods to echo the name of his girlfriend, Amaryllis (Virgil, Ecologue I, 4–5). Robert Frost's brilliant meditation on Milton's paradise in a sonnet called "Never Again Would Birds' Song Be the Same" has it that it was Eve's voice which imprinted the illusion of discourse—"Her tone of meaning, but without the words" on avian calls. Swift's great dialectic of the human and the humane, Houyhnhms and Yahoos: our animal and rational parts reversed—but a hint that too long a look outside of the Platonic cave of our own nature will make us inhumane in yet another way.

There is a wonderful revisionary return of all this in the great tutelary animal fiction, at least in English. I am thinking not only of the tutelary animals in both *Alice in Wonderland* and *Through the Looking-Glass* (and over a century of subsequent children's literature keeping alive the conventions of animal discourse) but even more of Kipling and T.H. White. In the *Jungle Book*, Mowgli, the *enfant sauvage* is raised by the wolf pack—a highly structured society—and tutored by others— panther, bear, python. In the great story called "Kaa's Hunting," a crucial dialectic distinction is drawn which reverses a popular taxonomic cliché: the *Bandar-log* who abduct Mowgli are the monkeys, primates, and thus closer to humans: but the closeness is only to the worst part of our nature—greed, irresponsibility, air-headedness, disorder. And it is the far more distant wolves who are metaphorically citizens, and the monkeys who constitute the grotesque metonymy, or the satiric distortion. This perhaps goes back to the Swiftian dialectic that reverses Plato's passionate horse and rational rider in persons of the Houyhnhms and Yahoos mentioned earlier. The remarkable education in the theory and practice of appropriateness and decency that graces the boyhood of the once and future king in T.H. White's great *The Sword in the Stone* entails Merlin, the young Arthur's donnish tutor, turning him into a series of creatures—a tench in the castle moat, a hawk in the mews, an owl, among others—in a complex reversal of the usual post-Aesopian tradition.

There is no time here to explore the world of animals who talk in fiction and, in particular, those who tell their own stories, like E.T.A. Hoffmann's wonderful *Kater Murr*, a tom-cat who scribbles his autobiography on the back of proof-sheets of a novel about the life of the musician Kreisler; Hoffmann's fiction is that a printer, not reading carefully, assembled a continuous intercalated narrative in which fragments of Murr's fly-on-the-wall view of uncertain events is apparently randomly interlarded with the tale of Kreisler. If Murr is perhaps the most inventive animal writer in the history of our literature it is because of Hoffmann's brilliant narrative invention which itself satirically revises and implicitly analyzes prior literary traditions of talking animals and beasts-as-humans and humans-as-beasts. Or there is Buck, the hero of Jack London's *The Call of the Wild* who, although he is not the narrator, has his consciousness and motives and inferences spoken for and of by an ordinary omniscient narrator. And I cannot resist mentioning the heartbreakingly beautiful and too little known story by James Agee called "A Mother's Tale," containing a marvelously framed miniature epic of a steer escaping from a slaughterhouse to journey westward again and tell the herd of what surely awaits them, narrated by a cow to a number of calves.

Even in earlier poetry we can see arising a whole tradition of poetic play with our ambivalence about household pets, including conventionalizing a cat/dog enmity that in domestic practice is frequently not there; and a little epigram in the *Greek Anthology* by Agathias of Byzantium that bemoans the death of a poet partridge at the claws of—guess who—the cat is the precursor of a considerable number of such texts over the years, whether of cat and bird (for example, Emily Dickinson's acute observations in "She sees a bird—she chuckles . . .") or Gray's famous archly moralizing verses on the favorite cat drowned in a tub of goldfish. In this tradition, which ends up with the dreadful cat-flattening Tom and Jerry cartoons, it is always the feline who is at fault. Another more interesting contemporary agenda has been raised by what I might call the contingent domesticity of television nature-film narratives. The remarkable technical skill involved in the making of these, and the mostly dignified reverence in which they seem to be framed, help to generate traps for our sentimentalities. We follow an animal mother or pair, watch the birth, growth, development, and education of their young, all speeded up across months of camera work and fictionalizing editing. The creatures are unceremoniously named by the narration, which under the complex visual circumstances makes them more familiar than any zoo-label name can do to its unfortunate bearer. But then, intercalated, is another *Bildungsroman*, that of a family, with its equally adorable and baptized young, of the first family's natural predators. And then we are shown the convergence of the twain, both of whom have been brought home to us. Many viewers look away as the maturing young Akbar chases and devours the graceful but unescaping Bonnie; but the moral of the story may really be that you do not deserve to cuddle up to the early spectacles unless you know the final unpleasantness is implicit in them and must not be denied.

Perhaps it takes the most profoundly imaginative of human faculties to scrutinize the very limits of those faculties themselves. With respect to our constructions of quasi-personhood for animals, we must always know that they are our constructions. As for animals, we can acknowledge that they must generate their own unrationalized taxonomies: the carnivores among them, for example, distinguish among wild and domestic prey, enemies, a few friends and specific signifiers, and what must be simply lumped as mobile landscape. But we also—with great complexity—acknowledge that our own taxonomies of all sorts and our own readings of the nature and behavior and variously experienced presence of all the animals continually breeds new modes of allegory, and moral fable, and mythopoeic representation.

With so many phases of poetic history interlayered with

zoological interpretation, our continuing process of assessment of our relations to the other creatures has re-emerged for different reasons. But we keep coming back to new versions of Milton's sense of the fact that in our fall the animal world was, through no fault of its own, inevitably engaged and condemned not only to be eaten, and kept, and put to work, and otherwise used, but to be studied and wondered at and contemplated and, if ever possible, understood. And yet at the same time we know that there will probably be thousands of creatures that—short of a total planetary demolition—will outlive us and our shortcomings, and who will evolve out of what we have done to ourselves and to their world. Animals may be thought of as sensing when they are near death, perhaps; but they do not know, as we do, that they will die. But perhaps these lines by the Scottish poet Edwin Muir put the matter best, with their glancing allusion to a Heideggerian formulation of human being as having been thrown onto it, and with their suggestion that whatever memories the beasts have, it is not like ours. I shall conclude with them. "The Animals" is from a book published in 1965 (pp. 207–8):

They do not live in the world,
Are not in time and space.
From birth to death hurled
No word do they have, not one
To plant a foot upon,
Were never in any place.

For with names the world was called
Out of the empty air,
With names was built and walled
Line and circle and square,
Dust and emerald;
Snatched from deceiving death
By the articulate breath.

But these have never trod
Twice the familiar track,
Never never turned back
Into the memoried day.

All is new and near
In the unchanging Here
Of the fifth great day of God,
That shall remain the same,
Never shall pass away.

On the sixth day we came.[9]

Notes

[1] He had a pocket edition. He seems to have taken Lyell's *Geology*, the Bible, and Humboldt. See Desmond and Moore, 1991, p. 129; Moorehead, 1969, p. 39.

[2] See also Philo, 1953 (Questions and Answers), Supplement I, 12–13.

[3] Actually, from a *Punch* cartoon of 1869 titled "Zoology," although my source (and that of others who remember the quotation but not where they read it) gives the line in the slightly but significantly different form which I quote. (In *Punch*: "Dogs is dogs and Cats is 'dogs' and rabbits is 'dogs' and so's parrots; but this 'ere 'tortis' is a insect, so there ain't no charge for it." See Williams, 1955, p. 116.)

[4] Quoted in Bialik and Ravnitzky, 1992, p. 12, where the rabbinic sources are given.

[5] Speciation [?] in the textual world of encyclopedias is an interesting matter. In Conrad Gesner's important *Historia Animalium* of the 1560's, there are twenty-odd varieties of dog, for example, more than of any other species.

[6] The story of Arachne is told in *Metamorphoses*, VI, 5–145.

[7] Even Gesner's progressive zoological agenda, which causes him to demystify a "supposed beast" (I quote from Edward Topsell's translation of a century later) like the Cacus, allows him to discuss the unicorn at some length, as well as a number of other mythical human-animal composites, many of which he discusses under a general class of apes.

[8] See the discussion by Withycombe, 1977, p. 257.

[9] From *Collected Poems* by Edwin Muir. Copyright © 1960 by Willa Muir. Reprinted by permission of Oxford University Press, Inc., and Faber and Faber Ltd.

References

Bialik, Hayim Nachman and Ravnitzky, Yehoshue Hana, eds., *The Book of Legends (Sefer Ha-Aggadah)*, reprinted by William G. Braude (New York: Schocken Books, 1992).

Borges, Jorge Luis, "The Analytical Language of John Wilkins," in *Other Inquisitions*, Ruth L.C. Simms, trans. (New York: Simon and Schuster, 1965).

Bulmer, R., "Why Is the Cassowary Not a Bird? A Problem of Zoological Taxonomy among the Karam of the New Guinea Highlands," in Mary Douglas, ed., *Rules and Meanings* (London and New York: Penguin Education, 1973), pp. 167–93.

Desmond, Adrian and Moore, James, *Darwin* (London: Michael Joseph, 1991).

Diaper, William, *Oppian's Helieuticks (1722)*, I, 1093–96, in *Complete Works*, Dorothy Broughton, ed. (Cambridge, MA: Harvard University Press, 1951).

Foucault, Michel, *The Order of Things*, anon. trans. (New York: Vintage Books, 1973).

Ginzberg, Louis, *The Legends of the Jews*, Henrietta Szold, trans. (Philadelphia: The Jewish Publications Society of America, 1909).

Hegel, G.W.F., *The Phenomenology of Mind*, J.B. Baillie, trans. (New York: Harper and Row, 1967).

Howe, Nicholas, "Fabling Beasts: Traces in Memory," *Social Research* 62:3 (Fall 1995).

Moorehead, Alan, *Darwin and the Beagle* (New York: Harper and Row, 1969).

Muir, Edwin, *Collected Poems* (New York: Oxford University Press, 1965). (Originally appeared in *One Foot in Eden*.)

Philo, *On the Account of the World's Creation as Given by Moses*, 148–150, G.H. Whitaker, trans., Loeb Edition, (Cambridge, MA: Harvard University Press, 1929).

Philo, *Questions and Answers on Genesis*, I, 20–22, Ralph Marcus, trans., Loeb Edition (Cambridge, MA: Harvard University Press, 1953).

Withycombe, E.G., *The Oxford Dictionary of English Christian Names, 3/e* (Oxford: Oxford University Press, 1977).

HISTORIES

Introduction

In his parable *Penguin Island* ([1908]1931), Anatole France relates how the old, blind monk Saint Mael inadvertently baptized a group of penguins, mistaking them for human beings. When the news reached heaven, it caused, so we are told, neither joy nor sorrow but extreme surprise. The Lord himself was embarrassed. He gathered an assembly of clerics and doctors, and asked them for an opinion on the delicate question of whether the birds must now be given souls. It was a matter of more than theoretical importance. "The Christian state," Saint Cornelius observed, "is not without serious inconveniences for a penguin. . . . The habits of birds are, in many points, contrary to the commandments of the Church." Who could deny that a penguin, once burdened with a soul, might get into unforseen—and undeserved—difficulties? After lengthy debate, the elders in heaven settled on a compromise. The baptized penguins were indeed to be granted souls, but, on Saint Catherine's recommendation, their souls were to be "of small size."

This section of the conference continues the discussion begun in the first section of the shifting boundaries between human and animal beings. Where shall we draw the line between ourselves and them? Who has a soul, and who has not? Who is in, and who is out? Who possesses those paradigmatically human qualities of consciousness, intelligence, language, free will, dignity, moral status, legal rights, and so on?

In most cultures, for most of human history, people's solution has been to opt—like Saint Catherine—for some less than absolute answer. While recognizing that no non-human animal can be in every way human, people have readily granted that particular animals may be more or less human in particular respects. The lion at times can be kingly, the fox can be sly, the ant industrious, the dog faithful. And equally—since the analogy runs both ways—a man or woman can be lion-like, or monkeyish, or foxy, or dogged. For Shakespeare, Mark Antony's "delights were dolphin-like, they show'd his back above the element they lived in" (*Antony and Cleopatra*, Act V, Scene ii).

This kind of recognition of a similarity that falls short of identity has not only been descriptively convenient and poetically creative, it arguably has been philosophically and scientifically sound. To the question, "Are animals people, and should they, therefore, be granted the respect and honor due from us to others of the human family," the correct answer, biologically and metaphysically, is, obviously, "No." To the question, "Do particular animals have human sides to them, and should we, therefore—for our own sakes as well as theirs—treat them from time to time as honorary humans," the answer is, obviously, "Yes."

Such a solution is not, of course, a tidy one. It can mean in practice that our relationships with animals become complex and thoroughly non-linear. We can hold multiple, even seemingly contradictory, attitudes to the very same animal—we may choose to enslave, worship, consume, abuse, befriend, hunt, play games with, grieve for it—all depending on which aspects of similarity or dissimilarity we choose to emphasize. And if there seems to be an inconsistency in our having these criss-crossing relationships, perhaps it exists only in the minds of observers who think we are intellectually and morally obliged to think about and behave toward animals in all-or-none ways: either admit them to the human club or bounce them from it for all time; either let them through the pearly gates or lock them out; either give them full-blown souls or none at all.

The three papers which follow address this issue of "in" or "out" from different perspectives. Harriet Ritvo, from her deep knowledge of Victorian culture, examines how classical ideas about the human-animal dichotomy were forced reluctantly to accommodate to the emerging proofs of evolutionary kinship. Stephen Glickman's essay on the myth and reality of the spotted hyena brilliantly uses a single case history to illustrate the tragi-comic side of human attempts to impose our own cultural frames on nature. Jerrold Tannenbaum addresses the issue of human rights, reviewing the history of protective legislation and arguing boldly against absolutist attitudes on either side. Perhaps the one thing these papers have in common, besides their focus on the relativity of human culture, is the lesson that human beings—some of them—have small souls too.

Nicholas Humphrey

References

France, Anatole, *Penguin Island*, A.W. Evans, trans. (London: Franklin Watts, [1908]1931).

Border Trouble: Shifting the Line between People and Other Animals

Harriet Ritvo

T HERE is not much doubt that human beings, a category that can be taken to include members of the modern species *Homo sapiens* as well, perhaps, as members of various extinct species of the same genus, are different from non-human animals. This crude, binary differentiation seems obvious, as it also seems obvious that hippopotami are different from non-hippopotamian animals or elephants from non-proboscideans. Humans appear reasonably distinctive physically, although such perceptions tend to be relative rather than absolute; it may be salutary, for example, to remember that any attentive modern zoo-goer can easily distinguish a chimpanzee from an orangutan, but most learned eighteenth-century naturalists perceived them as Asian and African populations of a single kind of animal. On the psychological or mental level, people have their characteristic human nature as bears have their ursine nature and sheep their ovine nature. What is far from generally obvious, however, is the kind and the degree of the difference between human beings and the rest of the animal kingdom. Dramatically varying answers have been proposed and, indeed, enjoy currency at the present time. This persistent, historically-grounded diversity of opinion reflects deep disagreement not merely about zoological taxonomy (significant as that is), but about a wide range of intellectual, social, and political issues.

Taking the long chronological perspective, it may seem odd that uncertainty about this question should persist. People have been thinking about it for a very long time. Western culture, for example, offers a standard or default account of the relationship between people and other animals. As part of the biblical creation myth, that account boasts a distinguished and ancient pedigree. To quote from *Genesis*, "And god said, Let us make man in our image, after our likeness: and let them have dominion over the fish of the sea, and over the fowl of the air, and over the cattle, and over all the earth, and over every creeping thing that creepeth upon the earth" (1:26). Implicit in this brief, dense passage is an absolute separation between people and all other creatures. Animals, thus, constitute a single group in some sense equivalent to the group that includes only humans, and the significant—indeed, the only specified—distinction between the groups is that humans resemble god, and animals do not. Or, to put it another way, humans occupy a middle station between god and the rest of the animate creation.

Thus, as so often, distinction is inevitably invidious; and separation is inextricable from hierarchy. Corollary to this fundamental human/animal dichotomy is the notion that what is good in people reflects their closeness to god (or their divine nature), and what is bad in them reflects their closeness to animals (or their bestial nature). This notion too has been well integrated into western literary and philosophical traditions. In ancient writings, as well as in those of the medieval period and the renaissance, "apes" (a term which probably referred primarily to monkeys) often figured as representatives of the lower side of human nature, whether lust or drunkenness or folly or baseness in general (Rowland, 1973, pp. 8–14). The evocative power of this dichotomy has persisted into the era of modern science. To cite just one Victorian example, Alfred Tennyson assumed it in the following exhortation to moral development, taken from his long meditative poem, *In Memoriam*:

    . . . Arise and fly
The reeling Faun, the sensual feast;
Move upward, working out the beast,
And let the ape and tiger die (CXVIII.11.25–28).

This horror of the beast within has its analog in the reiterated desire to understate or to deny altogether the connection between people and other animals, especially those—the chimpanzees, orangutans, and gorillas—that bear the closest resemblance.

It may also seem odd to introduce this issue with reference to literature and religion, rather than with reference to science. After all, at least from the perspective of the late twentieth century, the job of determining the systematic position of humans seems to fall squarely within the domain of biology. And certainly scientists have had a great deal to say about the matter. Since the renaissance, scientific consensus has gradually diverged from the traditional assertion of absolute, unbridgeable separation and shifted toward acknowledging relationship—and an ever closer relationship at that. The beginnings of this counter-traditional migration can be detected at least three centuries ago. For example, in 1699, Edward Tyson, often called the father of English comparative anatomy, published a treatise entitled *Orang-Outang, sive Homo Sylvestris. Or the Anatomy of a Pygmie compared with that of a Monkey, an Ape* [by which he meant a baboon] *and a Man*. It was the first account published in English of a scientific dissection of a chimpanzee. But as is usually the case, mere chronological priority was not the most important feature of this work. Tyson stated in his preface that the purpose of his exercise in comparative anatomy was to "observe *Nature's Gradation* in the Formation of *Animal* Bodies, and the Transitions made from one to another," thus implicitly including humanity in the animal series. And not only did Tyson present people as

anatomically continuous with animals, but his choice of
terminology further implied that the categories of "human"
and "orangutan" might not be completely distinct. Both of the
synonyms for orangutan (that is, chimpanzee) mentioned in
the title conflated it with people: the translation of "Homo
Sylvestris" is "wild man of the woods," and, conversely, the
humanity of the quasi-mythical Pygmies had long been the
subject of European speculation. Even at the end of the
eighteenth century, naturalists could claim that the "race of
men of diminutive stature" or "supposed nation of pygmies"
described by the ancients was "nothing more than a species of
apes . . . that resemble us but very imperfectly" (*An Historical
. . .*, ca. 1800, III, pp. 288–89) [see Figure 1].

The celebrated eighteenth-century systematizer Carolus
Linnaeus located humanity firmly within the animal kingdom,
constructing the primate order to accommodate humans, apes,
monkeys, prosimians, and bats. In what has become the
definitive (1758) edition of his *Systema Naturae*, he included two
species within the genus *Homo*. One was *Homo sapiens*,
subdivided into (mostly) geographical subspecies or races, such
as *H. sapiens Americanus* and *H. sapiens Europaeus*, and the other
was *Homo troglodytes*, which was also known, Linnaeus pointed
out, as *Homo sylvestris Orang Outang* (Linnaeus, [1758]1956, pp.
20–4). Thus, Linnaeus grouped Tyson's chimpanzee with
people, rather than including it in the crowded genus *Simia*
with all the monkeys and all other apes. This association was
visually reinforced by a tradition of natural history illustration
that made apes appear particularly human in their anatomy
and their behavior [see Figure 2].

At least in what has proved to be its most durable form, the
evolutionary theory that became increasingly respectable
during the nineteenth century cemented the connection. That
is, if eighteenth-century classification pointed out similarities
or affinities, as they were often called, between certain kinds of
animals, such as chimpanzees and humans, by the late
nineteenth century classification explained those similarities as

Figure 1. By permission of the Houghton Library, Harvard University.

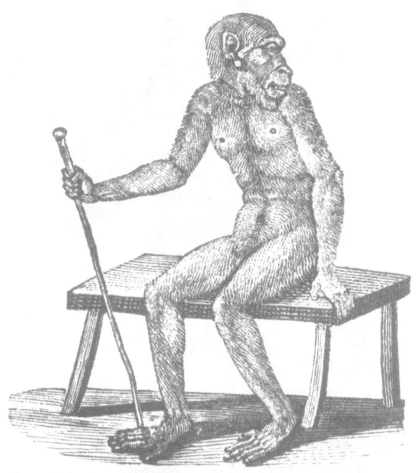

Figure 2. The Museum of Comparative Zoology, Harvard University.

the consequence of common genealogy or descent. As Charles Darwin put it in the *Descent of Man*,

> If the anthropomorphous apes be admitted to form a natural sub-group, then as man agrees with them, not only in all those characters which he possesses in common with the whole Catarhine group [old world monkeys], but in other peculiar characters, such as the absence of a tail and of callosities and in general appearance, we may infer that some ancient member of the anthropomorphous sub-group gave birth to man (Darwin, [1871]1981, I, p. 197).

The distinctive scientific developments of our own time have similarly emphasized the continuity between humans and our anthropoid relatives. DNA analyses have suggested that chimpanzees are more closely related to human beings than they are to gorillas, and that we are so similar genetically to the two recognized species of chimpanzee that, according to the biologist Jared Diamond, "in this respect as in most others, we are just a third species of chimpanzee" (Diamond, 1992, p. 23). Ethologists like Jane Goodall have observed

> the sometimes uncanny similarity between certain aspects of chimpanzee and human behavior: the long period of childhood dependency, the postures and gestures of the nonverbal communication system, the expressions of emotion, the importance of learning, the beginning of dependency on cultural tradition, and the startling resemblance of basic cognitive mechanisms (Goodall, 1986, p. 592).

More controversially, a number of researchers have claimed to have taught chimpanzees and other apes to talk, most frequently and effectively through the medium of American Sign Language.[1]

This story may seem like a model account of scientific progress, in which error yields to truth under the ineluctable pressure of fact. The impact of this accumulating evidence—incontrovertible as it seems to many people—has not been, however, uniformly persuasive. Many other people still find it problematic to consider themselves as "the third chimpanzee" or as any kind of ape at all, and a relatively restrained formulation like "people and other animals" can also seem startling, even provocative. Indeed, much less intimate connections—such as those proposed in the nineteenth or even the eighteenth century—continue to give offense in some quarters, as is most strikingly evidenced by the continuing resistance to the teaching of evolutionary theory in the public

schools. Such resistance is frequently caricatured as the result
of simple ignorance—an ostrich-like refusal to acknowledge
the intellectual realities of the contemporary world—but that
disparagement itself reflects a rather constrained sense of
chronology. Historically speaking, and even speaking of
relatively recent history, the passionate and committed refusal
to place people within the world of nature has not been
confined to the unlearned or even the unscientific.

That is, until times quite close to the present, most scientists
or naturalists were no more eager than their lay fellow citizens
to acknowledge a close kinship with apes and monkeys and,
thus, to narrow (let alone to close) the gap separating human
beings from animals. This philosophical or theological reluctance often informed and even shaped their scientific work. In
the early days of scientific classification, many naturalists
rejected Linnaeus's new primate order, despite his iconic status
as a systematizer. According to Thomas Pennant, one of the
preeminent British naturalists of the late eighteenth century,
"my vanity will not suffer me to rank mankind with *Apes*,
Monkies, *Maucaucos*, and *Bats*"(Pennant, 1793, p. iv); a
colleague similarly asserted that "we may perhaps be pardoned
for the repugnance we feel to place the monkey at the head of
the brute creation, and thus to associate him . . . with man"
(Wood, 1807, p. xvii). Other dissenters simply proposed their
own counter-taxonomies, which implicitly posited a much
wider separation. Thus, the anatomist William Lawrence
suggested that "the principles must be incorrect, which lead to
such an approximation" (that is, between humans, apes, and
monkeys in the primate order); instead, he argued that "the
peculiar characteristics of man appear to me so very strong,
that I not only deem him a distinct species, but also . . . a
separate order" (Lawrence, 1823, pp. 127, 131). This
exclusively human order was normally designated "Bimana,"
which stressed the erect posture and purpose-built feet
characteristic of people, in contrast with the four-handed apes
and monkeys of the order "Quadrumana."[2]

As evolutionary theory suggested a more concrete connection, it provoked more articulate resistance. The Lamarckian versions advocated by British zoological radicals in the 1820s and 1830s were countered by a scientific establishment self-described as conservative not only in technical matters but also in religion and politics.[3] Although not encumbered in the same way by incendiary political doctrines, Darwin's theory of evolution by natural selection was equally controversial when it appeared in 1859. Many of the first reviews of *On the Origin of Species* were vitriolic, and, as had been the case with Linnaeus's taxonomic approximation of humans and other primates, the critiques mixed scientific and extrascientific objections. For example, Adam Sedgwick, the Woodwardian Professor of Geology at the University of Cambridge and one of Darwin's early scientific mentors, anonymously expressed his "deep aversion to the theory: because of its unflinching materialism; . . . because it utterly repudiates final causes, and thereby indicates a demoralized understanding." He asserted that

> we cannot speculate on man's position in the actual world of nature, on his destinies, or on *his origin*, while we keep his highest faculties out of our sight. Strip him of these faculties, and he becomes entirely bestial; he he may well be (under such a false and narrow view) nothing better than the natural progeny of a beast, which has to live, to beget its likeness, and then die for ever" ([Sedgwick, [1860]1973, pp. 164–65).

Indeed, these two intellectual departments—that is, science and non-science—were not as apparently distinct in the Victorian period as they have subsequently become. Thus, one of the most celebrated early critics of *The Origin* was the Anglican bishop Samuel Wilberforce; his attack on Darwin's theory at the 1860 meeting of the British Association for the Advancement of Science has passed into legend.[4] Perhaps because the resistance to the human-animal connection was so broadly and variously grounded, the increasing familiarity of evolutionary ideas proved unable to dissipate or disarm it. As Darwin sadly noted at the end of *The Descent of Man*, written a decade after the appearance of *The Origin*, "The main conclusion arrived at in this work, namely that man is descended from some lowly-organised form, will, I regret to think, be highly distasteful to many persons" (Darwin, [1871]1981, II, 404).

That distaste has continued to shape scientific as well as popular formulations of the relation between humans and other animals, for which the relation betweens humans and other apes has remained the standard synecdoche. Although classification has long ceased to occupy the cutting edge of biological inquiry—no modern Linnaeus has anything like the prestige of the eighteenth-century master—its unambiguous hierarchies and diagrams still offer the most forceful and authoritative accounts of the nature and proximity of the connections between different kinds of animals. Despite the evidence of human similarity to chimpanzees—evidence that should, according to some scientists, not only make people one among several kinds of apes but put them squarely within the genus *Pan*, along with common chimpanzees and bonobos—most textbook taxonomical accounts construct a much greater distance. They tend to group gorillas, orangutans, chimpanzees, and sometimes gibbons in a single family (usually called *Pongidae*) and to isolate humans in another family (usually called *Hominidae*), the only other occupants of which are our certified fossil forebears.[5] The split in scientific opinion is on display at the American Museum of Natural History, where a placard in the newly opened hall of fossil mammals explains that humans are actually apes, while the primate hall downstairs retains *Hominidae* and *Pongidae*. To give a rough sense of the claim implicitly made by such a taxonomic structure, it may be useful to compare the degree of difference thus proposed with that represented by family distinctions within an order that does not engage human *amour propre* so directly. Among the carnivora, for example, the degree of difference recognized by distinct families is that between dogs (*Canidae*) and cats (*Felidae*).

But the story of the relationship between humans and animals is still more complicated. If the acknowledgement of human kinship with animals cannot be used to separate the scientific sheep from the vulgar goats—that is, if many experts have persistently been guided in their interpretation of the structure of the natural world by their sense of human uniqueness and dignity—it is also true that the connection between human worth and human uniqueness does not seem to have been universally acknowledged. Further, the opinions of scientists, theologians, and other elite intellectuals have not always been shared by the public at large, especially with regard to topics that have a direct and independent appeal. Animals, and especially large exotic animals, have traditionally exerted such a general fascination. The manifest similarity between humans and other primates— especially as evidenced in such non-functional details as the shape of ears and fingernails—has been as obvious to idle curiosity-seekers as to serious naturalists, but it has not always seemed to trouble them as much. On the contrary, it is possible to discern an alternative understanding, based as much in public spectacle and popular lore as in learned analysis, that not only acknowledged the connection between people and other animals but exaggerated and even reveled in it.

Often this fascination was expressed as credulity. For example, there was apparently an enthusiastic and enduring audience for anomalous creatures who exemplified the human/animal connection in their own heterogeneous bodies. The monster literature of the renaissance, which to a great extent simply repackaged the natural history of the medieval and classical periods, was full of old standbys like satyrs and minotaurs, as well as more modest—and perhaps therefore more plausible—combinations. Thus, in his *Historia Monstrorum*, the fifteenth-century Italian naturalist Ulisse Aldrovandi described a man with rabbit ears, a man with the tail and legs of a lion, and a fish with a human face, among other marvels (1642, pp. 10, 17, 21, 25, 353). The more sensational work of Pierre Boaistuau, *Certaine secrete wonders of Nature, containing a description of sundry strange things, seming* [sic] *monstrous in our eyes and judgement, bicause* [sic] *we are not privie to the reasons of them*, included such elaborate monstrosities as a man with an ass's head, a scaly body, and limbs borrowed from a variety of other animals (1569, p. 140).

Improbable as such creatures might seem, they did not simply fade away under the glare of the enlightenment. The hopeful rationalism of Boaistuau's title, implying that the inexplicability of the phenomena he described was only apparent and, therefore, only temporary, foreshadowed the learned discussions of mermaids that continued into the nineteenth century. These compound creatures were occasionally brought to Europe from the East Indies and displayed to the interested public for a fee. That public often included rather distinguished scientists. For the most part, their scrutiny resulted in predictable denunciation; for example, in 1858, Frank Buckland analyzed a mermaid he had seen at a London public house as part monkey and part hake. But even such dismissals had a grudgingly respectful subtext, since they implicitly acknowledged that the debunking was worth a certain amount of time and effort. And some naturalists were willing to entertain the possibility that mermaids actually existed. In his *New Dictionary of Natural History*, published in 1785, the popular naturalist William Frederic Martyn judged that "there seems to be sufficient evidence to establish its reality"; more than a generation later, the quinarian taxonomists William MacLeay and William Swainson found a space in their artificial and elaborately embedded system of categories for an as-yet-unknown amphibious primate which might, as MacLeay suggested in 1829, "explain why there is such a general feeling among mankind of all ages, in favor of the existence of mermaids and may indeed render the past or present existence of amphibious primates probable" (Martyn, 1785; MacLeay, 1829). Further, some allegedly half-human/ half-animal creatures were incontestably real. For example, a Laotian girl was exhibited in 1883 as "Darwin's missing link," not only because she was unusually hairy, but because she had prehensile feet and could pout like a chimpanzee.[6]

Another kind of posited combination expressed a still more intimate connection between people and other animals. Hybridization conventionally (although always problemati- cally) has been taken to indicate identity of species. Perhaps for that reason, the possibility of crossing distinct animal kinds— and thus in some measure usurping the divine creative role—has long fascinated people. The literature of eighteenth- and nineteenth-century natural history teemed with accounts of real if unusual hybrids, such as those between lions and tigers, as well as unreal and, therefore, still more unusual hybrids, such as those between horses and deer. And humans, too, could be the objects or the originators of passions that transcended or violated the species barrier, although accounts of this kind were carefully distanced by skepticism or censure. Indeed, the most common such breaches, those involving farmyard animals, were much more likely to figure in a legal context than in works devoted to natural history (Evans, [1906]1987, pp. 144–53). Zoological literature tended to confine its attention to more remarkable events. At the end of the eighteenth century, Charles White reported that orang utans "have been known to carry off negro-boys, girls and even women . . . as objects of brutal passion" (White, 1799, p. 34); more than sixty years later the Anthropological Society republished Johann Friedrich Blumenbach's summary of travelers' accounts that "lascivious male apes attack women" who "perish miserably in the brutal embraces of their ravishers" (Blumenbach, 1865, p. 73). Authors who reported human-ape encounters were circumspect about the possibility of progeny, although the idea clearly exerted a certain unacknowledged appeal. White recorded rumors "that women have had offspring from such connection" and proposed that "supposing it to be true, it would be an object of inquiry, whether such offspring would propagate, or prove to be mules" (White, 1799, p. 34). Blumenbach, still more cautious, asserted "that such a monstrous connection has any where ever been fruitful there is no well-established instance to prove" (Blumenbach, 1865, pp. 80–1). Along the same lines, in his account of chimpanzee anatomy, Edward Tyson had gone out of his way to assure his readers that "notwithstanding our *Pygmie* does so much resemble a *Man* . . . : yet by no means do I look upon it as the Product of a *mixt* generation" (Tyson, 1699, p. 2).

If few opportunities existed to view half human/half-animal creatures, whatever their alleged origins, it became increasingly easy to observe live specimens of animals that resembled humans in many respects. Although they often succumbed quickly to cold, damp northern climates because of their vulnerability to human respiratory infections, chimpanzees and orangutans were predictable features of nineteenth-century zoos and menageries. In designing the displays, their proprietors seem to have felt no fear of offending public sensibilities by suggesting the resemblance between exhibit and audience. On the contrary, the apes were encouraged to appear as human as possible. They ate with table utensils, sipped tea from cups, and slept under blankets. One orangutan displayed in London's Exeter Change Menagerie amused herself by carefully turning the pages of an illustrated book. At the Regent's Park Zoo, a chimpanzee named Jenny regularly appeared in a flannel nightgown and robe. A chimpanzee acquired by the Earl Fitzwilliam in 1849 was reported to walk "perfectly erect" and handle "everything like a human being"; in addition, its food was "choice, and wine a favorite beverage" (Bingley, 1804, I, 45–50; Jesse, 1834; Broderip, 1847, p. 250; "Importation . . . ," 1849, p. 2379). Consul, a young chimpanzee who lived in Manchester's Belle Vue Zoological Gardens at the end of the nineteenth century was considered highly "educated"; one photograph showed him fully clothed, including a hat, with a cigarette in one hand and the other on a table next to a bottle and a glass. When he died, an admirer was moved to ask: "Then who says/Thou'rt not immortal? That not mortal knows,/Not e'en the wisest—he can but *suppose*" (Peel, 1903, pp. 205–6; "In Memory") [see Figure 3].

It is possible to extract several kinds of conclusions from this rather inconsistent trajectory (if anything that so strikingly lacks a dominant direction can be called a trajectory). One has to do with the establishment of scientific consensus and the diffusion of scientific information, both of which are more complex and murky processes than they are sometimes considered to be. Those who survey the past simply to discover the roots of present orthodoxy are apt to caricature—or even to overlook altogether—alternative perspectives that were at least equally important to contemporaries. And if it is assumed that the authority of scientists is generally acknowledged by those who share their interests but not their expertise, many cultural manifestations are apt to appear not only surprising but inexplicable.

Another kind of conclusion leads in a political direction. Although scientific classification often seems like an arcane pursuit, the assumptions it embodies can influence the concrete experience as well as the abstract ranking or description of the creatures concerned. For example, scientists who use primates in American laboratories are officially encouraged to consider their happiness as well as their physical well-being, while no spiritual standards are applied to experimental rats, frogs, or zebrafish. And although genuine concern for the welfare of animals can derive from paternalistic or condescending attitudes, as well as from respectful ones, in general, the more distant from human beings other animals

Figure 3. The Museum of Comparative Zoology, Harvard University.

seem, the less obligation or inclination people feel to recognize claims that may interfere with their needs, their interests, or even their convenience. To put it another way, classifications that emphasize perceived similarity are apt to have gentler consequences than classifications that emphasize perceived difference.

Further, the implications of taxonomic generosity or ungenerosity have not been limited to non-human animals. The binary opposition between humans and animals has been historically parallel to a series of other oppositions, confined within the human species: for example, man/woman, upper-class/lower-class, European/non-European, adult/child. In each case, the occupants of the second category have routinely been disparaged by the more powerful occupants of the first. Such disparagement has often been expressed in terms of the hierarchy of nature, which is embodied in both the enlightenment (and pre-enlightenment) chain of being metaphor and in many versions of the tree or bush metaphor that has succeeded it. Both images suggest that the marked categories of people not only metaphorically resemble animals but actually approach them more closely than do other humans. Sometimes, as is always a possibility with figuration, analogy has slipped into identity. That is, groups who are compared to animals may also be treated like them. It was, for example, a truism of Victorian sexual science that women were closer to apes than were men, and that it was, therefore, appropriate to exclude them from a range of social responsibilities and privileges (Russett, 1989, pp. 14, 30; Poovey, 1988, p. 31). The anthropologist and physician Cesare Lombroso frequently argued for the existence of innate criminality in humans on the basis of the capacity—indeed the predilection—of animals to commit both crimes of passion and premeditated offenses.[7] Indeed, in some cases, borderline humans could seem less worthy than animals. Thus, Darwin ambiguously compared and contrasted human and animal others in an attempt to reconcile his readers to their own simian origins, pointing out that "there can hardly be a doubt that we are descended from barbarians," and then declaring that "I would as soon be descended from that heroic little monkey, who braved his dreaded enemy in order to save the life of his keeper . . . as from a savage who delights to torture his enemies, offers up bloody sacrifices, practises infanticide without remorse, treats his wives like slaves, knows no decency, and is haunted by the grossest superstitions" (Darwin, [1871]1981, II, 404–5).

This may seem like an ambiguous note on which to end. But that ambiguity is true to the subject under discussion. The issues raised in dividing people from other animals have not yet been resolved, either within the institution of science or from a broader philosophical or political perspective. And at a time when people seem increasingly inclined to define rigid boundaries and then to defend them, there is still a lot at stake in this resolution.

Notes

[1] See, for example, Cavalieri and Singer, 1993, which includes essays on the linguistic abilities of all the great apes: Fouts and Fouts, 1993; Miles, 1993; and Patterson and Gordon, 1993.

[2] See, for example, Owen, 1855, p. 41.

[3] See Desmond, 1989, especially pp. 288–295.

[4] On the early reception of On the Origin of Species, see Desmond and Moore, 1991, ch. 33.

[5] See, for example,Eisenberg, 1981, p. 460.

[6] Originally in Nature, 12 May 1882, cited in Howard, 1977, pp. 56–7.

[7] See Lombroso, 1895.

References

Aldrovandi, Ulisse, Monstrorum Historia (Bononiae: Nicolai Tebaldini, 1642).

Bingley, William, Animal Biography (London: Richard Phillips, 1804).

Blumenbach, Johann Friedrich, The Anthropological Treatises . . . , Thomas Bendyshe, ed. and trans. (London: Longman, Green, Longman, Roberts and Green/The Anthropological Society, 1865).

Boaistuau, Pierre, Certaine Secrete wonders of Nature, containing a description of sundry strange things, seming [sic] monstrous in our eyes *and judgement, bicause* [sic] *we are not privie to the reasons of them,* Edward E. Fenton, trans. (London: Henry Bynneman, 1569).

Broderip, William, *Zoological Recreations* (London: Henry Colburn, 1847).

Cavalieri, Paola and Singer, Peter, eds., *The Great Ape Project: Equality Beyond Humanity* (New York: St. Martin's Press, 1993).

Darwin, Charles, *The Descent of Man, and Selection in Relation to Sex* (1871) (Princeton NJ: Princeton University Press, 1981).

Desmond, Adrian, *The Politics of Evolution: Morphology, Medicine, and Reform in Radical London* (Chicago: University of Chicago Press, 1989).

Desmond, Adrian and Moore, James, *Darwin* (London: Michael Joseph, 1991).

Diamond, Jared, *The Third Chimpanzee: The Evolution and Future of the Human Animal* (New York: HarperCollins, 1992).

Eisenberg, John E., *The Mammalian Radiations: An Analysis of Trends in Evolution, Adaptation, and Behavior* (Chicago: University of Chicago Press, 1981).

Evans, E.P., *The Criminal Prosecution and Capital Punishment of Animals: The Lost History of Europe's Animal Trials* (1906) (London: Faber and Faber, 1987).

Fouts, Roger S. and Fouts, Deborah H., "Chimpanzees' Use of Sign Language," in P. Cavalieri and P. Singer, eds., *The Great Ape Project: Equality Beyond Humanity* (St. Martin's Press, 1994), pp. 28–41.

Goodall, Jane, *The Chimpanzees of Gombe: Patterns of Behavior* (Cambridge MA: Harvard University Press, 1986).

An Historical Miscellany of the Curiosities and Rarities in Nature and Art . . . (London: Champante and Whitrow, ca. 1800).

Howard, Martin, *Victorian Grotesque: An Illustrated Excursion into Medical Curiosities, Freaks and Abnormalities—Principally of the Victorian Age* (London: Jupiter Books, 1977).

"Importation of Another Specimen of the Chimpanzee," *Zoologist* 7 (1849).

"In Memory of Consul," pamphlet in the Belle Vue Collection, Chetham's Library.

Jesse, Edward, *Gleanings in Natural History, Second Series* (London: John Murray, 1834).

Lawrence, William, *Lectures on Comparative Anatomy, Physiology, Zoology, and the Natural History of Man; delivered at the Royal College of Surgeons in the Years 1816, 1817, and 1818* (London: R. Carlile, 1823).

Linnaeus, Carolus, *Systema Naturae: Regnum Animale* (1758) (London: British Museum of Natural History, 1956).

Lombroso, Cesare, *L'Homme Criminel* (Paris: Ancienne Librairie Germer Ballière, 1895).

MacLeay, William, "Draft Classification of Mammals on Quinary Principles," Linnean Society Archives, 1829.

Martyn, William Frederic, *A New Dictionary of Natural History; or, Compleat Universal Display of Animated Nature* (London: Harrison, 1785).

Miles, H. Lyn White, "Language and the Orang-utan: The Old 'Person' of the Forest," in P. Cavalieri and P. Singer, eds., *The Great Ape Project: Equality Beyond Humanity* (St. Martin's Press, 1993), pp. 42–57.

Owen, Richard, "On the Anthropoid Apes and their relations to Man," *Proceedings of the Royal Institution of Great Britain* 2 (1855).

Patterson, Francine and Gordon, Wendy, "The Case for the Personhood of Gorillas," in P. Cavalieri and P. Singer, eds., *The Great Ape Project: Equality Beyond Humanity* (St. Martin's Press, 1993), pp. 58–77

Peel, C.V.A., *The Zoological Gardens of Europe: Their History and Chief Features* (London: F. E. Robinson, 1903).

Pennant, Thomas, *History of Quadrupeds* (London: B. and J. White, 1793).

Poovey, Mary, *Uneven Developments: The Ideological Work of Gender in Mid-Victorian England* (Chicago: University of Chicago Press, 1988).

Rowland, Beryl, *Animals with Human Faces: A Guide to Animal Symbolism* (Knoxville: University of Tennessee Press, 1973).

Russett, Cynthia Eagle, *Sexual Science: The Victorian Construction of Womanhood* (Cambridge MA: Harvard University Press, 1989).

[Sedgwick, Adam], "Objections to Mr. Darwin's Theory of the Origin of Species," *Spectator* March 14, 1860/April 7, 1860, in David L. Hull, *Darwin and His Critics: The Reception of Darwin's Theory of Evolution by the Scientific Community* (Chicago: University of Chicago Press, 1973).

Tyson, Edward, *Orang-Outang, sive Homo Sylvestris. Or, the Anatomy of a Pygmie compared with that of a Monkey, an Ape, and a Man* (London: Thomas Bennet, 1699).

White, Charles, *An Account of the Regular Gradation in Man, and in Different Animals and Vegetables; and from the Former to the Latter* (London: C. Dilly, 1799).

Wood, William, *Zoography; or the Beauties of Nature Displayed* (London: Cadell and Davies, 1807).

The Spotted Hyena from Aristotle to the Lion King: Reputation Is Everything*

Stephen E. Glickman

As several distinguished authors of the present age have undertaken to reconcile the world to the Great Man-Killer of Modern times; as Aaron Burr has found an apologist, and almost a eulogist; and as learned commentators have recently discovered that even Judas Iscariot was a true disciple, we are rather surprised to find that some one has not undertaken to render the family of Hyenas popular and amiable in the eyes of mankind. Certain it is, that few marked characters in history have suffered more from the malign inventions of prejudice (Goodrich and Winchell, 1885, p. 283).

Hyenas: To Know Them is to Love Them

T HE situation that Goodrich and Winchell described in 1885 did not improve during the next century. Hyenas have a

* I am grateful to my wife, Krista, for helping me in all the phases of research and writing involved in this paper. I am also indebted to Laurence G. Frank, my friend and collaborator, who introduced me to hyenas; and to Kay Holekamp and Laura Smale who have taught me a great deal about these animals from their own field studies. In that regard, I thank the Office of the President, the Wildlife Conservation and Management Department, and the Narok County Council, Republic of Kenya, for facilitating our hyena research. Finally, I am appreciative of the efforts of many friends and colleagues who called my attention to relevant material that I otherwise might have missed. These include Carol Christ, Carol Clover, Helen Ettlinger, Don Friedman, Sam Gosling (and his parents), Diane and Fred Leavitt, Roger Short, and, particularly, Nicholas Howe. The material may not have been included in this article, but I am still thankful for their interest. Portions of the research involved in this paper were supported by a grant from the National Institute of Mental Health [MH-39917].

87

terrible reputation in Western culture. However, scientists who study these animals at close range have a different view. Hans Kruuk (1972), whose field studies of the 1960s completely changed our understanding of the life of the spotted hyena, has written:

> . . . there is a magic about hyaenas which can only be understood by those of us who have watched them, for some time. There is now a growing band of us, who came to the African bush with all our prejudices, with all that 'common knowledge' about hyaenas which proved so totally wrong, and who just fell for the spell of animals which were so totally different" (Kruuk, 1990, p. xiii).

This chapter had its origins in an effort to understand this discrepancy: why the positive attitudes and feelings about hyenas, held by the scientists who knew them best, were so at odds with the general culture? This, in turn, led to other questions. First, what are the essential elements of the hyena reputation in contemporary Anglo-American culture? Second, has that reputation changed across time or from one cultural setting to the next? And, finally, how do reputations relate to biological realities?

The family Hyenidae contains four extant species: striped hyenas (*Hyaena hyaena*), brown hyenas (*Parahyaena brunnea*), spotted hyenas (*Crocuta crocuta*), and a small, termite-eating animal, the aardwolf (*Proteles cristatus*). As we shall see, the public image of hyenas has largely been created on the basis of real, exaggerated, or imagined traits of striped and spotted hyenas. It is the latter species, the "laughing hyena" of popular lore, which forms a primary focus of this article, although we shall also be drawn to its somewhat smaller relative, the striped hyena.

By virtue of their size and abundance, spotted hyenas are among the most significant predators on the African Savannah.[1] In terms of tonnage of meat consumed, they are, perhaps, the most significant terrestrial carnivore on the planet. Often living in large female-dominated social groups, or "clans," they display intriguing behavioral complexities that parallel those of many venerated old-world primates. Every adult female spotted hyena breeds and the clan is organized in terms of subgroups of mothers, daughters, and their offspring. Long-lasting dominance hierarchies exist among these clan families. Adult males appear in the group as immigrants from other clans. They are totally subservient to the resident females and their older offspring. Spotted hyenas defend their kills against much larger marauding lions and their hunting territories against other clans. They are remarkable animals. However, spotted hyenas, or their immediate relatives, the brown and striped hyenas, are rarely found in zoological parks in this country, despite a self-conscious shift among contemporary zoos from functioning as entertainment parks to institutions dedicated to public education and preservation of endangered species. According to one recent count, spotted hyenas were to be found in but 11 of 164 North American zoological parks.[2] One major American zoo that I visited several years ago had Asiatic tigers housed within the boundaries of an African savannah exhibit but no spotted hyenas.

I believe that this serious miscarriage of biological justice can be traced to the poor public reputation of the hyena. There are no "save-the-hyenas" committees, and their persistent public relations problems could have very serious consequences for the preservation of hyena habitats and the long-term prospects for these species. This has been an issue of concern to the scientists who study these animals. Mills concludes his monograph: "If this book has helped in any way to convince people that hyaenas are worth conserving, not only because of their intrinsic value, but because of their beauty and fascinating behaviour, it will have been worth the many hours of toil that it has taken me to try to convert the wonderful experiences of watching them, into some sort of coherent and scientifically meaningful form" (Mills, 1990, p. 273).

We may begin with a brief survey of the portrayal of hyenas in twentieth-century American culture and then consider the shaping of the hyena image over a span of several thousand years, contrasting European views with those of contemporary Africans. Along the way, I will occasionally digress to compare views of the hyena with those of other animals, particularly the lion. It is one of the great ironies of spotted hyena existence that, over a span of several million years, they have held their own in direct competition with lions, only to lose (what may be their most critical battle) in the court of public esteem.

Hyenas and Anglo-American Culture in the Twentieth Century

A trio of spotted hyenas appear in the recent Disney movie, *The Lion King*. Several artists from the Disney studios spent two days in the hills above the Berkeley campus of the University of California observing and sketching hyenas maintained in our colony at the Field Station for Behavioral Research. Their visit was arranged by Dr. Laurence Frank, who has followed a clan of hyenas in the Masai Mara Reserve in Kenya for more than 15 years, and who was instrumental in establishing the Berkeley colony. Frank, and other scientists working at the colony, asked how hyenas were to be portrayed in the film and expressed a strong request that it be positive. The artists explained that the script was written and that a trio of hyenas (Banzai, Shenzi, and Ed) were to be the allies of an evil older lion, who would eventually lose to the hero, a noble young lion. However, they seemed very appreciative of the animals and said they would do their best to make them appear comical instead of evil.

It is difficult for me to judge how successful the artists

were in their efforts. Banzai and Shenzi are not particularly appealing, at least in the ordinary sense of the term. They are portrayed as evil, gluttonous characters who would do anything for food. Ed is the the third member of the trio. According to one text accompanying a childrens' computer game based on *The Lion King*, Banzai and Shenzi are "relatively intelligent," but Ed is "a little slow" and is " . . . known for his maniacal laughter. He's been known to gnaw on his own leg by mistake" (*The Lion King* . . . , 1994, p. 91). These are not new images of hyenas. Their portrayal in contemporary media as weird, dangerous, repulsive animals, with a few recent exceptions, has been quite consistant.

The world of cartoons is particularly relevant to understanding attitudes toward hyenas, since their appreciation depends upon shared perspectives in the audience. Consider this cartoon from the *New Scientist*, in which we find an array of the more serious prejudices wrapped in a single caption [Fig. 1]: "We're scavengers, we're ugly and we smell bad, if we didn't laugh we'd crack."

Hyenas have fared no better in serious literature. In his account of travelling on safari in Africa, Hemingway claimed that his African guide M'Cola found Hemingway's killing of many animals humorous, hyenas most of all:

> It was funny to M'Cola to see a hyena shot at close range. There was that comic slap of the bullet and the hyena's agitated surprise to find death inside of him. It was funnier to see a hyena shot at a great distance, in the heat shimmer of the plain, to see him go over backwards, to see him start that frantic circle, to see that electric speed that meant he was racing the nickelled death inside him. But . . . the pinnacle of hyenic humor, was the hyena, the classic hyena, that hit too far back while running, would circle madly, snapping and tearing at himself until he pulled his own intestines out, and then stood there, jerking them out and eating with relish.
>
> Fisi, the hyena, hermaphroditic self-eating devourer of the dead, trailer of calving cows, ham-stringer, potential biter-off of your face at night while you slept, sad yowler, camp-follower, stinking, fowl, with jaws that crack the bones that the lion leaves,

"WE'RE SCAVENGERS, WE'RE UGLY AND WE
SMELL BAD. IF WE DIDN'T LAUGH, WE'D CRACK."

Fig. 1. Cartoon from the *New Scientist.*

belly dragging, loping away on the brown plain, looking back,
mongrel dog-smart in the face; wack from the little Mannlicher
and then the horrid circle starting (Hemingway, 1935, pp.
37–8).

In this painful passage, Hemingway raises many of the
myths that have followed hyenas and contributed to their
sorry reputation: hermaphrodites, scavengers, singers of sad
songs, smelly, ugly, and, ultimately, comical in their failure to
comprehend the "realities" of our perceptions.

Perhaps Hemingway was following in the footsteps of
another Great White Hunter, Theodore Roosevelt, who had
made a similar trek to Africa several decades before
Hemingway. According to Roosevelt and Heller: " . . . the
hyena is a singular mixture of abject cowardice and the utmost

ferocity. Usually feeding on carrion, and often hesitating to attack even the weakest animal if it is unhurt and on its guard, the ravenous beast will, on occasions, even when single but especially when in troops, assail very formidable creatures" (Roosevelt and Heller, 1915, p. 259).

Roosevelt and Heller add accounts of hyenas attacking sheep, goats, donkeys, mules, cattle, dogs, men, women, and especially children. But perhaps most surprising to these authors was that " . . . under certain circumstances they do attack lions. . . . " To appreciate their surprise, one has to realize the extraordinary reverence that Roosevelt and Heller display toward lions as brave, powerful killers and predators and their conviction that hyenas are lowly scavengers. They write that: "Ordinarily the hyenas merely attend the lion at a respectful distance, eager to get whatever he leaves, and they occasionally pay with their lives if they grow too impatient" (Roosevelt and Heller, 1915, p. 260). However, Roosevelt and Heller then describe a number of incidents, which they find very puzzling, in which hyenas confronted healthy lions and even killed old or crippled lions.

Some Biological Realty

Zoologist Hans Kruuk, following hyenas in the Serengeti, reports that in cases where lions and hyenas were "sharing" a carcass, 53 percent had been killed by hyenas and 33 percent by lions. He reports that George Shaller, following lions in the same ecosystem, found that 54 percent of the prey involved in "shared" feeding had been killed by hyenas and 34 percent by lions (Kruuk, 1972, p. 129). It is clear that the noble, much-admired king of beasts scavenges more often on hyena kills than hyenas scavenge on lion kills. It should be added that from a biologist's perspective, scavenging is an extremely honorable, essential profession. I doubt that the affection of biologists for brown and striped hyenas is lessened by the greater reliance of these latter species on food killed by others. However, it seems likely that the image of spotted hyenas as "cowardly scavengers" is fundamental to their reputational problems, as it has been for vultures and jackals—other favorite subjects for human metaphor. It is also of particular interest that the prejudice was so powerful it clouded the perception of otherwise knowledgeable people (like Roosevelt and Heller) to what was happening in front of their eyes. When they came upon hyenas feeding at a carcass, they merely assumed that the animal had originally been killed by lions or other predators. When they found a cluster of distressed hyenas surrounding some feeding lions, they assumed that the prey had been dispatched by lions and that the hyenas were awaiting their turn to scavenge. When they actually witnessed kills, or observed hyenas fighting to hold their prey against intruding lions, these were assumed to be exceptions, and the reputation of hyenas as cowardly scavengers was maintained. As we shall see, these charges have a long history. In the 1990s, several decades after Kruuk's and Schaller's studies, the educated public and many scientists have been slow to recognize the complexity of feeding among hyenids.

There was another problem that has caused endless difficulties for the hyena reputation: the charges of sexual ambiguity noted by Hemingway. As all of these issues were discussed by Aristotle, his writing becomes a logical place to search for the origins of the hyena reputation.

Aristotle

More than 2300 years ago, Aristotle considered almost all of the major contemporary charges against hyenas: first, that they are hermaphrodites; second, that they are scavengers; third, that they are naturally cowardly or, at least, "timorous"; and, fourth, that they are potentially treacherous and threatening for people. We may begin where Aristotle begins, in his *History of Animals* (*HA*) with a consideration of their presumptive hermaphroditism.[3]

What is recounted concerning its genital organs, to the effect that every hyena is furnished with the organ both of the male and the female, is untrue. The fact is that the sexual organ of the male hyena resembles the same organ in the wolf and in the dog; the part resembling the female genital organ lies underneath the tail, and does to some extent resemble the female organ, but it is unprovided with duct or passage, and the passage for the residuum comes underneath it. The female hyena has the part that resembles the organ of the male, and, as in the case of the male, has it underneath her tail, unprovided with duct or passage; and after it the passage for the residuum, and underneath this the true female genital organ. The female hyena has a womb, like all other female mammals of the same kind (*HA*, VI, 32, 579b, 16ff).

The topic reappears in the text of Aristotle's *Generation of Animals* (*GA*), at the conclusion of a section dealing with various reports of fish and certain birds being fertilized via their mouths. He castigates other natural scientists promoting such ideas for " . . . speaking too superficially and without consideration" (*GA*, III, 6, 756b, 18), noting that orally ingested semen could not reach the uterus and, in any case, would be destroyed in the intestine. Aristotle then returns to rumors of hyena hermaphroditism:

Much deceived also are those who make a foolish statement about the trochus and the hyena. Many say that the hyena, and Herodorus the Heracleot says that the trochus, has two pudenda, those of the male and of the female, and that the trochus impregnates itself but the hyena mounts and is mounted in alternate years. This is untrue, for the hyena has been seen to have only one pudendum, there being no lack of opportunity for observation in some districts, but hyenas have under the tail a line like the pudendum of the female. Both male and female have such a mark, but the males are taken more frequently; this casual observation has given rise to this opinion (*GA*, III, 6, 757a, 2ff).[4]

He then concludes, almost irritably, "But enough has been said of this." Aristotle obviously believed that he had disposed of the hyenas-as-hermaphrodites rumor by observation and had even been able to provide a logical explanation for the false belief, namely, the line under the tail that concealed a duct, albeit that the duct had a blind ending. Aristotle was correct; hyenas are not hermaphrodites. (In this regard, he was ahead of Hemingway.) However, there was a problem. As has been noted by a number of biologists (Watson, 1878; Platt, 1958; Matthews, 1939), Aristotle was surely dealing with striped hyenas, when the rumor must have originated with spotted hyenas. In his initial general description of the hyena, Aristotle writes, "The hyena in colour resembles the wolf, but is more shaggy, and is furnished with a mane running all along the spine" (*HA*, VI, 30, 579b, 15).[5] Such a mane is characteristic of the striped hyena, not the spotted hyena. From this single anatomical characteristic, it is clear that Aristotle was not considering the species on which the original rumor was based. The description of a "line beneath the tail" might also fit the striped hyena better than the spotted. Although all hyenids have anal scent glands in this position, both brown and striped hyenas " . . . possess a large glandular pouch below the tail which largely obscures the external genitalia . . . " (Walker, 1968, p. 1267). Finally, in Aristotle's time, striped hyenas would have been common in North Africa and the Middle East, while the range of the spotted hyena had largely, if not completely, retreated to sub-Saharan Africa. D'Arcy Thompson concluded that Aristotle did much of his writing on biological subjects while residing on the Isle of Lesbos, off the coast of Turkey, and did some travelling on the mainland as well (Thompson, 1916). Turkey could have been prime striped hyena habitat in those times, and, as recently as 1877, British naturalists reported that striped hyenas "were not uncommon near Smyrna and in the southern districts" (Danford and Alston, 1877, p. 270). Smyrna is less than 75 miles southeast of Lesbos. It is likely that the hunters that

Aristotle consulted would have been familiar with striped hyenas, and, perhaps, he even had the opportunity for direct examination of this animal. His anatomical descriptions are certainly quite precise.

If Aristotle had only examined a spotted hyena, he would have appreciated the source of the rumor; for in this species the external genitalia of the female are virtually identical to those of the male. The vaginal labia have fused to form a pseudoscrotum, containing two pads of fatty tissue which appear superficially to be testes; there is no external vagina. The clitoris, which is traversed by a central urogenital canal, has developed until it is the size and approximate shape of the male penis, and female hyenas display erections similar to those of the male. Female spotted hyenas urinate, copulate, and give birth through this organ [Figure 2]. Internally, they are ordinary female mammals. It is only within the last few years that scientists have been able to distinguish between males and females without internal examination or very close inspection (Neaves, Griffin, and Wilson, 1980; Frank, Glickman, and Pouch, 1990; Frank and Glickman, 1994). The anomalous "masculinization" of the female spotted hyena promoted the maintenance of hermaphroditic rumors well into the present century, and their ambiguous sexual anatomy has contributed substantially to the bad reputation of this species, although it was more troubling in some eras than others.

In another section of his *History of Animals,* Aristotle discusses threats posed by the hyena: "It will lie in wait for a man and chase him, and will inveigle a dog within its reach by making a noise that resembles the retching noise of a man vomiting. It is exceedingly fond of putrefied flesh, and will burrow in a graveyard to gratify this propensity" (*HA*, VIII, 5, 594b, 2). Included within this brief passage are four separable concerns that people have had of hyenas: (a) they are active predators of particular danger to us; (b) they are crafty or deceitful and capable of using their vocal skills to lure an

Fig. 2. Drawing of a female hyena by Dr. Christine Drea. The drawing depicts a female spotted hyena with an erect clitoris.

animal into trouble; (c) they are scavengers who prefer rotting meat; and (d) they will disturb human gravesites. All of these themes were troubling, and all would be picked up by later writers.

At several points in the *Parts of Animals* (*PA*), Aristotle advanced a general theory of bravery and cowardice. Courage was viewed as proportional to the heat of the blood, and he believed that there was a negative correlation between the relative size of the heart and qualities of bravery among various species of mammals. "The heart is of large size in the hare, the deer, the mouse, the hyena, the ass, the leopard, the weasel, and in pretty nearly all other animals that either are manifestly timorous, or betray their cowardice by their spitefulness" (*PA*, III, 4, 667a, 20ff).[6]

He moved on to explain this correlation with an interesting analogy, noting that " . . . a fire of equal size gives less heat in a large room than in a small one" (*PA*, III, 4, 667a, 25). Therefore, since the heart is the source of heat, the blood of cowardly animals would be cooler than the blood of brave ones as the result of being heated in a larger chamber. Aristotle's inclusion of the hyena in his list of timorous/cowardly animals again did little to enhance their reputation.

The final mention of hyenas is contained in a section of *On Marvellous Things Heard* (*OMTH*). According to Barnes (1984), although this material is traditionally included in Aristotle's works, it is almost surely not written by Aristotle. Here we get the taste of magic and power which will appear in much more expanded form in Pliny the Elder. "In Arabia they say there is a certain kind of hyaena, which, when it sees some wild beast, before itself being seen, or steps on the shadow of a man, produces speechlessness, and fixes them to the spot in such a way that they cannot move their body; and it is said that they do this in the case of dogs also" (*OMTH*, 145, 845a, 24–8).

From Natural History to Morality and Religion: Pliny the Elder, and the Author of Physiologus

The next stop in our journey occurs approximately 400 years after Aristotle's biological works. It involves a man whose books on natural history were to influence peoples' attitudes toward animals for at least the next 1600 years: Gaius Plinius Secundus, also known as Pliny the Elder.[7] He claimed that his Natural History dealt with 20,000 matters of importance, drawn from 100 selected authors, among whom Aristotle figures prominently. However, as Rackham has observed, "In selecting from these he has shown scanty judgment and discrimination, including the false with the true at random; his selection is coloured by his love of the marvellous, by his low estimate of human ability and his consciousness of human wickedness, and by his mistrust of Providence" (*NH*, Volume I, p. ix). Rackham concludes that: "Nevertheless it is a mistake to underrate the value of his work. He is diligent, accurate, and free from prejudice."

Pliny's writings about the hyena fall into two categories: the first dealing with "traditional" natural history, the second with accounts of practical results to be gained by using fractions of hyenas for medical purposes. As a natural historian, Pliny recounts Aristotle's description of the hyena's neck and mane, his denial of hyena hermaphroditism, and the magical powers of hyenas to freeze other animals in place. However, after repeating Aristotle's account of hyenas using vomiting sounds to attract dogs, he elaborates on the possibilities of hyena vocalization, observing that " . . . among the shepherds homesteads it simulates human speech, and picks up the name of one of them so as to call him to come out of doors and tear him in pieces, . . . " (*NH*, Volume 3, Book 8, Paragraph 45, 107–9). We then come upon one of the few semi-positive references to hyenas in 2300 years: "The Magi have held in the highest admiration the hyaena of all animals, seeing that they have attributed even to an animal; magical skill and power, by which it takes away the senses and entices men to itself" (*NH*, Volume 8, Book 28, Paragraph 25, 92).

He then devotes five pages (of the English translation) to describing the hyena as a walking pharmacopoeia, including the "fact" that a person carrying anything made of hyaena leather is not attacked, while the skin of the hyena head, when tied on, cures headaches; their teeth (when touched to the corresponding human tooth) relieve toothaches, and the gall, if applied to the forehead, cures opthalmia. Among other cures, Pliny observes " . . . that barrenness in women is cured by an eye taken in food with liquorice and dill, conception being guaranteed in three days" (*NH*, Volume 8, Book 28, Paragraph 27, 98–9). Finally, in a somewhat confusing passage, Pliny notes that, "A hyaena's genitals taken in honey stimulate desire . . . , even when men hate intercourse with women . . . " and adds that " . . . the peace of the whole household is assured by keeping in the home these genitals and a vertebra with the hide still adhering to them" (*NH*, Volume 8, Book 28, Paragraph 27, 99).

The *Physiologus* is generally attributed to a single author, who lived somewhere between the second and fifth centuries AD. Curley has observed that Pliny transmitted Indian, Hebrew, and Egyptian legends that had reached him through varied routes to the early Christian world. He then adds that: "The anonymous author of *Physiologus* infused these venerable pagan tales with the spirit of Christian moral and mystical teaching, and thereafter they occupied a place of special importance in the symbolism of the Christian world" (Curley, 1979, p. xi). Whereas Aristotle and Pliny had presented their work in a secular context, Physiologus draws Christian moral implications from these animal stories, occasionally making drastic changes in the story to accommodate the moral lesson. Some species, particularly the panther or leopard, were the beneficiaries of such changes. Hyenas did not benefit from the new perspective. Their ambiguous sexuality, rejected by Aristotle and Pliny, was reactivated and became the focus in the *Physiologus*. "There is an animal which is called the hyena in Greek and the brute in Latin. The Law said, 'Thou shalt not eat the brute, nor anything similar to it' [cf Lev. 11:27]. This animal is an *arenotelicon*, that is, an alternating male-female. At one time it becomes male, at another female, and it is unclean because it has two natures, Therefore, Jeremiah said, 'Never will my heritage be to me like the cave of the brute' [cf Jer. 12:9]" (Curley, 1979, pp. 52–3).

Physiologus, adds that "double-minded men are compared to the brute . . . being men at the signal for gathering the congregation" but taking on a "womanly nature" when the congregation is dismissed. He then adds an anti-Semitic twist to the story of the hyena which would persist throughout the bestiaries of the middle ages: "The sons of Israel are like this animal since in the beginning they served the living God but later, given over to pleasure and lust, they adored idols. For this reason, the Prophet likens the synagogue to an unclean animal. Whoever among us is eager for pleasure and greed is compared to this unclean brute since he is neither man nor woman, that is, neither faithful nor unfaithful" (Curley, 1979, p. 53).

Hyenas in Medieval Times: The Bestiaries

In the Middle Ages, bestiaries appeared which were heavily influenced by *Physiologus*. A twelfth-century Latin bestiary translated by T. H. White begins with a variant of the portrait presented in the *Physiologus*. "This is an animal called the YENA, which is accustomed to live in the sepulchres of the dead and to devour their bodies. Its nature is that at one moment it is masculine and at another moment feminine, and hence it is a dirty brute" (White, 1960, pp. 30–1).

The compiler of this bestiary then adds the now familiar tales of hyaenas imitating human voices or the sounds of human vomiting with the goal of luring prey to their death, and reasserts that hyaenas dig up graves in quest of buried corpses. The bestiary then turns to the analogy between the Sons of Israel and the brute, concluding that: "Since they are neither male nor female, they are neither faithful nor pagan, but are obviously the people whom Solomon said: 'A man of double mind is inconstant in all his ways.' About whom also the Lord said: 'Thou canst not serve God and Mammon.'" The final passage in this bestiary repeats selected magical tales from Pliny and others about using the stone from a hyaena's eye, placed under the tongue, to forsee the future and the ability of hyenas to freeze animals in place by circling them three times.

Several later bestiaries I have examined appear to follow this text fairly closely in terms of content (Porion, 1988; Barber, 1993). There is also a common theme to the illustrations. The hyena is always portrayed in the act of devouring a human corpse; most commonly the body is in a coffin, often with a church in the background [Fig. 3]. This form of scavenging involves the disturbance of the peaceful rest of the interred body, and any such disturbance by external agents has been a subject of concern in human societies.[8] The hyenas are all portrayed with a highly exaggerated mane running the length of the spine—as described originally by Aristotle. With the passage of time, the illustrations grow more elaborate, moving, as Barber observes, from simple line drawings to beautifully

Fig. 3. Drawing from twelfth-century Latin bestiary portraying hyenas scavenging on human corpses.

colored portrayals of animals in some English thirteenth-century bestiaries (Barber, 1993, pp. 11–12). However, there are no stripes or spots on the hyenas in these bestiaries. The bodies are uniformly colored, and it seems doubtful that the artists had ever seen a hyena.

Thoughts of Hyenas in Britain: Sixteenth-Seventeenth Centuries

In the fifteenth and sixteenth centuries, travelers to Africa and the Middle East provided new eyewitness (and second-hand) accounts of animals in remote areas, stimulating intense interest in Europe (Lloyd, 1971)]. Of particular relevance were the travels of a Moor, born in Granada but raised in Morocco, who ultimately converted to Christianity. Best known by the name of Leo Africanus, he wrote about his travels in Arabic and then contributed to the translation of that manuscript into Italian in approximately 1526 A.D. (Brown, 1896, V. 92–94).[9] John Pory, in turn, translated Africanus' *History and Description of Africa* into English in 1600. New, somewhat fanciful material on hyenas was brought to the attention of the English public, and some old stories were repeated:.

> This beast . . . in bignes and shape resembleth a woolfe, saving that his legges and feete are like to the legs and feete of a man. It is not hurtful to any other beast, but will rake the carkeises of men out of their graves, and will devour them, being other wise an abject and silly creature. The hunters being acquainted with his denne, come before it singing and playing upon a drum, by which melodie being allured foorth, his legs are intrapped in a strong rope and so he is drawn out and slaine (Brown, 1986, V. 94, p. 947).

The writings of these travelers presaged the return of a new, more secular natural history. In 1551, the Swiss naturalist Conrad Gesner compiled material on natural history from many sources and devoted particular attention to providing the best possible illustrations in the published work (Gesneri,

1551). Portions of this work, with associated illustrations, were translated from Latin and edited by Edward Topsell early in the seventeenth century. The material on hyenas was both interesting and influential. "The first and vulgar kind of Hyaena is bred in Affricke and Arabia, being in quantity of body like a wolfe, but much rougher haird, for it hath bristles like a horsses mane all along his back, and the middle of his backe is a little crooked or dented, the colour yellowish, but bespeckled on the sides with blew spots, which make him looke more terrible, as if it had so many eies" (Topsell, 1608, p. 340).

The preceding description combines the mane of the striped hyena with the spots of *Crocuta* in an account calculated to inspire fear. Confusion still reigned over the existence of multiple species of hyena. The drawing that Gesner presented of the "firste" kind of hyena reflected that confusion, containing spots composed of parallel stripes [Fig. 4]. However, merely filling in the details of the body was an "advance" over the mediaeval bestiaries. There were also

Fig. 4. Drawing from Gesner's *Natural History* (1551), portraying a hyena eating a dog.

important changes in attitude reflected in text. Following Aristotle, Gesner/Topsell now denies the hermaphroditism of hyenas but adds some new and interesting details (describing ". . . in the femal a bunch like the stones of the male," which could refer to the fatty tissues found in the pseudoscrotum of the spotted hyena) (Topsell, 1608, p. 340). He then goes on to suggest that confusion over hyena sexuality might have arisen from observations of fish with the same name, " . . . which turneth sexe, and peradventure some men hearing so much of the fish, might mistake it more easilye for the four-footed beast, and applye it thereunto" (Topsell, 1608, p. 340). An extensive recital of familiar and unfamiliar magical properties of hyenas is then described. Finally, Topsell concludes the discussion of the "firste" hyena: "Such is the folly of the Magitians, that they beleeve the transmigration of soules, not only of one man into another, but also of man into Beasts. And therefore they affirm, that their men Symis and religous votaries departing life send their souls into Lyons, and their religous women into Hyaenaes" (Topsell, 1608, p. 342). Topsell adds three other animals within the category of hyenas: the Papio or Dabuh (which is clearly a baboon), the Crocuta, " . . . an Aethiopian four-footed beast . . . ingendered betwixt a Lyoness and an Hyaena," and the Mantichora, a fanciful monster with the body of a lion and a human head (Topsell, 1608, p. 342–45).[10]

A new skepticism was emphasized in the writings of Sir Thomas Browne, a bit later in the seventeenth century. In his *Pseudodoxia Epidemica* he dismisses the "annual mutation of the sexes in the hyaena"[11] and provides a rational explanation for the supposedly hypnotic powers of the hyena over dogs, attributing a variety of such effects to " . . . a vehement fear which naturally produceth obmutescence; and sometimes irrevocable silence (*PE*, V. 1, p. 339). It was well for the hyena reputation that their sexual "double-mindedness" was rejected at this time. The very rational and liberal Browne makes clear that animals, as well as people, were to follow a single "Law of

their Coition," that " . . . they observe and transgress not . . . ";
it is only man that " . . . hath in his own kind run thorow the
Anomalies of venery; and been so bold, not only to act, but
represent to view, the irregular ways of lust" (*PE*, V. 2, p. 41).
Species were obligated to follow their particular law of coition,
and a beast that changed sex from year-to-year would surely
have transgressed.

Hyenas also occupied the attention of writers whose major
focus was not on natural history. Sir Walter Raleigh, in his
History of the World (1614, Book I), dealt with hyenas in the
context of Noah's Ark.[12] In a delightfully "rational" passage,
he grapples with the problem of fitting all of the existent
animal species within the limited space provided by the ark.
Detailed calculations are provided encompassing the space
required by both sexes of each species and the requisite animal
food, as well as space to house Noah, his family, and necessary
supplies. Space is tight, and Raleigh relieves some of the
pressure by noting that: "For those Beasts which are of mixt
natures . . . it was not needful to preserve them, seeing that
they might be generated again by others: as the Mules, the
Hyaena's, & the like; the one begotten by Asses and Mares, and
the other by Foxes and Wolves" (Raleigh, 1614, Book I, pp.
94–5).

References to hyenas also appear with some frequency in
English literature of the time. Although these writers were not
clear about the various species of hyena, the focus was on
vocalizations, and this means the spotted hyena (striped and
brown hyenas have a much more limited vocal repertoire). In
As You Like It, Shakespeare has Rosalind telling Orlando how
she will change after they are wed, using a number of animal
analogies: "I will be more jealous of thee than a Barbary
cock-pigeon over his hen; more clamorous than a parrot
against rain; more new-fangled than an ape; more giddy in my
desires than a monkey: I will weep for nothing like Diana in
the fountain, and I will do that when you are supposed to be

merry; I will laugh like a hyen, and that when thou are inclined to sleep" (Act IV, Scene One).[13]

Hyena vocalizations are also the focus of metaphor in George Chapman's *Eastward Ho* ("I will neither yield to the song of the siren nor the voice of the hyena" [Act V, Scene 1]) and, much less directly, in Milton's *Samson Agonistes*, as Samson speaks of Dehlila: "Out, out hyaena; these are thy wonted arts, and arts of every women false like thee" (Milton, 1971, p. 369). John Carey attributes the latter reference to the stories in Pliny concerning the deceitful nature of hyena vocalizations used to lure innocents to their death. He also calls attention to Ben Jonson's reference to hyenas in *Volpone* ("Now, thine eies / Vie teares with the hyaena" [IV vi 3]) as continuing this use of hyenas to represent deceit (Milton, 1971, pp. 369–70).

Toward the end of the seventeenth century, John Dryden translated selections from Ovid's *Metamorphosis*, written between A.D. 2 and A.D. 8, into rhymed English verse. The bisexuality of the hyena was resurrected: "A wonder more amazing wou'd we find? / Th' Hyaena shows it, of a double kind, / Varying the sexes in alternate years, / In one begets, and in another bears" (Dryden, 1961, Book XV, p. 502).

Modern Natural Histories: Eighteenth–Nineteenth– Twentieth Centuries

In the eighteenth and nineteenth century there was a notable advance in the biological accuracy of books concerned with natural history. Travelling was sufficiently frequent, and communication was common enough so that an author could compare recent accounts of wildlife in remote areas, examine specimens, and perhaps even observe wild animals from far-away places in European zoological parks. This both constrained the more fabulous tales of earlier centuries and provided a personal perspective that had often been lacking since the times of ancient Greece and Rome. The three species of hyena were finally recognized and named. Scientist-writers had already rejected stories of hermaphroditism, but the publication of a set of papers by Watson (1877, 1878) provided the actual anatomical details of spotted hyena genitalia and settled (or should have settled) that issue.

Natural historians continued to focus on the scavenging habits of the hyenids and their potential as grave-robbers, even though there was obvious awareness of the predatory possibilities of spotted hyenas. In addition, one old theme was resurrected, and a new dimension emerged. The old theme was bravery and cowardice; the new dimension involved the possibility of "taming" hyenas as pets.

Bravery and Cowardice

Natural historians of the eighteenth and nineteenth century as well as a number of writers in the present century appear to have been preoccupied with the dimensions of bravery and cowardice as applied to animals. Moreover, the criterion was not how the animals related to one another within their habitat, but how they related to us, and particularly how they related to hunters with guns. Lions are generally much admired: "Accustomed to measure their strength with that of every animal they meet, the habit of conquering renders them terrible and intrepid. . . . Wounds enrage, without terrifying them. They are not disconcerted even by the appearance of numbers. A single lion of the desert often attacks a whole caravan; and if . . . he finds himself fatigued, instead of flying, he retreats fighting, always opposing his face to the enemy" (Buffon, 1830, pp. 3–4).

The contrast with views of the hyena is clear and to the latter's disadvantage. For example, Nott stated that, "All writers agree that the hyaena lacks courage, and is only ferocious when he himself is free from harm" (Nott, 1886, p. 106). He goes on to note that hunting hyaenas is not considered good sport, and relates an incident in which a "well-known Indian sportsman" by the name of McMaster found himself face to face with a terrified hyaena. Previously, he had not "wished to waste a shot," but now " . . . although I had spared it before, I could not resist taking his worthless life as he stood" (Nott, 1886, p. 106). Flower and Lyddeker characterize the striped hyena as "essentialy a cowardly animal," although they do distinguish it from the spotted hyena, which is described as "larger and bolder . . . hunting in packs, and uttering very frequently its unearthly cry" (Flower and Lydekker, 1891, pp. 542–43). Mention of the "cowardice" of striped hyenas emerges in the otherwise sedate passages of the eleventh edition of the *Encyclopaedia Britannica* (1910–1911, V. 13, p. 174), is repeated as part of a more general commentary on hyenids in the fifteenth edition published in 1976 (V. 3, p. 935), and is even found in Walker's authoritative *Mammals of the World*. There it is asserted that spotted hyenas " . . . are cowardly and will not fight if their prospective victim defends itself" (Walker et al., 1968, V. 11, p. 1265).

What have hyenas done to earn such human scorn? First, as regards the striped hyena, there are many older stories suggesting that striped hyenas could be captured alive by human beings, by hand, without resistance (for example, Leo Africanus); second, that even spotted hyenas, as predators, select the sick and the weak as their natural prey. By extension, when attacking people, their targets would be the smaller, weaker humans, particulary children. That this is still an issue is indicated by a recent report in the journal, *Tropical Doctor*, identifying snake and hyena bites as a significant cause of amputations among the children of Senegal (Loro, Franceschi, Dal Lago, 1994, p. 99). Finally, spotted hyenas are very wary animals, whose style when hunting large, dangerous prey involves a great deal of "testing," that is, of approach and withdrawal. This would fit well with our notions of human cowardice, that is, they do not come out and fight-like-real-men, face-to-face.

Hyenas as Pets

There is evidence from paintings and bas-reliefs in tombs that in ancient Egypt hyenas were tamed and kept as pets, as well as being artificially fattened as food (Zeuner, 1963). That striped hyenas could be tamed and kept as pets was a matter of fascination for authors in the eighteenth (Buffon, 1830, p. 95), nineteenth (Nott, 1886, p. 114), and twentieth (Ognev, 1962, p. 317; Roosevelt and Heller, 1915, p. 260) centuries. In earlier editions of his *Natural History*, Buffon denied the possibility of taming hyenas: "His disposition is extremely ferocious and, though taken young, can never be tamed" (Buffon, 1830, p. 94). He corrects this in a supplement inserted in later editions, describing personal observation of a male hyena who, " . . . having been tamed when young, was remarkably gentle; for though his master often provoked him with a cudgel, in order to erect his mane, . . . instantly afterwards seemed to forget the affront" (Buffon, 1830, p. 95). I will return to the question of humans forming reciprocal, affectionate relationships with hyenas toward the conclusion of this paper.

African Attitudes

It is apparent that the reputation of hyenas in Africa is no better than their reputation in Anglo-American culture. Beidelman (1975, p. 183) states that the Kaguru of Tanzania, East Africa, speak of spotted hyenas with a combination of disgust, fear, and derision but also as funny or comical. They are viewed as "incredibly greedy," with "boundless appetites," and as being "so stupid or perverse that they even prefer bones to good meat." The Kaguru believe that hyenas are hermaphrodites, and "were horrified" when Biedelman asked whether they would eat hyena flesh. The Kaguru also believe that hyenas dig up the graves of the dead and attribute the "exceedingly unpleasant odour" of hyenas to their disgusting diet. In Kaguru folklore, tales of hyenas are generally associated with anti-social behavior. Sapir, discussing the attitudes of the Kujamaat Diola of Southern Senegal in West Africa, notes that in their folktales " . . . the Hyena becomes a vehicle for a very definite character type which combines the perceived characteristics of greediness (especially with food), aggressiveness, trickiness of a particularly crude sort . . . and often plain stupidness." He concludes that: "The characteristics that Beidelman . . . has inferred for Hyena in Kaguru folklore are essentially those developed in Kujamaat tales. In fact, one can safely say that its role is fairly uniform across Africa" (Sapir, 1981, pp. 532–33).

Allen F. Roberts is the author of a new book on *Animals in African Art* (1995).[14] Roberts reviews African attitudes toward hyenas in the context of correlated artistic productions. I recently received a postcard printed in Benin from Professor Roberts. He noted that to the best of his knowledge, this is the only representation of a "whole" hyena in African Art. It depicts a hyena regurgitating a stolen goat, certainly a portrayal of gluttony and anti-social behavior. There is also the stupidity of trying to swallow something so outsized, and, as in many of the folk tales recounted by Beidelman, there may be a moral lesson regarding the futility of theft. Although full-body depictions of hyenas may be rare in Africa, Roberts decribes and offers photographs of hyena masks that are comonly worn by dancers in a number of African cultures. He observes that through wearing such masks, " . . . people can "become" hyenas to play dramatic roles that usually run counter to social harmony or the tenets of human civilization" (Roberts, 1995, p. 75).

African and Western Attitudes

Over the years, African and Western attitudes have converged on a negative vision of hyenas. Similar characteristics, that is, sexual ambiguity, scavenging, disturbing gravesites, and eating the bones of our dead, have been the focus of disapproval in widely different cultural settings. However, there are also significant cultural differences. The judgement that hyenas are ugly and cowardly is much more prominent in our culture, while the greed, gluttony, stupidity, and comical foolishness of hyenas is a more integrated part of African lore, as is an appreciation of these animals as powerful and potentially dangerous. Although these cultural differences reflect the closer contact that Africans have had with hyenas, the quote from Hemingway presented earlier in this paper and the general portrayal of hyenas in *The Lion King* are much more closely tied to the African view of hyenas than traditional Western views. In both Hemingway and *The Lion King* there is an emphasis on greed, gluttony, and stupidity that is ultimately designed to be comical. This reaches its "pinnacle" when a hyena feeds on its own body, as described in *The Green Hills of Africa* and in the American children's computer game based on the movie (*The Lion King . . .*, 1994)

Hyena Reputations Across Time in Western Culture

Most of the negative characteristics associated with hyenas can be found in Aristotle but in relatively non-judgemental form: the animals are what they are. Pliny, with his jumble of stories both true and fabulous, is also relatively unjudgemental, although there is certainly an element of wonder. It is in the *Physiologus* that we first come upon explicit, negative judgements. Early in the Christian era and continuing through the middle ages, two themes emerged: that hyenas changed sex from year to year—a morally unacceptable practice—and that they preyed upon human corpses, digging up graves. The former was linked metaphorically to the Jews and reflected the anti-semitism of this extended period, while the latter was threatening to many human traditions and persists to the present day.

In the post medieval world, the hyena reputation was reexamined. After an early period of uncritical enthusiasm for animal stories, a set of new, somewhat more reality-based attitudes gradually emerged, and hyenas were "appreciated" for characteristics, such as their vocalizations, in seventeenth-century English literature. Finally, as we move into the eighteenth and nineteenth centuries, writers come to understand hyena speciation. However, this did not help the hyena reputation. Their ambiguous sexuality has receded as a moral concern in Europe, but several old concerns moved to center stage: in particular, that they were cowardly, but dangerous, scavengers.

The Biology of Reputation

Reputation is related to biology, but it is clearly not a simple relationship. The fact that spotted hyenas were misperceived as scavengers undoubtedly contributed directly to their poor reputation. Perhaps there is some aspect of their eating food that is unappetizing, or would even be unhealthy for us, that encourages us to reject the "ingestors" of such items. Possibly, there is even a lapse of the work ethic that we find offensive. Why are they not out there working hard and "earning" their food, like an admirable predator? The fact that hyenas are dangerous to humans, to the peaceful rest of our buried friends and relatives and to pets and livestock, must also have contributed to their persistently negative image, as has the misperception of their sexual ambiguity and the strange perception of their "cowardice." However, it is important to recognize the extent of cultural variation.

The Kaguru may have been horrified by the thought of eating hyena flesh (Beidelman, 1975), but the Kujamaat "may eat hyena" (Sapir, 1981, p. 531)[15] and ancient Egyptians evidently found hyenas to be worth fattening in captivity and eating (Zeuner, 1963). People in Western Europe were often preoccupied with thoughts of hyenas disturbing human gravesites, but others, in Africa, were taking the ecologically reasonable step of deliberately setting their dead in the open, to be disposed of by hyenas (Schwabe, 1978). The European Natural Histories of the eighteenth-twentieth centuries are filled with admiration for predators, but as Beidelman observes, "When selecting animals to use metaphorically, Kaguru choose carnivores for negative qualities and herbivores for more positive ones. All larger carnivores (lions, leopards, hyenas, pythons) are described negatively and are a threat to man and his livestock" (Beidelman, 1975, p. 192). The European admiration for predators may depend on distance and weaponry.

As to the supposed sexual ambiguity of hyenas, it is clear that this was much more disturbing to Europeans during medieval times than before or since. Just as there has been great cultural variation in human response to sexual variations in our own species, so there has been variation in attitudes toward presumptive sexual variation in animals. We have seen similar fluctuations across time in European views of cowardice, and this trait has not been a prominent aspect of the African portrait of hyenas described above.

Finally, hyenas have also been castigated for their vocalizations, supposedly foul odors, and ugly appearance. Spotted hyenas do have an incredible array of vocalizations, as one might expect in a social, nocturnal predator (Kruuk, 1972, pp. 220–22; Mills, 1990, pp. 179–87). They are not particularly "smelly," certainly not in the class of a lion. They do have anal scent glands which produce a somewhat soapy/musky smelling paste. However, as the surgeon-anatomist John Hunter observed in the late eighteenth century, in the course of dissecting a striped hyena, these glands " . . . contain a yellow substance somewhat harder than common butter, which smells like musk. They are always distended with this matter, and but very little of it can escape in substance. I would seem as if it was only meant as smelling in a bottle" (Hunter, 1861, p. 58). It is unfortunate that Dr. Hunter never had the opportunity of observing the scent-marking behavior of this species or of the other hyenas. They do extrude these anal glands and deposit paste on stones, tree trunks, grass, or other objects (Reiger, 1981). However, in the ordinary course of events, you are not likely to come upon any concentrated scent-marking site. Visitors to our colony often comment on the lack of odor.

In regard to their "ugly" appearance, they do have large heads and short hindlegs, in comparison with a generalized mammal. Perhaps their appearance does touch some primitive aesthetic template. However, bears also have short hindlegs and that has not prevented them from being adopted as positive cultural icons in the form of childrens' toys. In addition, large heads are characteristic of babies and presumed to be one aspect of their charm (Tinbergen, 1951). This is an area where the media portrayal of hyenas has clearly contributed to the general view.

After spending a great deal of time with spotted hyenas during the past decade, I have come to the conclusion that peoples attitudes often tell us little about hyenas but reveal a great deal about our own needs and anxieties, as well as our readiness to perceive the natural world from a very personal perspective and to make our ideas and images cohere. In the long run, it is important to understand the routes through which we have constructed these cohesive views. What are the essential elements that initiated the process, and what are the conditioned qualities that complete the picture?

Learning to Love Hyenas: A Personal Journey

I believe that there is a special attraction of dangerous carnivores that invites the attention of other animals, including humans. In part, this is just practical fear reduction.

Thompson's gazelles gather on the African savannah, clustering about their primary predator, the cheetah (Kruuk, 1972). Presumably, they are attempting to deal with the danger presented by this predator by gathering information about its location, behavior, and intentions.

Our species also has a set of mechanisms for dealing with dangerous predators. We gather information and learn about them, a bit like gazelles. We also attempt to conquer and control them, as in circus acts, or, sometimes, we denigrate them as vile, repulsive, ugly, comical creatures. There is another strategy. On occasion, we attempt to bond with them.

I believe that my own view positive of hyenas has been constructed, in part, intellectually, by studying them. Knowing the nuances of their lifestyle and the difficulties that they face on a daily basis and being familiar with sights and sounds of hyenas changes my view of them. They begin to look attractive and appealing. I hear their vocalizations, not as eerie, but as a remarkable and necessary trait of a nocturnal mammal with a rich social life. And I find that their foul smell is truly mythical.

But something else has happened that is less cognitive and more emotional; my colleagues and I have bonded with these animals, as Hans Kruuk did with a pet hyena (Kruuk, 1972).[16] As the result of rearing them and watching their individual lives and personalities unfold, we have formed attachments [Fig. 5]. It is a potentially dangerous process for the scientist, although it may be inevitable and can be beneficial to all involved (Davis and Balfour, 1992). It is important to make sure we are aware of what has occurred and try to utilize our familiarity and comfort with the animals to enrich our understanding of their life. Hyena researchers are not alone. Gordon Burghardt, an ethologist at the University of Tennessee who has worked with bears, has recently written an article describing the bonds between humans and bears and the consequences for his research

Fig. 5. Krista Glickman with an adult female hyena.

(Burghardt, 1992). The people I know who work with wolves have certainly had to grapple with similar problems (Fentress, 1992).

At the beginning of this paper, I raised the question of why we were so out of step with the rest of Western culture in our attitudes toward hyenas. I think these are the reasons: a combination of knowledge, familiarity, and bonding. The animals look and sound different to us then they do to others, particularly as hyenas are portrayed in the popular media. We carry a different set of images in our heads.

Hopefully, we will be able to communicate our understanding of these animals beyond the traditional scientific community. I think we owe them that much. And there are some encouraging signs. More and better documenteries have appeared that reveal the complexities of hyena life (for example, *The Sisterhood*; *Lions and Hyenas: Eternal Enemies*). Articles have been published that increase public familiarity and convey a more appealing visual image (Stevens, 1993), and, in a recent edition of *Road and Track*

(47, October 1995, p. 40), it was announced that a new high performance Italian sports car has been built: the Lancia Hyena Zagato. There is even an alternative rock group called "The Laughing Hyenas," and a newsgroup on USENET devoted to people who are fans of hyenas (address: alt.fan.hyena). Hyenas on MTV and the information super-highway! Now there would have been a thought with which Aristotle could grapple.

Notes

[1] Information about the behavior of spotted hyenas in nature can be found in the books by Kruuk and Mills, the field reports of Frank (1986a, p. 1500; 1986b, p. 1510), Hofer and East (1993), and Holekamp and Smale (1991, p. 306).

[2] Personal communication from Mr. Gary Noble of the St. Louis Zoo, who has undertaken to develop a "studbook" of all spotted and striped hyenas in North American zoological parks.

[3] Throughout this paper, references to Aristotle's *History of Animals*, *Parts of Animals*, and *Generation of Animals* are based on the 1912 Oxford edition, which has extensive footnotes for these biological works (Smith and Ross, 1958). References to *On Marvellous Things Heard* are from the most recent Oxford compilation (Barnes, 1984).

[4] In a footnote to this section, the translator (A. Platt) observes that it is impossible to guess what the "Trochus" might have been.

[5] This feature is also described in *HA*, VIII, 5, 594b, 1–2, with the hyena having " . . . a mane like a horse, only that the hair is stiffer and longer and extends over the entire length of the chine."

[6] By spitefulness, Aristotle may be referring to the phenomenon of "surplus killing," that is, killing in excess of what can be consumed. Lions were much admired for their supposed restraint in not killing more than could be eaten.

[7] All references to Pliny's *Natural History* (*NH*) contained in this article refer to the Loeb Classical Library edition (Rackham, 1958).

[8] See Richardson (1987) for a discussion of the dread of dismemberment in a very different context.

[9] Facts of the life of Leo Africanus are considered in the introduction to the first volume, written by Dr. Brown.

[10] The critical distinction between spotted and striped hyenas was still causing confusion. Spotted hyenas, in the modern genus *Crocuta*, were probably present in Ethiopia in those times as they are today. In the next century, the great French naturalist, George Louis LeClerc, Comte de Buffon (1707–1788), would begin his consideration of hyenas with an extensive discourse on the many names that had been applied to these species, but he still did not appreciate the nature of their speciation.

[11] Pseudodoxia Epidemica (PE) was first published in 1646. The quotations provided in this article are from the sixth edition, published in 1672 (Sayle, 1904, V. 1, p. 174).

[12] Sir Walter Raleigh's *The History of the World* was originally published in 1614, while Raleigh was in prison. Quotations provided in this article are taken from an edition printed in London in 1652.

[13] The "laugh" of the hyena is not a joyous sound. At best it is associated with excitement, at worst with a very negative event. For example, hyenas would commonly "laugh" when displaced from a kill by lions. Shakespeare's use is raw anthropomorphism.

[14] Professor Roberts was curator of the exhibition at the Museum for African Art that was held in conjunction with the "In the Company of Animals" conference.

[15] Sapir adds that " . . . it is hardly a preferred food."

[16] Kruuk concludes his acknowledgements with a salute to his tame hyena, Solomon, "who gave us so much fun under the guise of scientific information."

References

Barber, R., *Bestiary* (Woodbridge, Suffolk: The Boydell Press, 1993).

Barnes, Jonathon, ed., *The Complete Works of Aristotle* (Princeton: Princeton University Press, 1984).

Beidelman, T.O., "Ambiguous Animals: Two Theriomorphic Metaphors in Kaguru Folklore," *Africa* 45 (1975).

Brown, R., ed., *Leo Africanus, The History and Description of Africa*, John Pory, trans. (London: The Hakluyt Society, 1986).

Buffon, Comte de, *The Natural History of Quadrupeds* (Edinburgh: Thomas Nelson and Peter Brown, 1830).

Burghardt, G., "Human-Bear Bonding in Research on Black Bear

Behavior," in H. Davis and D. Balfour, eds., *The Inevitable Bond: Examining Scientist-Animal Interactions* (Cambridge: Cambridge University Press, 1992).

Curley, M.J., *Physiologus* (Austin, TX: University of Texas Press, 1979).

Danford, C.G. and Alston, E.R., "On the Mammals of Asia Minor," *Proceedings of the Zoological Society of London* (1877).

Davis, H. and Balfour, D., eds., *The Inevitable Bond: Examining Scientist-Animal Interactions* (Cambridge: Cambridge University Press, 1992).

Dryden, J., in *Ovid's Metamorphoses*, translations supervised by S. Garth (New York: Heritage Press, 1961).

Fentress, J., "The covalent animal: on bonds and their boundaries in behavioral research," in H. Davis and D. Balfour, eds., *The Inevitable Bond: Examining Scientist-Animal Interactions* (Cambridge: Cambridge University Press, 1992).

Flower, W.H. and Lydekker, R., *An Introduction to the Study of Mammals Living and Extinct* (London: Adam and Charles Black, 1891).

Frank, L.G., "Social Organization of the Spotted Hyena (*Crocuta crocuta*). I. Demography," *Animal Behavior* 34 (1986a).

Frank, L.G., "Social Organization of the Spotted Hyena (*Crocuta crocuta*). II. Dominance and Reproduction," *Animal Behavior* 34 (1986b).

Frank, L.G. and Glickman, S.E., "Giving birth through a penile clitoris: parturition and dysocia in the spotted hyaena (*Crocuta crocuta*)," *Journal of Zoology* 234 (1994): 659.

Frank, L.G., Glickman, S.E., and Powch, I., "Sexual dimorphism in the spotted hyaena (*Crocuta crocuta*)," *Journal of Zoology* 221 (1990): 308.

Gesneri, C., *Historiae Animalium* (Zurich: Tiquri, Apud. Christ. Froschoverum, 1551).

Goodrich, S.G. and Winchell, A., *Johnson's Natural History* (New York: A.J. Johnson & Company, 1885).

Hemingway, E., *Green Hills of Africa* (New York: Scribners, 1935). Reprinted with permission of Scribner, a division of Simon & Schuster, from *Green Hills of Africa* by Ernest Hemingway. Copyright 1935 by Charles Scribner's Sons. Copyright renewed © 1963 by Mary Hemingway.

Hofer, H. and East, M.L., "The commuting system of Serengeti

spotted hyaenas: how a predator copes with migratrory prey,"
Animal Behavior 46 (1993): 547.

Holekamp, K.E. and Smale, L., "Dominance Acquisition During
Mammalian Social Development: The Inheritance of Maternal
Rank," *American Zoologist* 31 (1991): 306.

Hunter, J., *Essays and Observations on Natural History, Anatomy,
Physiology, Psychology, and Geology*, Posthumous papers arranged
by Richard Owen (London: John Van Voorst, 1861).

Kruuk, H., *The Spotted Hyena* (Chicago: University of Chicago Press,
1972).

Kruuk, H. in M.G.L. Mills *Kalahari Hyaenas* (London: Unwin Hyman,
1990).

Lloyd, J.B., *African Animals in Renaissance Literature and Art* (Oxford:
Clarendon Press, 1971).

Loro, A., Franceschi, F., and Dal Lago, A., "The Reasons for
Amputations in Children (0–18 years) in a Developing Country,"
Tropical Doctor 24 (1994).

Matthews, L.H., "Reproduction in the Spotted Hyaena, *Crocuta
crocuta (Erxleben)*," *Philosophical Transactions, Series B* 230 (1939):
1.

Milton, J., *Complete Shorter Poems*, J. Carey, ed. (London: Longman,
1971).

Neaves, W.B., Griffin, J.E., and Weilson, J.D., "Sexual dimorphism
in the phallus in spotted hyaena (*Crocuta crocuta*)," *Journal of
Reproduction and Fertility* 59 (1980): 509.

Nott, J.F., *Wild Animals: Photographed and Described* (London:
Sampson, Low, Marston, Searle and Rivington, 1886).

Ognev, S.I., *Mammals of Eastern Europe and Northern Asia* (Jerusalem:
Israel Program for Scientific Translations, 1962).

Poirion, D. *Le Bestiaire* (Paris: Phillipe Lebaud, 1988).

Rackham, H., ed. and trans., *Pliny's Natural History* (Cambridge, MA:
Harvard University Press, 1958).

Reiger, I., "*Hyaena hyaena*," *Mammalian Species, American Society of
Mammalogists*, No. 150 (1981).

Richardson, R., *Death and the Destitute* (London: Routledge & Kegan
Paul, 1987).

Roberts, A.F., *Animals in African Art* (New York: The Museum for
African Art, 1995).

Roosevelt, T. and Heller, E., *Life-Histories of African Game Animals*
(London: John Murray, 1915).

Sapir, J.D., "Leper, Hyena, and Blacksmith in Kujamaat Diola Thought," *American Ethnologist* 8 (1981).

Sayle, C., *The Works of Sir Thomas Browne* (London: Grant Richards, 1904).

Schwabe, C.W., Cattle, Priests, and Progress in Medicine (Minneapolis: University of Minnesota Press, 1978).

Smith, J.A. and Ross, W.D., eds., *The Works of Aristotle* (London: Oxford University Press, 1958).

Stevens, J.E., "Secrets of the Spotted Hyena," *Image Magazine, San Francisco Examiner* (August 8, 1993).

The Lion King: Official Game Book (Indianapolis: Brady Publishing, 1994).

Thompson, D., "On Aristotle as a Biologist," in *Herbert Spencer Lectures* (Oxford: Clarendon Press, 1916).

Tinbergen, N., *The Study of Instinct* (Oxford: Oxford University Press, 1951).

Topsell, E., *The History of Four-Footed Beasts and Serpents* (London: E. Cotes, 1608).

Walker, E.P. et al., *Mammals of the World* (Baltimore: Johns Hopkins Press, 1968).

Watson, M., "On the Female Generative Organs of *Hyaena crocuta*," *Proceedings of the Zoological Society of London* (1877).

Watson, M., "On the Male Generative Organs of *Hyaena crocuta*," *Proceeding of the Zoological Society of London* (1878): 416.

White, T.H., *The Bestiary* (New York: Capricorn Books, 1960).

Zeuner, F.E., *A History of Domesticated Animals* (New York: Harper and Row, 1963).

Animals and the Law: Property, Cruelty, Rights

Jerrold Tannenbaum

In American society, the legal system is the predominant engine of public policy and, perhaps, moral suasion. It is therefore important for all people who are affected by animals—which is, of course, virtually everyone—to know something about how the law treats animals. If the law precludes appropriate ways of thinking about animals, we ought to know about it so that the law can be changed. On the other hand, if the law's general approach to animals is sound, the existing legal apparatus may provide a useful foundation for treating animals fairly.

This paper offers some basic information about how the American legal system deals with animals. This information is placed in historical context. Some readers will find this history surprising, even if they (wisely) begin with the view that the law can work in strange ways and does not always reflect reason, morality, or common sense. However, the main interest of this paper is not historical, and I will refer readers to other sources for more detailed historical discussions. My primary aim is to use historical data to address the following question: Does the law treat animals as essentially inanimate objects, not possessed of important interests, and unworthy of protection in their own right? This question is important because many who view themselves as animal advocates answer it affirmatively. Some of these people believe that the law stands in the way of decent treatment of animals. Some blame the law for much of what they consider to be unethical behavior toward animals.

I will argue that, in fact, the basic conceptual apparatus American law has fashioned to deal with animals is sensible and serviceable. The law's view of animals is not hopelessly manipulative and heartless. Indeed, fundamental legal concepts relating to animals provide considerable moral space.[1] This space allows for some inappropriate treatment of animals, but it is also consistent with excellent treatment of animals. If people have too often treated animals poorly, that is usually not the law's fault, but their own. To blame fundamental legal concepts relating to animals for bad treatment of animals will therefore prove as unproductive as it is historically inaccurate.

THE LEGAL STATUS OF ANIMALS: THE "ACTIVIST" VIEW

It is useful to begin with what I shall call the *activist view of the legal status of animals*. In using the term "activist" I do not mean to disparage those who espouse the view. Many people who advocate this view call themselves animal activists. Indeed, some who espouse the view use much stronger terms to describe themselves, such as animal use "abolitionists" or animal "liberationists."

I will elaborate on the activist view as its contentions are explored further. Initially, it can be characterized as making the following assertions:

(1) The law's fundamental approach to animals is historically outmoded and ethically unacceptable.
(2) This fundamental approach views most animals as the property of human beings, and claims to protect animals through so-called "cruelty-to-animals" laws.
(3) The classification of animals as property and cruelty laws
    (a) do not afford due regard to the interests of animals;
    (b) do not, and indeed cannot, afford them legal "rights;" and
    (c) are in fact not even intended to protect animals and their interests, but to serve human interests.

(4) Therefore, substantial and fundamental changes in the law's conception of and approach to animals are required.

ANIMALS AS PROPERTY

The Popular Definition of Property

In general, animals are classified by the law as property. However, what the law means by "property" is not what many people appear to understand by the term in ordinary discourse.

Many people seem to conceive of property as something that belongs to one and over which one has complete control, to the absolute exclusion of anyone else. Recently, I observed an altercation between a twelve-year-old boy and his next door neighbor that illustrates this common usage of the term "property" beautifully. The boy was playing soccer with his friends in his back yard. One of the children inadvertently kicked the ball over the fence separating the boy's yard from his neighbor's. The boy ran onto the neighbor's lawn to retrieve the ball. The neighbor was standing behind her kitchen window awaiting the impending trespass. She rushed out her back door. Running toward the ball with the obvious intention of seizing it, she uttered words that children throughout the ages have heard from so-called neighbors: "This is my *property*!," she screamed. "Get off of my *property*!" Not to be outdone, the boy responded, as he quickly grabbed the ball and jumped over the fence back into friendly territory, "Well, this ball is *my* property!"

Here, both the boy and neighbor viewed "their property" as something over which they have the right of control. And the assertion "This is my *property*!" is commonly meant to convey that no one else may infringe on such control. One's property, on this view, is something that belongs to one, something the use or destiny of which one has the sole right to determine. If someone else is to use, or have any affect on, one's "property," this can happen only if one gives such a person permission to do so.

In his eighteenth century survey of English law, William Blackstone provided a classic statement of this conception of property. Blackstone also believed (incorrectly) that this conception was to be found deep in the historical foundations of English law. "There is nothing," he wrote, "which so generally strikes the imagination, and engages the affections of mankind, as the right of property; or that sole and despotic dominion which one man claims and exercises over the external view of the world, in total exclusion of the right of any individual in the universe."[2]

Blackstone applied this notion of absolute and exclusive control to the concept of "chattel" or personal property (as distinguished from "real property," that is, land or real estate). He listed typical forms of such chattel over which their owners supposedly have absolute dominion. Included in this list are animals.

> Chattels *personal* are, strictly speaking, things *movable*; which may be annexed to or attendant on the person of the owner, and carried about with him from one part of the world to another. Such are animals, household stuff, money, jewels, corn, garments, and every thing else that can properly be put in motion and transferred from place to place.[3]

> [Regarding property in chattels personal, there is] property in *possession absolute*, which is where a man hath, solely and exclusively, the right, and also the occupation, of any movable chattels; so that they cannot be transferred from him, or cease to be his, without his own act or default.[4]

Blackstonian Absolute Possession and the Activist View of the Legal Status of Animals

If Blackstone's analysis of the concept of the law's concept of property were correct, it would be obvious that the law must afford animals very little protection and, indeed, must have a very low regard for them. For, according to this analysis, animal owners would have "sole and despotic dominion" over their animals. Owners could, at their option, treat their animals well, but they could also treat them quite miserably, indeed, could do with them as they wish, without interference from others. Moreover, to say that something can freely be the object of "despotism" is to place very little value upon it. A thing or being over which a despot exercises despotic control is far less important to the despot than the despot is to himself. If such a being's interests are to be respected or furthered, this will not be something to which the being has a *right*. Such protection will only come at the sufferance of the despot, and can be turned on or off at the despot's will or whimsy.

The Blackstonian conception of property has played an important role in what I have termed the "activist view" of the legal status of animals. Many proponents of this view accept the analysis of property offered by Blackstone, and then conclude that the law should no longer classify animals as property because people ought not to exercise sole and despotic dominion over other sentient beings. Such an argument is presented by Professor Gary Francione in a recent critique of current legal principles governing the treatment of animals. Citing Blackstone and the philosopher John Locke, upon whom Blackstone relied in formulating his views regarding property, Francione considers how the law deals with situations in which the interests of animals conflict with the interests of their owners.

As far as the law is concerned, it is as if we were resolving a conflict between a person and a lamp, or some other piece of personal property. The winner of the dispute is predetermined by the way in which the debate is conceptualized in the first place. The human interest in regarding animals as property is so strong that even when people do not want to consider animals as mere "property," and, instead, view animals as members of their family, as in the case of dogs, cats, and other companion animals, the law generally refuses to recognize that relationship. For example, if one person negligently kills the dog of another, most courts refuse to recognize the status of the animal as family member and will limit the owner to the same sort of recovery that would be allowed if the property were inanimate.[5]

In one important respect Francione's apparent view of the nature of property differs sharply from Blackstone's. Central to Francione's notion of property—and to the activist view of the legal status of animals—is that property, or at the very least, paradigm forms of property, are in an important sense or respect inanimate objects such lamps or other personal or household goods. This explains Francione's statement that animal owners who consider their animals as members of their families do not treat these animals as "mere" property. Francione appears to think that only something which is an inanimate object, and indeed is considered (somehow) as no more valuable than an inanimate object is "mere" or *pure* property.[6] In contrast, in the passage quoted above, Blackstone not only included animate beings—animals—in his list of kinds of personal property, he placed them first on the list. Locke too not only included animate beings—animals—among kinds of personal property, he placed animals first on his list of kinds of property. Locke thought that *historically* the "original" property right was the "right a man has to use any of the inferior creatures, for the subsistence and comfort of his life."[7] Locke believed that this was the first property right because, he believed, God bestowed this right on humankind on the fifth day of Creation.

For the historical record, it is important to note that Locke's theory of property did not, as Francione asserts, have "an extraordinary influence on the common law."[8] Common law principles relating to possessory rights in land and chattel, as well as the forms of action in which such rights were pursued, were largely in place well before Locke's time.[9] Moreover, as will be discussed below, Blackstone's principle of sole and despotic dominion over personal property was never a feature of the common law.[10] However, at least Locke and Blackstone never suggested that inanimate objects such as lamps constituted the paradigm forms of personal property. Blackstone included animals in his list of kinds of chattel without qualification. He did not distinguish between animals as one fundamental kind of property and inanimate objects as another, although he believed that the law treated wild animals differently from domestic species because the former tended to stray and therefore were more difficult to possess.[11] He never would have suggested that people who viewed animals as members of the family would thereby cease to view them as "mere" or true property. To Blackstone and Locke animals were every bit as much property as "household stuff, money, jewels, corn, garments, and every thing else that can properly be put in motion and transferred from place to place."

As we shall see, the actual history of the legal concept of property provides *absolutely no support* for the claim that property, "true" property, or property properly speaking is or should be inanimate. However, at this stage in the discussion the claim that "mere" or true property is inanimate provides the opportunity to further elaborate on the activist view of the legal status of animals. Regarding animals as property, the activist view holds as follows:

(1) The law regards animals as a form of property.
(2) Historically as well as conceptually, the law began with a category of "property" and placed animals in this category.
(3) Animals can be removed from the category of property just as they were placed in it.
(4) Personal property, by its very nature, involves "sole and despotic dominion" over it by its owners.
(5) Therefore, if animals are property their interests are not important. Animals must give way to their owners. They certainly cannot have legal "rights."
(6) If the law is to accord animal interests due regard, or to afford them rights, it must remove them from the category of property.
(7) The paradigm of personal "property" is an inanimate object, or to use Blackstone's examples (other than animals)

"household stuff, money, jewels, corn, garments, and every thing else that can properly be put in motion and transferred from place to place."

(8) Inanimate things cannot have rights.

(9) Therefore, as property, animals cannot have legal rights.

(10) If the law is to accord animal interests due regard, or to afford them rights, it must remove them from the category of property.

Property in Post-Conquest English Law: Land

To understand how American law conceives of property in general and animals as property in particular one must look to the beginnings of our law—which is to say one must start with the victory of William the Conqueror over King Harold in 1066. The early history of Anglo-American law is fascinating in its own right, and I cannot even begin here to do justice to it or the considerable body of scholarship it has engendered.[12] It can be said that shortly after the Conquest, and proceeding in earnest beginning with the reign of Henry II (1154–1189) foundations of what lawyers call the "common law" were enunciated by judges and were modified by various royal charters (such as the Magna Carta), acts of Parliament, and litigation in the common law courts. This common law was brought to the American colonies and became part of the law of the first 13 states upon the creation of the United States. Later states (with the exception of Louisiana, which had been a French colony and retained much of its old civil rather than the common law tradition) also adopted the common law. This incorporation of the common law proceeded sometimes by legislative edict but more often through decisions of judges, who accepted and built upon fundamental legal causes of actions and concepts which had been enunciated in the early English common law courts.

The origins of English law were feudal,[13] and Anglo-American property law property still shows signs of its feudal roots. The first, and initially the most important form of "property" (this word must be placed in quotation marks because it arrived relatively late) in English law was land. To William the Conqueror and his men, nothing was more important than the land. They had to conquer the indigenous population for it, and for some time had to defend their new territory against foreign invaders. The feudal system of land distribution and possession, which William brought with him from the Continent, was well suited for these difficult times. Figure 1 illustrates how the system worked. There was only one person who could properly be called an "owner" of land, and that was the King. The King owned all the land in the kingdom. In return for service (which involved an oath of fealty and supplying a certain number of men for the King's army) the King granted his knights the right to possess land. In turn, these men swore fealty to the King's knights, and were then granted land in return for their service. This system worked well as long as there were wars or skirmishes, but was not suited for times of peace. As is illustrated in the figure, if A (one of the King's immediate knights) died or failed in his duties of service to the King, all of the land possessed by A would revert to the King. This would displace all of A's men (and their families) if the successor chosen by the King did not choose to keep them on. Likewise, when one of A's men (B, for example) died or failed in his duties of service to A, the land he possessed would revert to A, displacing B's son or sons and their families unless A chose to keep them on in return for their service.

This system reflected the principle that all rights to possess any land derived from the King's control over all the land, and from the linkage of possession of land to the King's need for military service. Possession of land was in a real sense personal. It was as transitory and temporary as the personal relations between the King and his immediate grantees, and the grantees of the King's grantees. It is therefore not surprising that the terms "owner" and "ownership" did not appear with any frequency in the English law books until the late 1500s—because no one except the King really "owned" any land. Indeed, even the term "property" did not appear with frequency in statutes or case reports until the seventeenth century. Rather, people spoke of their estates (in land) and their possessions (of movable things).[14] When the word "property" was used, it did not denote Blackstonian absolute and despotic dominion, because English law did not recognize any such dominion except by the Sovereign. Rather, when it was used (which, initially, was not very often) the term "property" meant the *best* right of possession.[14] This remains its core meaning in Anglo-American law.

For many years following the Conquest, English law struggled to effect something that seems absolutely unproblematic to us today: how to "own" land in the modern sense. How could men on the level of B through K in Figure 1 be permitted to hold onto their land after A died? How could one of these pass on his land to an eldest son? Other sons? A daughter? The Church? How could someone sell one's right to possess land (later viewed as one's land) to someone other than a blood relative? The law instituted a number of fictions and procedures, sometimes bordering on the bizarre,[15] as it fashioned rules that resemble contemporary land law. All these developments presupposed, at first in actuality and later more in theory, the principle that the King owns all the land, and that everyone who has title to land derives it ultimately from the King.

Property in Post-Conquest English Law: Movable Goods

A second major category of what was later called "property" comprised what the law called "movable goods" or "chattel." For a number of years, some legal historians thought that the common law had taken an entirely different approach to movable things from that which it had taken to land. While the

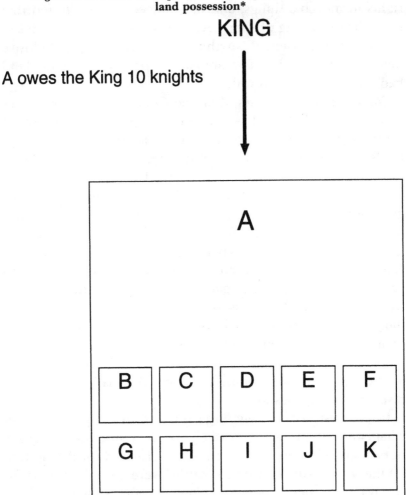

Figure 1. A schematic representation of post-Conquest feudal land possession*

* This diagram was used by the late Professor Samuel Thorne in his lectures on English legal history at Harvard Law School.

King was said to "own" all the land and to condition possession of land on satisfactory service to him, movable things, some historians thought, were different. Movable things such as animals, books, and clothing were capable of absolute ownership in a way land was not.

Modern scholarship has debunked this view of property rights in movable things, and has, indeed, demonstrated that from the beginning of Anglo-American law, property rights in movable things were if anything less absolute and exclusive than rights in land.[16] The animals of post-conquest England had a great deal to do with this.

For a number of years after the Conquest, there was very little of value that most people (even knights in possession of substantial amounts of land) could possess. Therefore, few kinds of movable things were the subject of litigation. (From the beginning, the common law courts developed principles of law out of actual litigated cases, and not consideration of general or abstract possible situations. This remains a central feature of English and American jurisprudence, as distinguished from European civil law codes that derived from Roman law.) Knights and men-in-service had their armor; this could be the subject of litigation if there was a dispute as to who had the right to possess it. However, except for armor most valuable inanimate objects consisted of books (which few people could read), embroidered vestments and crucifixes, and jeweled crowns. Such things were secured in Church sanctuary or the King's treasure house and were thus safely away from disputes over possession.[17]

In contrast, a relatively large number of people possessed what were then called "cattle": oxen, cows, sheep, goats, horses, and chickens. These animals were used for draught, to till the soil, or for food. Such "cattle" were important, first for survival, and then as wealth. When someone stole one's cattle or claimed the right to possess it, this was a serious matter, and litigation would often ensue.

Thus, the development of Anglo-American law of personal property began to a large extent with litigation over the rights of possession of such cattle. This is why the law came to use the term "chattel" for movable goods over which there were disputes as to rights of possession. The similarity in the words "cattle" and "chattel" is not coincidental. Among the first chattel were cattle. The word "chattel" derived from the Latin *catalla*.[18]

This fact bears repeating. It was not the case, as some proponents of the activist view of the legal status of animals assert, that animals were just one kind of personal property. Animals were a seminal kind of personal property. Animals—animate beings—came early, perhaps even first, and inanimate objects were included as they too became the subject matter of litigation.

The Nature of Post-Conquest Cattle: Chattel as Money

Several features of the cattle of post-Conquest England played an important role in the development of the law of personal property. First, such cattle were *movable*. They could be transported from place to place and more importantly from possessor to possessor. Second, cattle were extremely *valuable*. Indeed, some were so valuable to their possessors that they were worth a great deal in barter or trade—they could, singly or in groups, be used to obtain other movable goods of value. Third, these cattle were usually not considered unique and irreplaceable. Most oxen or sheep were pretty much the same as any others. Today's breeds of domestic animals, and uniquely valuable animals, were as yet unknown.[19] These "cattle" were, to use a contemporary legal term for essentially similar and interchangeable objects, *fungible*. Without such fungibility, they could not be used in trade, and passed from possessor to possessor.

Because of their transportability, value, and fungibility, cattle became an early form of money.[20] For money, be it made of gold, mineral stones, beads, paper, or something else, is essentially something of value that can be used to obtain goods or services, but is not considered (except by the miser) to be valuable in itself. The Domesday Book, William the Conqueror's massive catalogue of the lands and goods in his new kingdom, used the Latin term *pecunia* (money) for cattle.[17]

Cattle, Chattel, and the Nature of Personal Property

The use of cattle as money, and the fact that cattle were generally fungible, helps to explain a feature of English and American property law that some people find surprising and inexplicable. If someone steals or destroys your property, or destroys its value to you, the law allows you to bring a civil action against the wrongdoer. However, except in circumstances in which the property can be characterized as unique, if you prevail you will not be entitled to your property back or a piece of property just like it. You will be awarded the monetary value of your property at the time of its loss. For example, if a negligent driver renders your car a total loss, he or his insurance company will not be obligated to replace your car with a model of like age and quality; you will receive money supposedly representing the fair market value of your vehicle at the time of the accident. As the great historians of English law Pollock and Maitland explain,

> Our common law in modern times has refused, except in rare cases, to compel the restitution of a chattel. Having decided that the chattel belongs to the plaintiff and that the defendant's possession is wrongful, it nevertheless stopped short of taking the thing by force from the defendant and handing it over to the plaintiff. Its judgment was that the plaintiff should recover from the defendant the chattel or a sum of money that a jury had assessed as its value. This left the defendant the choice between delivering up the thing and paying a sum of money, and if he would do neither, then goods of his were seized and sold, and the plaintiff in the end had to take money instead of the very thing he demanded.[21]

This feature of English and American property law derives directly from the cattle of post-Conquest England. The refusal of the common law to compel restitution of a chattel might seem mysterious or bizarre until one considers what the objects of litigation concerning cattle really were. If you stole or destroyed someone's sheep or ox, you deprived him of something of value to be sure, but something which in itself could generally be used to purchase other things. Therefore, it was no major problem to compensate a plaintiff wrongly deprived of cattle with money (whatever form it might take: at first cattle itself and later coin of the realm) so that he could then replace it with another, equivalent, animal. He thus received equal *value*, which seemed to the law entirely fair.[22]

"Personal" Versus "Real" Property

The fungibility of chattel, and the law's concentration on its value to its possessors and not its inherent nature, is underscored by another historical fact about which few people (and even few lawyers) are aware. Today, lawyers speak about two *kinds* of property, "real" property and "personal" property. Real property, or real estate, consists of land and buildings. Personal property, or chattel, consists of movable goods. Many lawyers are under the impression that the term "real" property refers to land because the term derives from the Latin word *res* for "thing" and because land is a tangible thing. In contrast, the term "personal" property (or "personalty") is thought to have something to do with the fact that most such property is of direct personal value or of value in personal life, as are, for example, such things as household goods, clothing, and books.

Such a historical account of the terms "real" and "person" is not even intuitively plausible, because chattels are things and because land can be of great personal value. More importantly, this interpretation of the distinction between real and personal property is historically incorrect. The interpretation also reinforces a view that is in part responsible for misunderstandings about the nature of the legal status of animals. As commonly understood, real and personal property are two kinds of property. The terms "real" and "personal" always referred, it is said, to the property itself—to some characteristic of *it*. This is part of the approach that insists that the law has always viewed, and still views, animals (like other chattel) as a certain kind of *object*, and real estate as a different kind of *object*.

Historically, the terms "real" and "personal" did not refer to two different kinds of property. The terms referred to two different kinds of legal action. A real action was a lawsuit in which someone who had been deprived of something or had the right to something could obtain *that thing*, hence the term "real." A court could order transferal of possession of that thing to him. Legal actions for land were "real," because from the very beginning of English law courts could award or ratify possession of land directly to a plaintiff who was entitled to it. Land was sufficiently unique in the eyes of the law that monetary compensation was insufficient. In contrast to real actions, personal actions were those in which possession of a thing need not be given or restored but in which the courts could compensate the *person* for his loss. Such compensation was measured in terms of monetary value. Movable goods— chattel—such as animals, household goods, and other possessions were the subject of personal actions because, as has already been explained, the law did not generally consider them unique. Actions for chattel were personal because one had a right not to the thing but to its value. Over time, the terms "real" and "personal" came to be applied to land on the one hand and chattel on the other because the former had been the subject of real actions and the latter generally of personal actions.[23]

*Property Not as Things But as a Relation between Owner
and Property*

The origins of the distinction between "real" and "personal" property—and more importantly, the reluctance of the law to grant or restore possession of all but the most unusual kinds of chattel—are of immense importance in understanding how our legal system actually views personal property in general, and animals as property in particular.

No sole and despotic dominion. First, we have already seen why regarding at least most forms of chattel, property rights do not involve sole or despotic dominion over one's property. If one had such dominion, one would always be entitled to the return of one's property, or if it were no longer available, to a piece of property just like it. Instead, except in the case of unique property one is entitled to its monetary value. "Personal property" is not a category that marks out or characterizes those things that are property. "Personal property" is more a characteristic of the *rights* of *people* regarding that which is property. The law has traditionally entitled people to the value of chattel and to repossession of land because the law has declared what it believes to be the value of various kinds of possession to their possessors. In most cases, for chattel this value is monetary.

Rights of possession subject to qualification by government. That property rights do not involve sole and despotic dominion is also illustrated by the fact that governments (local, state, and sometimes federal) can place constraints upon one's ability to possess, use, or transfer one's property. Limitations can be placed on both chattel and real estate. One cannot own an automobile without obtaining the state's permission in the form of a certificate of registration, and one cannot put such property to its ordinary use without further permission in the form of a driver's license. Many localities do not permit the possession of certain kinds of domestic animals (such as chickens or pigs). Many communities require owners to obtain a license to keep a dog, demand that dogs away from their owner's premises be controlled on a leash, prohibit the possession of dogs or cats that have not been licensed or vaccinated for rabies, or prohibit the possession of more than a specified number of dogs or cats on one's premises. Most localities do not allow people to construct or modify their own buildings without obtaining a building permit, and many sharply restrict the kinds or style of buildings they allow. These are but a few of the constraints on the possession and use of personal and real property.

At the time of this writing, there is considerable litigation dealing with the difficult question of the extent to which the government should be allowed place constraints upon the use, enjoyment, and alienability of various kinds of property.[24] As is discussed below, the concept of property does place limits on how far governments may go in constraining owners regarding its use. This point implies two quite different principles. First, pursuant to the Fourteenth Amendment of the United States Constitution, if something is property, certain kinds of interference with it (including outright seizure) will be considered a "taking." Due process requires that people be given notice of the government's intention to take and the opportunity to be heard, and they must be compensated fairly for the value of their property if such taking is valid. Second, government can sometimes intrude into the use of land or things to such an extent that it becomes inappropriate to say that these things can be anyone's property in the first place. For example, as is noted below federal and state governments have the authority to prevent anyone from possessing or selling species of wild animals, but can allow people to possess or sell them to the extent that people are allowed to possess and sell ordinary kinds of chattel such as household goods. If the government prohibits possession of a certain kind of animal, that kind of animal cannot be the property of private citizens. However, as a government expands the ability of people to do certain things with these animals, these people come closer to having "property" rights in them, and eventually do have such rights. Thus, if the government allows land owners to feed certain species of birds that fly onto their property, but does not permit capture or restraint of these animals, such permission does not grant property rights. In contrast, if the government allows people to purchase certain species of wild birds in pet stores, to keep these animals as their own, and to sell them if they wish, property rights have been conferred.

Priority of possession or use. It is beyond the scope of this discussion to offer a detailed analysis of the legal concept of property. However, certain things can usefully be said here. Items classified as "property" typically do have certain characteristics. These characteristics are so frequently associated with property that they have come to be viewed as features of property itself. Among the typical characteristics of "property" both real and personal is the ability of its owner to sell it, to pass it to his heirs after his death, to use it as collateral for various kinds of obligation, to use or consume it, and to prevent others from using or consuming it.[25]

However, both the history and practice of Anglo-American law show that it is a mistake to define property merely as things which have such attributes, or even in terms of these attributes themselves. In essence, the term "property" refers to *priority* of certain kinds of rights of possession or use. To say that I own a piece of land, an automobile, or an animal, is not to say that I have Blackstonian "sole and despotic dominion" over it. It is to say that I am ordinarily *more* entitled to possess or use it than anyone else, and that if someone else attempts to take it from me or to deprive me of my use or enjoyment of it, the law will recognize my priority—by giving it back to me in the case of unique and irreplaceable property or by compensating me for its value in the case of property that is not unique or that cannot be returned. The features we generally associate with property itself can be associated with it because of priority of possession and use. For example, I can sell a piece of land, a car, or an animal, because I have priority of possession against others, and the law allows me to transfer this priority for a price.

Put another way, the legal concept of property signifies relationships not just between an owner and his or her property but between an owner and other people. As Constitutional lawyer Bruce Ackerman observes

> instead of defining the relationship between a person and "his" things, property law discusses the relationships that arise *between people* with respect to things. More precisely, the law of property considers the way rights to use things may be parceled out amongst a host of competing resource users. Each resource user is conceived as holding a bundle of rights vis-à-vis other potential users . . . And it is probably never true that the law assigns to any single person the right to use anything in absolutely *any* way he pleases. Hence, it risks serious confusion to identify any single individual as the owner of any particular thing. At best, this locution may sometimes serve as a shorthand for identifying the holder of that bundle of rights which contains a range of entitlements more numerous or more valuable than the bundle held by any other person with respect to the thing in question. . . . For the fact (or is it the law?) of the matter is that property is not a thing, but a set of legal relations between persons governing the use of things.[26]

An Example of the Value of Property Status for Animals

At the panel discussion following the presentation of the papers collected in this volume, philosopher Colin McGinn offered an example which illustrates how the concept of property connotes rights of priority of possession. To appreciate the significance of the example one must know that McGinn opposes, among other things, the killing of animals for food. McGinn's example is important because it shows that the concept of property is so powerful that it is likely to be applied to animals even by many people who consider themselves animal activists.

McGinn related that he owns a cat. Imagine, he said, someone who is convinced that he can provide McGinn's cat a better life than McGinn. This person comes along and takes or demands the cat. McGinn observed that his reaction would be to claim that the cat is his *property*. He would not be interested in a dispute (in a court of law or somewhere else) regarding what is in the best interests of this cat. He would respond, and hope that the law would agree, that this is just *his* cat, his property, and that is all there is to it. In the absence of the most compelling kinds of abuse that would be prohibited by law people should, McGinn observed, be allowed to keep their pets, their property.

THE LEGAL STATUS OF ANIMALS FROM THE TWELFTH TO THE EARLY NINETEENTH CENTURY

Once animals were established as chattel, which occurred by the early 1100s, their fundamental legal status remained essentially unchanged in Anglo-American law until the eighteenth century. The major development during this period was the early fashioning of a distinction between wild and domestic animals and of associated principles of property rights. King John acquiesced formally to these principles in the Magna Carta of 1215. They were incorporated into the laws of the various colonies and then the states of the United States as part of the common law.[27]

Animals were classified into two categories, wild (*ferae naturae*) and domestic (*domitae naturae*). To determine into which of these classes an animal belongs one looked not at the characteristics of the animal itself, but rather whether the law classified its kind as wild or domestic. Thus, an extremely ornery and dangerous dairy bull was (and is still) classified as domestic because it belongs to a kind or species of animal the law considers domestic. The law began with a very small list of domestic kinds or species. Everything else was wild. Domestic animals were, roughly, those that had been tamed, bred, and used for farming, food, and draught—what the courts termed for quite some time "useful" animals. Crucial to the legal concept of domesticity was economic use and benefit. Thus, household pets were not classified as domestic for several centuries because they tended, as a group, to have little economic usefulness or value. Domestic animals were those to which the generic term "cattle" had been applied: oxen, cows, horses, sheep, goats, donkeys, mules, and chickens.

Wild Animals: Ownership by the State

Regarding wild animals, the Magna Carta affirmed the principle that the King had the right to possess all such animals. However, this did not allow him to do whatever he wanted to these animals because others (at first the nobility and later the general populace) had an interest in possessing and using them as well. People could possess or own wild animals in several ways. They could be given permission or license by the King to do so, which Blackstone described as possession *propter privilegium*.[28] Most of the licenses granted by the English kings were to hunt and fish. If the King said nothing about who could take, hunt, or possess certain animals or kinds of wild animals, ownership was still possible under various circumstances. One could, for example, own a wild animal by taming and using it (possession per *industriam*);[29] however, in general, if such an animal strayed from one's land it ceased to be one's own and reverted to its "natural" legal state, in which it was either owned by the King or subject to taming and use by someone else. One could also be said to have the right of possession *propter impotentiam* of animals that were so young or old that they were too weak to move by themselves from one's land.[30]

Most of these principles remain in effect in the United States today. The basic rule of American law relating to wildlife is that the state and federal governments own all wild animals in their respective lands in trust for the people. These governments may, and frequently do, give private citizens permission to possess or own members of wild species. Such permission can take the form of a general law, regulation, or policy decision allowing people to own certain species or exempting them from the general requirement of seeking permission to own a wild animal. (States typically provide blanket permission to own exotic birds commonly kept as pets or typical home aquarium fish.) Permission can also be granted by the issuance of written licenses or permits, which allow people to possess or deal with identified animals or species in specified ways. The authority of the federal or state governments to control the possession of wildlife extends to exotic as well as indigenous wild species. For example, a Northeastern state can dictate whether, or under what conditions, anyone within its territory may possess a boa constrictor just as it can dictate whether, or under what conditions, skunks or squirrels native to the state may be possessed.

The power of government to restrict or impose conditions on the ownership or possession of wildlife is extremely broad, and courts will not overturn such government decisions unless fundamental constitutional or legal principles are violated. (For example, a state could not prohibit only members of a racial or ethnic group from owning certain kinds of wildlife, because this would constitute unconstitutional discrimination.) This explains why one state can prohibit the possession of ferrets by private citizens, another state can condition such possession on the issuance of a license, and a third may allow anyone to possess a ferret. If a state decides that it shall allow ownership or veterinary treatment of certain wild species, or certain species under certain conditions, its decision will stand as long as it can articulate a defensible reason related to its interest in controlling wildlife for the public benefit.

Domestic Animals: Ordinary Chattel

As we have seen, domestic animals were not only chattel or personal property. They were among the first clearly recognized such forms of property. It is therefore not surprising that domestic animals were treated in fundamental respects as other kinds of personal property. They could generally be possessed (later "owned") by people without permission from the Sovereign. To be sure, specific principles relating to the keeping of animals developed that related to certain particular characteristics of domestic animals, such as rules prohibiting certain kinds of domestic animals in certain localities or kinds of localities. However, regarding their status as property, the principles that applied to the ownership of property applied to them. Like other kinds of property, they could be owned, sold, taxed, bailed (given to others temporarily for care or improvement), inherited, and stolen. Constitutional provisions applicable to property applied as well to animals. Thus, the Fourteenth Amendment of the U.S. Constitution, which prohibits the states from taking one's property without due process of law applies to domestic animals as it applies to other forms of chattel.

The Emergence of Dogs and Cats as Domestic Animals

Today pets such as household dogs and cats are classified by the law as domestic animals. This may seem obvious, but it is in fact a relatively recent development. Indeed, for some time the law was uncertain about how to categorize dogs and cats. They were not wild animals because these species had been tamed and kept in human habitations for centuries. But neither did they seem like domestic animals, for the paradigm domestic animals were farm animals that were of direct economic use. The uncertain status which dogs and cats occupied in American law until the twentieth century is illustrated by an 1897 decision of the U.S. Supreme Court. The case involved a challenge to a Louisiana statute that conditioned the recognition of dogs as full-fledged personal property upon registration and declaration of their value with one's local government. If this was not done, a dog "owner" could not sue someone who wrongfully destroyed or harmed his dog, because the dog was not property the destruction of which entitled one to compensation. The particular case involved a dog possessor who had failed to register his valuable Newfoundland bitch and was denied the opportunity to sue when it was run over by a train. The dog's "owner" claimed that the law was unconstitutional because dogs were property and that he was therefore denied due process in being prevented by the State from suing to recover the value of the animal.

The Supreme Court held that dogs were not as a matter of fundamental or common law domestic animals and that a state could therefore decide to grant them full property status only upon fulfillment of specified conditions. This approach was justified, according to the Court, because dogs cannot be

> considered as being upon the same plane with horses, cattle, sheep and other domesticated animals, but rather in the category of cats, monkeys, singing birds and similar animals kept for pleasure, curiosity or caprice. . . . Unlike other domestic animals, they are useful neither as beasts of burden, for draught (except for a limited extent), nor for food. They are peculiar in the fact that they differ among themselves more widely than any class of animals, and can hardly be said to have a characteristic common to the entire race. While the higher breeds rank among the noblest representatives of the animal kingdom, and are justly esteemed for their intelligence, sagacity, fidelity, watchfulness, affection, and, above all for their natural companionship with man, others are afflicted with such serious infirmities of temper as to be little better than a public nuisance. All are more or less subject to attacks of hydrophobic madness.[31]

Although the Court has never specifically overruled this holding, it is undoubtedly not the law today. Monkeys and "singing birds" are wild animals that can be possessed only with permission of the state or federal government. However, because of their importance and value to their owners, courts now consider dogs and cats domestic animals.

The Early Legal Status of Animals: Property Without Interests

Although the legal status of animals as property did not give
their owners sole and despotic dominion over them, until the
emergence of animal cruelty laws in the 1800s the law did not
recognize that animals had any interests which their owners, or
anyone else, were legally obligated to respect. All conditions
placed on the ability of people to possess, use, or dispose of
their animal property were based on human interests. What
was good or bad for the animals themselves was irrelevant. It
was not a crime to subject one's own animal to considerable
suffering, just as it was not a crime to mutilate one's own book
or blanket. One could be prosecuted for injuring or killing an
animal belonging to someone else, but the injury the law considered to be done in such cases was not to the animal but to its
owner. Several states enacted so-called "malicious mischief" laws,
which prohibited maliciously injuring or destroying another
person's animals. Some of these laws spoke specifically of animals, and in others animals were included because they were
simply one kind of property that could be maliciously damaged
or destroyed. However, all such laws applied only when the
miscreant evidenced malice toward the animal's owner. These
laws were not intended to protect the animals themselves.[32]

CRUELTY

Development of Cruelty Statutes

Proponents of what I have termed the "activist" view of the
legal status of animals believe that there is something inherent
in the legal concept of property that prevents the law from
considering the interests of animals—and from affording them
legal rights—while at the same time considering them
property. This position incorrectly interprets a historical phase
in the development of the legal status of animals as a feature of
property itself. The law did not consider animal interests

important or afford them legal rights not because it considered them property, but because it did not think protecting them for their own sakes was important. With the emergence of cruelty laws animals eventually became sufficiently important to become property with legally cognizable interests.

The English common law did not make it a crime to harm the interests of animals themselves, nor did it use the term "cruelty" to apply to treatment of animals. The first cruelty-to-animals law in Anglo-American jurisprudence was contained in the Body of Liberties adopted by the Massachusetts Bay Colony in 1641. Liberties 92 and 93 provided as follows:

> *Liberty 92* No man shall exercise any Tyranny or Crueltie towards any bruite Creature which are usuallie kept for man's use.[33]

> *Liberty 93* If any man shall have occasion to leade or drive Cattel from place to place that is far of, so that they be weary, or hungry, or fall sick, or lambe, It shall be lawful to rest or refresh them, for a competent time, in any open place that is not Corne, meadow, or inclosed for some peculiar use.[33]

Although there is a record of at least one prosecution under these statutes,[34] they did not become a lasting part of the laws of the Bay Colony or of Massachusetts after it became a state. No other American colony adopted this prohibition of "tyranny" or "crueltie."

It was not until 1822 that an enduring cruelty law appeared in Anglo-American law. After heated debate, the English Parliament passed Martin's Act, an "Act to prevent the cruel and improper treatment of Cattle." The Act authorized a fine of from 10 shillings to 5 pounds or imprisonment not exceeding 3 months for persons convicted of cruelly treating "Horses, Mares, Geldings, Mules, Asses, Cows, Heifers, Steers, Oxen, Sheep and other Cattle."[35] In 1835 Martin's Act was extended to all domestic animals.[36]

The first modern American cruelty law was enacted in New York State in 1828. This statute declared that "Every person who shall maliciously kill, maim, or wound any horse, ox, or other cattle, or sheep, belonging to another, or shall maliciously and cruelly beat or torture any such animal, whether belonging to himself or another, shall, upon conviction, be adjudged guilty of a misdemeanor."[37] This law, like the original Martin's Act, applied only to what English law characterized generically as "cattle." It also required a malicious intent on the part of the defendant.

In 1866 Henry Bergh, founder of the American Society for the Prevention of Cruelty to Animals, persuaded the New York State Legislature to amend the cruelty law to include all animals, and to allow for prosecution for negligent as distinguished from intentional or malicious infliction of cruelty. The law now provided that:

> Every person who shall by his act or neglect, maliciously kill, maim, wound, injure, torture or cruelly beat any horse, mule, ox, cattle, sheep, or other animal belonging to himself or another, shall, upon conviction, be adjudged guilty of a misdemeanor.[38]
> Every owner, driver or possessor of an old, maimed, or diseased horse or mule, turned loose or left disabled in any street, lane or place of any city of this state, who shall allow such horse or mule to lie in any street, lane or public place for more than three hours after knowledge of such disability, shall, on conviction, be adjudged guilty of a misdemeanor.[39]

In 1867 Bergh wrote and the New York Legislature passed yet another cruelty law. This statute has proved to be one of the most remarkable pieces of legislation in the history of American law. Not only has much of it remained on the books in New York; its basic language still forms the nucleus of the majority of state cruelty laws in this country. One section prohibited participation in or assisting the "fighting or baiting [of] any bull, bear, dog, cock, or other creature."[40] Other provisions required persons who kept animals in a pound to provide "a sufficient quantity of good and wholesome food and water."[41] A section that remains in place today provided that "If any person shall carry or cause to be carried, in or upon any vehicle or otherwise, any creature, in a cruel or inhuman manner, he shall be guilty of a misdemeanor. . . ."[42] The statute exempted from its application "properly conducted scientific experiments or investigations, which experiments shall be performed only under the authority of the faculty of some regularly incorporated medical college or university of the state of New York."[43]

The first section of the 1867 statute contained language that can be found in many contemporary American state cruelty-to-animals laws.

> If any person shall over-drive, over-load, torture, torment, deprive of necessary sustenance, or unnecessarily or cruelly beat, or needlessly mutilate or kill, or cause or procure to be overdriven, over-loaded, tortured, tormented or deprived of necessary sustenance, or to be unnecessarily or cruelly beaten, or needlessly mutilated or killed, as aforesaid, any living creature, every such offender shall, for every such offense, be guilty of a misdemeanor.[44]

The similarity of the 1867 statute to the current New York law is apparent. The present statute provides that

> a person who overdrives, overloads, tortures or cruelly beats or unjustifiably injures, maims, mutilates or kills any animal, whether wild or tame, and whether belonging to himself or to another, or deprives any animal of necessary sustenance, food or drink, or neglects or refuses to furnish it such sustenance or drink, or causes, procures or permits any animal to be overdriven, overloaded, tortured, cruelly beaten, or unjustifiably injured, maimed, mutilated or killed, or to be deprived of necessary food or drink, or who willfully sets on foot, instigates, engages in, or in any way furthers any act of cruelty to any animal, or any act tending to produce such cruelty, is guilty of a misdemeanor . . .
>
> Nothing herein contained shall be construed to prohibit or interfere with any properly conducted scientific tests, experiments, or investigations, involving the use of living animals, performed or conducted in laboratories or institutions, which are approved for these purposes by the state commissioner of health . . .[45]

Structure of Contemporary American Cruelty Laws

Cruelty Laws as Criminal Statutes

Several important features of cruelty laws follow from their status as criminal statutes. First, the crime of cruelty, like most crimes, is a matter of state and not federal law. Each state has its own animal cruelty law or laws, and because cruelty laws are creations of state legislatures, these laws can and do vary from state to state. The legislature of each state determines what the crime (or in some states a number of different crimes) are called; although many states call their laws "cruelty" statutes, some states do not use this term in the title of these laws. The legislature also determines through the words of the statute what constitutes the crime of cruelty; therefore, what is cruelty in one state might not constitute the crime of cruelty or any crime in another. The legislature also sets potential punishments for those convicted of cruelty. Because the concept of cruelty is a matter of statutory and not fundamental common law, state legislatures have virtually unbridled discretion to define what cruelty is, what animals are covered by the cruelty law, and what sorts of persons may or may not be prosecuted for cruelty. Thus, if a state legislature decides that only domestic animals are covered by the statute, or that farmers or veterinarians cannot be prosecuted for cruelty under all or specified conditions, that is the end of the matter. Indeed, a state's legislature could decide to do away with any crime of cruelty or any criminal offenses whose purpose is to protect animals.

Because cruelty is a crime, cruelty actions are criminal and not civil in nature. The plaintiff is the state, and, as in all criminal cases, in order to convict the defendant the state must prove beyond a reasonable doubt that the defendant committed acts specified by the legislature as elements of the crime.

Differences in Cruelty Laws

Cruelty laws can differ from state to state regarding virtually any part of the definition of the crime, the mechanism of enforcement, or the penalty.[46] Among the more general kinds of differences are the following.

States can take a different approach regarding which animals are protected. In most states, one can commit cruelty upon any animal. However, some states limit the crime to domestic animals or owned animals. In all states, one can be prosecuted for cruelty for an animal that generally is covered by the statute even if one is the owner of that animal. This is a critical point because it demonstrates once again that the legal status of animals as property does not give animal owners despotic dominion over this property.

In some states, the crime of cruelty requires a malicious intent to harm an animal or harming an animal intentionally in the sense of knowing that one is harming the animal and still intending to do so. However, in the great majority of states, the crime of cruelty does not require knowing that one is harming an animal or causing it pain. In most states, it is sufficient that the infliction of such pain be unnecessary or unjustifiable— although in such states intentionally causing unjustifiable pain will also be cruelty, and in some of these states such intentional cruelty is subject to more severe punishment than inflicting unnecessary pain negligently.

Some cruelty statutes specifically exempt certain kinds of animals or persons from their purview. At the time of this writing, 26 states and the District of Columbia exempt from prosecution persons conducting research on animals; some of these states condition this exemption on the approval of the facility in which such research is conducted by a state agency, or the regulation of such research by federal or state authorities. Some states exempt ordinary and accepted agricultural practices. Some specifically exempt hunting done in accordance with state laws and regulations. In some states, veterinarians are exempted from prosecution. The power of the state legislature to craft the crime of cruelty is most apparent in its authority to exempt whole classes of animals and people from the purview of the law. Even the most egregious acts of abuse are immune from prosecution if exempted.[47]

Until recently, cruelty was almost universally a misdemeanor, which is defined in most states as a crime the
maximum imprisonment for which is one year or less.
However, a number of states have made cruelty under certain
circumstances a felony, which is punishable by a term of
imprisonment exceeding one year.[48]

Supplementary "Cruelty" or Animal Protection Statutes

In many states, what is called "the" cruelty law—the statute
setting forth the basic offenses relating to improper treatment
of animals—is part of a larger set of laws. This whole set of
laws is sometimes referred to as the state's "animal cruelty
laws" or "animal protection laws." A state's legislature may add
to the section of its penal code defining the basic crime of
cruelty a number of statutes dealing with more specific
problems. A typical addition to the basic cruelty offense is a law
specifically prohibiting and setting penalties for organizing or
participating in animal fighting enterprises such as dog and
cock fights. (Some states add such specific offenses as sections
included in one encompassing cruelty law.) Table 1 provides
an example of how the basic cruelty law can be supplemented.
It sets forth the titles of specific animal protection laws that
have been added to New York's cruelty law over the years.
Many of these sections provide specific possible punishments
which (as in the case of animal fighting, for example) can
exceed those of the basic cruelty provision.

Meaning of the Legal Concept of "Cruelty"

"Cruelty" a Misnomer

Historically, these laws have been called "cruelty" laws
because the word "cruelty" has been prominent in them. This

TABLE 1. Additions to the New York State cruelty statue

351	Prohibition of animal fighting
353	[Basic cruelty statute]
354	Sale of baby chicks and baby rabbits
355	Abandonment of animals
356	Failing to provide proper food and drink to impounded animal
357	Selling or offering to sell or exposing diseased animal
358	Selling disabled horses
358-a	Live animals as prize prohibited
359	Carrying animal in a cruel manner
359-a	Transportation of horses
360	Poisoning or attempting to poison animals
361	Interference with or injury to certain domestic animals
362	Throwing substance injurious to animals in public place
364	Running horses on highway

is unfortunate, because in ordinary usage the term connotes an intentional or malicious mistreatment of animals. Although the cruelty laws of many states once required such mistreatment, and some still do, this was not the import of other early cruelty laws, and today the vast majority of states do not require intentional or malicious infliction of injury for "cruelty." It can be argued that state statutes would do better to distinguish a crime of intentional "cruelty" from, say, animal "neglect." However, use of the word "cruelty" is probably too ingrained to make such a reform possible, although the laws of some states now use different terms to distinguish between the crime of intentional, malicious, or reckless abuse of animals on the one hand and the crime of inadvertent or negligent mistreatment on the other.[49]

"Cruelty" and Pain

In some states, the mere unjustified killing of an animal is included within the offense of cruelty. However, this is not the essence of the crime of cruelty as it has developed, and states that have included wrongful killing of an animal in their definitions of cruelty appear to have done so because the existing cruelty law may have been the most convenient place to put such an offense.

As it is typically understood, cruelty involves the infliction of pain, distress, or discomfort upon an animal. It is typically not cruelty to kill one's own or another person's animal painlessly. The latter kind of situation may be unlawful, but as a kind of unlawful deprivation of another's property. The legal concept of cruelty has not included the failure to provide pleasures or other positive goods. Thus, one can be prosecuted for failing to provide one's dog or cat sufficient food or water or for leaving it outside in the cold. One cannot be prosecuted for failing to play with it as much as one might or for failing to make it happy. Cruelty laws reflect an ethical principle once almost universally held in Western societies but now increasingly felt to be insufficiently respectful of animals: that people's only fundamental ethical obligation is to avoid causing animals they use pain, or unjustifiable pain.[50]

Cruelty: Unnecessary Or Unjustifiable Pain

As the term has been defined explicitly in state statutes or by courts interpreting these statutes, the essence of the crime of cruelty is the *unnecessary* or *unjustified* infliction of pain upon an animal. However, by "unnecessary" or "unjustified" the law does not mean what ethical deliberation would demonstrate to be unnecessary or unjustified. The cruelty laws are intended to prohibit what is regarded by the great majority of people in society as the infliction of unnecessary or unjustifiable pain. This point was made brilliantly by a New York City Magistrate in 1911. A prosecution had been brought pursuant to the state's law (part of the second Bergh statute of 1867 and still law in New York) prohibiting the transportation of animals in a "cruel manner." The case involved turtles that had been transported into the Port and then to the City's Chinatown where they were to be made into turtle soup. The animals had been placed on their backs; they were tied together by means of a rope which was passed through a hole in a fin of each turtle. The Magistrate explained the meaning of the legal concept of "cruelty" as follows:

... The infliction of pain alone is insufficient for the purpose of a prosecution [for cruelty]; but the question is: Was unjustifiable pain inflicted? The statute itself contemplates and permits the infliction of a certain amount of pain. Certain physical pain may be necessary and justifiable in given cases. I would call it a legal license permitting the infliction of unavoidable pain. Many are the cases where animals suffer or are permitted to suffer physical pain, but it is insufficient in law to warrant a holding by a committing magistrate. By biblical mandate man was given "dominion over the fish of the sea and every creeping creature that moveth upon the earth." Man is superior to the animals, and some of them he uses for food and in such cases the incidental pain and suffering is treated as necessary and justifiable. It must have come to the attention of many that the treatment of "animals" to be used for food while in transit to a stockyard or market is sometimes not short of cruel and, in some instances, torturable. Hogs sometimes have the nose perforated and a ring placed in it; ears of calves are similarly treated; chickens are crowded into freight cars; codfish is taken out of the waters and thrown into barrels of ice and sold on the market as "live cod"; eels have been known to squirm in the frying pan; and snails, lobsters and crabs are thrown into boiling water.

Irrespective of the devious means that might be adopted to destroy life before these cruelties are perpetrated upon them, still no one has raised a voice in protest. These practices have been tolerated on the theory, I assume, that in the cases where these living dull and cold-blooded organisms are for food consumption, the pain, if any, would be classified as "justifiable" and necessary.

The question as to whether the pain caused to such creatures, often classed as dull nervous organisms, is "justifiable" or not cannot be easily answered. Public opinion at different times among different races has swung from one extreme to another. The Emperor Augustus nearly exterminated peacocks to regale himself in Rome with their brains. Today the world would hold their death unjustifiable. Then again, juries and magistrates of different localities, races, or education, with varying ideas of taste and opinion, may hold widely divergent ideas as to whether the improved flavor of lobster boiled alive makes such torture "justifiable". . .[51]

Legal "cruelty" refers to what society in general would consider unnecessary or unjustifiable infliction of pain on animals. Thus, legal cruelty is not synonymous with what we might call ethical cruelty, what sufficient and persuasive ethical deliberation might demonstrate to be the infliction of unnecessary or unjustifiable pain. For example, let us suppose, as I have argued,[52] that it is morally wrong to raise veal calves in so-called "crates," in which they are not free to walk or turn around, socialize with other animals, and are fed a liquid diet (in order to produce white, anemic flesh). Assuming that this husbandry practice is wrong, and *ethically* cruel in the sense of inflicting unjustifiable discomfort or distress, it would not follow that treating veal calves in this way constitutes legal cruelty. It might be difficult to persuade a jury that such animal treatment is unjustifiable; in some states raising calves in this way is exempted from the purview of the cruelty law as an "accepted" animal agricultural practice.

What Kinds of Animal Pain are "Unnecessary" or "Unjustifiable?"

A state's legislature can indicate what kinds of infliction of animal pain (and what kinds of pain) are unnecessary or unjustifiable by including in the cruelty law categories or examples of behavior that will not be tolerated, or by exempting certain kinds of activities from the statute. Courts that have interpreted and applied cruelty laws have also indicated more general kinds of behavior that are, and are not, typically considered unnecessary or unjustifiable. For example, it is clear that attacking an animal and causing it pain in order to defend oneself, another person, or one's property would not be classified as "cruel." (Of course, using or causing a patently excessive amount of pain in defending a person or property could well constitute legal cruelty. A jury might well find that attempting to get a dog off one's property by shooting or stabbing it is excessive, unnecessary, and therefore "cruel.")

Neither statutes nor court decisions define with rigorous precision standards that are to be applied by juries or judges in determining whether a given activity constitutes legal cruelty. However, it is possible to focus on several general criteria that are used frequently to determine whether the infliction of animal pain is legally "unnecessary" or "unjustifiable."

The severity and duration of pain. Other things being equal, severe and long-lasting pain is more likely to trigger a jury determination of cruelty than trivial or brief pain. Likewise, an experience appropriately described as "pain" or "suffering" is more likely to result in a determination of cruelty than a less intense mental state, such as discomfort.

Perceived legitimacy (by society as a whole) of the particular activity involving animals. An important component of legal cruelty is the generally perceived justification or legitimacy of the behavior in which animal pain is inflicted. The more justified the activity is considered, the more animal pain that will be tolerated. For example, the search for pain-killing drugs for human cancer patients is generally considered a legitimate endeavor. Therefore, even in states that allow cruelty prosecutions of animal researchers, causing laboratory animals severe pain in order to determine whether a given drug stops such pain would not constitute cruelty, provided the animals do not suffer such pain longer than is necessary to make the determination and provided the entire experiment is scientifically sound. However, someone who causes dogs or cats even brief pain by smashing their heads against a brick wall would be engaged in legally cruel behavior, because this activity would not generally be perceived to have any justification.

Avoidability of the pain given the activity or aim. Also critical to whether a certain amount or kind of pain will be considered unnecessary or unjustifiable is whether 1) any pain is required for the (legitimate) use of animals in question and 2) whether, if some pain is required, more pain than necessary is inflicted. If, for example, there is a completely painless way of doing a certain kind of experiment on animals, and a researcher does this work in a way that causes the animals pain, then the infliction of any pain may well be unjustifiable and therefore legally "cruel."

Motivation of the defendant. Most states do not require a malicious intent or enjoyment of the infliction of animal pain. However, even in jurisdictions that do not have such a requirement, cruelty is more likely to be found where there is such a motivation. This may be so because enjoyment of animal suffering is not viewed as a legitimate purpose, or because the intentional infliction of animal suffering is something prosecutors and juries feel is especially deserving of punishment.

Perceived value or moral status (by society as a whole) of the animal or species. It is impossible to look at the application of cruelty laws and not conclude that some animals are given more favorable and others less favorable status because of their perceived value by society as a whole. Most cruelty prosecutions involve dogs and cats—our preferred household pets. There may be an increasing number of cruelty prosecutions involving horses, probably because many horses are being kept as companion animals and because horses are admired and respected by an apparently growing number of people.[53] At the bottom of the list are so-called animal "pests" such as mice or rats, the treatment or mistreatment of which rarely result in cruelty prosecutions.[54]

In general, a successful cruelty prosecution is most likely when a defendant has done something that runs afoul of all the above criteria, that is when an animal about which society feels especially protective has been caused severe and avoidable pain for a reason that is generally considered illegitimate.

"Necessity"

As has been noted, the legal concept of cruelty is commonly defined as the infliction of *unnecessary* or unjustified pain. Use of the word "necessity" (like the word "cruelty" itself) is so commonplace and historically ingrained that objections to it are probably fruitless. However, speaking of "necessity" in the context of cruelty prosecutions is not terribly helpful. It is not strictly speaking necessary for any animals to experience pain because of human activity because it is not strictly speaking necessary for humans to use animals for any purposes. Typically, to claim that a given amount or kind of animal pain or distress is "necessary" is to make two judgments: (1) that a human aim for which the pain (distress, and so on) is imposed is legitimate or is sufficiently important to justify the animal pain; and (2) that the amount or kind of pain in question is in fact required for the achievement of that aim. Thus, cattle ranchers and slaughterhouses would not be subject to cruelty prosecutions on the grounds that it is not necessary for people to eat meat. Cruelty laws operate against the background of many human uses of animals that are *assumed* appropriate. Cruelty laws then require that given these uses, no more pain should be inflicted than is "necessary." But even here, the notion of "necessity" is used loosely. A person ordinarily does not commit cruelty if he causes just a slight amount of pain that need not be caused. The law is not so precise. Generally, *substantially* more pain than is actually required for a legitimate animal use is required.

The Underlying Purposes of Cruelty Laws

Professor Francione asserts that cruelty statutes are archaic and unsatisfactory because they are not intended to protect animals. Rather, he claims, when cruelty laws were first written and interpreted their stated intention was to protect people. By prohibiting the infliction of certain kinds of pain upon animals, it was said, the law would encourage people to treat each other better. Cruelty laws, it was said, created no direct legal duties to animals because animals were not the ones considered harmed by cruelty. According to Francione even today "most courts agree these statutes are intended to prevent humans from acting in cruel ways toward each other,"[55] and a "close examination of these statutes indicates quite clearly that they have an *exclusively* humanocentric focus, and the duties they impose give no corresponding rights for animals."[56]

The claim that cruelty laws are not intended to protect animals is important to what I have termed the activist view of the legal status of animals. For if cruelty laws do not protect animals, they would not create duties to animals. In general, duties and rights are correlative: if you have a duty to someone, typically that person has a right to whatever you have a duty to do (or not to do) to him. Therefore, if the cruelty laws protected animals and created duties to animals, they would provide rights for animals. This is something the activist view of the legal status of animals cannot accept, because the activist view argues that the law does not *now* give legal rights to animals, and must be changed in certain ways to provide rights.

There is ample evidence that many legislators who enacted and judges who interpreted cruelty to animals laws did, at least during the early history of these laws, believe that their purpose was not to protect animals against people but to protect people against themselves. Many cruelty laws were initially placed in sections of state penal codes dealing with "offenses against the public." One can still find courts that hearken back to the rationale of the cruelty statutes as essentially people and not animal protection measures. In 1931, the Supreme Judicial Court of Massachusetts declared that the cruelty laws "are directed against acts which may be thought to have a tendency to dull humanitarian feelings and to corrupt the morals of those who observe or who have knowledge of these acts."[57] This rationale for the cruelty laws was quoted by a Massachusetts intermediate appellate court as late as 1981.[58]

It is clear that many state legislatures and courts once did view the protection of people from themselves as the underlying purpose of cruelty to animals laws. However, this was not how everyone who wrote or interpreted these laws saw them. It is altogether possible that some advocates of cruelty laws who really believed in animal protection did not express their underlying purpose explicitly because it was the standard and virtually universally accepted position of theologians and philosophers that people could not have direct legal or ethical duties to animals.[59] In any event, many people who wrote and campaigned for the passage of the early animal cruelty laws in England and the United States clearly did believe that these laws were necessary for the protection of the animals themselves. John Lawrence, the co-author of Martin's Act, wrote the following about the beings his efforts were intended to protect.

> Justice, in which are included mercy, or compassion, obviously refers to sense and feeling. Now is the essence of justice divisible? Can there be one kind of justice for men, and another for brutes? Or is feeling in them a different thing to what it is in ourselves? Is not a beast produced by the same rule, and in the same order of generation with ourselves? Is not his body nourished by the same food, hurt by the same injuries; his mind actuated by the same passions and affections which animate the human breast; and does not he, also, at last mingle his dust with ours, and in like manner surrender up the vital spark to the aggregate, or fountain of intelligence? Is this spark, or soul, to perish because it chanced to belong to a beast? Is it to become annihilate? Tell me, learned philosophers, how that may possibly happen.[60]

George Angell, a prominent Boston attorney and founder of the Massachusetts Society for the Prevention of Cruelty to Animals, was a leading American proponent of animal cruelty laws. He expressed with simple eloquence the motivation of many in the animal protection movement of the late 19th century. When asked why animals deserve the special protection of laws, he responded, "*First* for their own sake; second because protection to animals is protection to men."[61] In his famous "Ten Lessons on Kindness to Animals," which was distributed to primary school students throughout the United States, Angell summarized why he believed we humans do indeed have direct obligations to animals.

> Hosts of animals there are that make the world more pleasant, and our lives happier. They add a charm to the meadows by their bleating and browsing, and peaceful existence. They make the woods pleasant with their sweet songs and fair plumage. They peep with bright eyes from tree, and bank, and bush. They cover and fill the streams, and ponds, and lakes, and ocean with their brilliant forms and abundance of life. Since they do so much for us, we should do all we can to make their lives comfortable and happy, and always give them proper food, and shelter, and care, and kind word, and kind treatment.[62]

As will be illustrated below, a number of early court decisions interpreting cruelty laws indicated that among the purposes of these laws was the protection of animals and the creation of legal duties to animals. Most importantly, *today*, if one were to ask legislators, prosecutors, judges, and employees of humane societies who enforce cruelty laws, they would say, virtually *universally*, that the primary purpose of these laws is to protect animals. The following words of a New York State judge justifying the cruelty conviction of the owner of a carriage horse are typical.

> History prior to the middle of the 19th Century is devoid of any laws as to cruelty to animals. Only recently has there been codification in our legal system forbidding cruelty to animals. But, the moral obligation of man toward the domestic animal is well documented in the Bible. "A righteous man regardeth the life of his beast" (Proverbs 12:10). He has consideration for its feelings and needs.
> The Bible also states that if you see an animal hurt or

over-burdened, one should not look away but help it (Deuteronomy 22:4.)

It is truly a humanitarian sentiment that domestic animals are in fact considered part of the human community. Thus, they should be treated with respect and given proper care.[63]

Cruelty and Animal Rights

Whatever they might originally have been intended to do, cruelty laws today clearly are intended at the very least to protect animals. They create legal duties to animals. They therefore afford legal rights for animals. This fact was acknowledged explicitly even by some early court decisions. According to a 1898 opinion of the Mississippi Supreme Court,

> the common law recognized no rights in [domestic animals] and punished no cruelty to them, except in so far as it affected the rights of individuals to such property. [Cruelty] statutes remedy this defect, and exhibit the spirit of that divine law which is so mindful of dumb brutes as to teach and command, not to muzzle the ox when he treadeth on the corn; not to plow with an ox and an ass together; not to take the bird that sitteth on its young or its eggs; and not to seethe a kid in its mother's milk. To disregard the rights and feelings of equals, is unjust and ungenerous, but to willfully or wantonly injure or oppress the weak and helpless, is mean and cowardly.[64]

The following words of the Louisiana Supreme Court are especially noteworthy.

> The [cruelty] statute relating to animals is based on "the theory, unknown to the common law, that animals have rights which, like those of human beings, are to be protected. A horse, under its master's hands, stands in the relation of the master analogous to that of the child to the parent." Reasoning from that basis, we feel certain that the [cruelty laws in question] do not interfere with the private right of property . . .[65]

Both these decisions interpret the cruelty laws as creating legal duties upon people not to treat animals cruelly, and the correlative legal right for animals not to be treated cruelly. The latter decison is significant. It recognizes that there is nothing in the legal status of animals as property, or in cruelty laws, that precludes the ascription of legal rights to animals. As has been noted above, from its very inception the concept of property (which, in fact, began and remains in essence the concept of lawful priority of possession) does not imply an owner's despotic dominion over his property. The concept of property is sufficiently flexible to allow some property to be animate, sentient beings, and to allow that owners of certain kinds or items of property have legal obligations to this property, and for this property to have legal *rights*.

Legal Rights

An exhaustive consideration of the concept of a legal right is not possible here. However, because misinterpretations of the history and core meaning of this concept are an important part of the activist view of the legal status of animals, some discussion of the term "right" as it is used in the law is appropriate. As the philosopher Joel Feinberg explains, "to have a right is to have a claim *to* something and *against* someone, the recognition of which is called for by legal rules or, in the case of moral rights, by principles of an enlightened conscience."[66] Legal rights presuppose that the holders of rights have interests, in the sense of something that is a good for them even if they cannot articulate or fully comprehend that something is in their interests. As Feinberg explains, a right is something to which a right holder can be entitled, even if recognizing that right does not satisfy or benefit people who must act in accordance with the right. Thus, to say that people have the legal right to freedom of speech is to say that they ordinarily may speak their mind if they wish to do so, even if their doing so is displeasing or inconvenient to others. Legal rights, Feinberg explains, are exceedingly strong protections, which is why they are so valuable. A right is

an extremely valuable possession, neither dependent on or derivative from the compassionate feelings, propriety, conscientiousness, or sense of *noblesse oblige* of others. It is a claim against another party in no way dependent on the love of the other party or the loveableness of its possessor. Hence, wicked, wretched, and odious human beings maintain certain rights against others, and the duties of others based on those rights are incumbent even on those who hate the claimant, and hate with good reason. A right is a matter of justice, and justice, while perhaps no more valuable than love, sympathy, and compassion, is nevertheless a moral notion distinct from them.[67]

As Feinberg observes, cruelty laws give animals covered by these laws the legal *right* not to be treated in violation of the statutes, for it is, he notes, the purpose of these laws to protect *animal* interests, and to protect such interests even when doing so prevents people from treating animals in ways they would prefer.

Legal Rights and the Activist View

In recent years, a number of animal activists, some of them lawyers and philosophers, have insisted that neither the cruelty laws, nor any laws now existing, give animals rights.[68] This is so, it is claimed, because to have a legal right is to have what the law calls "standing" to sue on one's own behalf and for one's own benefit. Children can have such standing because under certain circumstances a legal guardian can be appointed by a court to pursue a legal action in the child's behalf. However, as activists correctly note, cruelty-to-animals laws do not give animals legal standing. Prosecutions for cruelty are not undertaken by an attorney appointed to represent animals but by public prosecutors. Nor does the legal system afford the opportunity for an animal to sue for compensation or other relief, someone such as its owner or veterinarian.

The claim that the possession of a legal right requires legal standing has been repeated so often in the literature that even some people who do not accept the goals of animal activists assume that the claim is correct. It is incorrect. State and federal criminal laws, for example, do not confer legal standing upon private citizens to sue criminals. Prosecutions are undertaken by public prosecutors, and in most instances cannot result in monetary compensation for victims. Yet we say that people have the right not to be treated in ways prohibited by the criminal laws. They would have this right even in the absence of the ability to institute civil actions in their own behalf against criminals. Generally only a government authority, and not a private citizen, can sue to abate a nuisance (such as a polluting factory) that affects the public at large.[69] However, most lawyers and laymen would say that individual members of the public have a legal right not to be harmed by such a nuisance.

The claim that animals now have no legal rights, and that legal standing is necessary for their having rights, is understandably important to the proponents of the activist view of the legal status of animals. Recent surveys show that most adult Americans believe that animals have moral, and should have some legal rights.[70] Some activists want animals to be able to sue their owners, veterinarians, and others. If legal rights entailed legal standing, it would be much easier to convince the public that animals should be given such standing.

Giving animals standing to sue would, to be sure, give them a right they do not now have. Lawsuits undertaken by animals in their own behalf could also result in their being afforded more, and more extensive legal rights than they now have, if, for example, courts order that animals be provided certain kinds of care to which they are not now legally entitled. Animals can also be given a wider range of legal rights than are now afforded by cruelty laws without giving them legal standing to sue. However, it is not the case that, without standing, animals do not have rights or have "rights" in some corrupted or improper sense.[71]

Criticisms of Cruelty Laws

In recent years proponents of the activist view of the legal status of animals and others have strongly criticized cruelty laws. I want to consider the more common criticisms briefly.

1. According to Professor Francione, cruelty laws are archaic because they are intended not to protect animals but to foster better behavior by people toward other people.[72] As I have indicated, this claim is incorrect as a contemporary description of the intent of cruelty laws. This is an important point. Cruelty laws, and the legal status of animals of property, are sufficiently flexible to allow for quite different underlying rationales. One rationale can (and once may have been) the protection of people from each other. But another rationale can be (and now is) protecting the animals as worthy of protection in their own right.

2. It is also claimed that cruelty laws cannot really protect animals because in requiring a determination of "justifiable" pain or suffering, they allow human interests to be balanced against animal interests. The inclusion of human interests, it is said, means that these laws are not intended to protect animals. Cruelty laws do require a balancing of human and animal interests. But this does not imply that animal interests are not important, any more than it would follow from such a balancing that human interests are not important.

3. According to Francione, the problem with cruelty and other so-called animal protection laws is not just that they require balancing of human and animal interests, but that they afford animal interests little if any weight.

> The problem is that animals do not have rights under the law. There are, of course, many laws on the federal and state levels that purport to protect animals from "in humane" treatment, but these law do not really confer rights in the sense that we usually use that term. Rather, these laws concern animal welfare, or the notion that animals may be exploited by humans as long as this exploitation does not result in the infliction of "unnecessary" pain, suffering, or death. . . . Legal welfarism requires that we "balance" the interests of humans and animals in order to decide what constitutes "humane" treatment and "unnecessary" suffering. The problem is that the framework of legal welfarism contains numerous normative considerations that render any attempt to "balance" as an empty exercise for the most part—at least as far as animal interests are concerned. The result of legal welfarism is that in many instances, a relatively trivial human interest is balanced against an animal's most fundamental interests in not experiencing pain or death, and that the human interest nevertheless prevails. We all reject "unnecessary" cruelty, but we still allow bow hunting, pigeon shoots, rodeos, and all sorts of activities that are very difficult to justify on any coherent moral ground . . .[73]

There is nothing inherent in the principle that human and animal interests should be balanced which dictates that such balancing is an "empty exercise" destined to go against the animals. In fact, cruelty laws have been used to prohibit a number of activities in which some people may still prefer to engage, such as dog fighting. If Francione believes, as do some activists, that it is inherently immoral to balance human against animal interests because it is inherently wrong to use animals for any purposes, then his objection to cruelty laws is indeed fundamental—and is destined to be ignored by the legal system. For, I submit, our legal system will never prohibit the balancing of human and animal interests. If Francione's problem with cruelty laws is that they have been used in ways that are not always sufficiently respectful of animal interests, he is surely correct. But there is nothing in cruelty laws that prohibits the legal system from giving certain animal interests greater weight than has been done in the past. This might be accomplished either by adding specific prohibitions or requirements to cruelty statutes, or by convincing juries that certain kinds of treatment of animals have no redeeming justification.

4. The cruelty laws are also attacked on the grounds that their emphasis on "unnecessary" or "unjustifiable" infliction of pain does not stimulate careful ethical deliberation about appropriate treatment of animals because (a) the notions of "necessity" and "justifiability" are inherently vague; and (b) the application of these notions begins with the assumption that certain practices (such as meat-eating and animal research) are acceptable.

There is some justice to these criticisms. In fact, as Francione observes, cruelty laws often begin by regarding as sacrosanct certain activities which, at the very least should be open to careful ethical thought, and then shift our attention to the issue of whether, given the acceptance of these activities, more pain is inflicted than is required for them.

However, in my view this criticism is ultimately unconvincing because it reflects either a misunderstanding of how the cruelty laws are supposed to work, or attempts to substitute for their underlying rationale substantive policies regarding animals that will never be accepted by the American or any other legal system.

In condemning "unnecessary" or "unjustifiable" pain or suffering, cruelty laws do call upon courts and juries to make ethical determinations of inappropriate treatment of animals. Certainly, these laws are not intended to condemn behavior simply because the great majority of people condemn it, but because, it is felt, application of the cruelty laws reflects a correct appreciation that certain ways of treating animals are ethically unacceptable. However, cruelty laws are not intended to raise to the level of legal prohibition and obligation all ethical standards applicable to our relations with animals. Put another way, cruelty laws were fashioned at a time when it was believed (wisely, in my view) that it is not the purpose of the law to prohibit everything that is morally bad and to require everything that is morally obligatory. The law, it was believed, should embody only those core, central, and most fundamental ethical principles the violation of which are intolerable in a civilized society. Cruelty laws do not attempt to prohibit all kinds of behavior that ethical deliberation might demonstrate are inappropriate, but only the worst kinds of treatment of animals. And they do this by reflecting the general values and views of society—which, as the New York City Magistrate quoted above observed are certainly capable of change and development. That cruelty laws reflect only a core element of ethical principles relating to animals is demonstrated by the fact that they do not require that one be good to animals, but only that we not treat them extremely badly.

The ability of cruelty laws to expand their range of prohibited activities as society's views about appropriate treatment of animals changes allows for vigorous ethical debate about how animals should be treated. It also allows for the inclusion within the class of legally prohibited behavior activities that come, over time, to be generally seen as inappropriate. In Europe some cruelty-to-animals laws now prohibit intensive rearing of "milk-fed" veal.[74] If this husbandry practice becomes generally disfavored in the United States it may some day also be viewed as imposing "unnecessary" or "unjustifiable" pain or discomfort on the animals. In 1988 Sweden enacted a comprehensive law to prevent "cruelty" that gives agricultural animals specified "rights," among which are the right of cattle to be grazed outdoors, of poultry "to be let out of cramped battery cages," and of cows and pigs "to have access to straw and litter in stalls and boxes."[75] The Swedish law neither states nor implies that such animals are not the *property* of their owners.

The activist view of the legal status of animals is also activist in the sense that it seeks to involve the law more actively in the prohibition of a wide range of kinds of use and treatment of animals. For some activists, this approach may stem from a general view that the law should vigorously enforce ethical standards. For others, a more pervasive role of the law in human-animal relations may derive from the position that so many terrible things are done to animals that greater legal intervention is necessary. Activists find patently immoral practices involving animals that many people find acceptable. Some would prohibit by law not just rodeo (calls for the abolition of which do not appear to upset some Easterners or animal welfare advocates who have an immediate revulsion to this sport[76]) but also meat-eating, horse racing, and animal research aimed at helping people and other animals. There are serious ethical issues relating to the human use and treatment of animals. The approach of the cruelty laws allows discussion of these issues to develop. An approach which would prohibit, by law, the balancing of human and animal interests does not.

5. It is also argued that cruelty laws are not vigorously enforced.[77] This is true, but is a function not of the intent of these laws but of the terrible burdens that must be borne by the criminal justice system. In most major metropolitan areas prosecutors must devote already scarce resources to murders, rapes, arsons, robberies, and assaults. Animal cruelty prosecutions must be placed within this context, and they must often assume a lower priority. The solution to this problem is to give prosecutors resources they need to punish and prevent all kinds of crimes. Some activists would replace or add to public prosecution the appointment of legal guardians to bring private civil lawsuits against alleged violators of animal interests.[78] Among the predictable results of such an approach would be the influx of still more lawsuits—now by animals against pet owners, farmers, animal researchers, and veterinarians—into overburdened courts that already must usually delay criminal cases for months and civil cases for years.

6. It is also objected that the variety in cruelty statutes among jurisdictions makes no ethical sense. In one state, certain practices (for example, cockfighting) may be cruel, and in others completely exempt from the cruelty statutes. But if a certain behavior toward animals is wrong, how can it not be so wherever this behavior occurs?

In my view there are cruelty laws that are not sufficiently comprehensive and therefore need revision. However, objecting to cruelty laws on the grounds that they can vary misses much of the point of the approach of these laws. Cruelty laws are not expected to remain the final word on the ethics of our relations with animals. That a given state's law does not prohibit certain behavior does not make such behavior appropriate. Nor does it prevent the statute from being amended to prohibit it. Like other statutes passed by legislatures, cruelty laws result from a political process in which different points of view are expressed. This is especially valuable in the animal area, where there are different points of view and it is not always obvious where and to what extent the law should intervene in protecting animals. (For example, while some people argue that state cruelty laws should not exempt animal researchers from prosecution, a strong argument can also be made that regulation of research animal welfare ought to reside with federal and state authorities who, unlike overburdened local prosecutors, have resources to articulate and enforce appropriate standards for the care and use of these animals.)

AN ACTIVIST PICTURE OF HUMAN-ANIMAL RELATIONS

I have found it virtually impossible to convince proponents of the activist view of the legal status of animals that the historical account of the concept of property offered here is correct. During the conference at which this paper was presented, several speakers and members of the audience insisted that people have complete dominion and control over *property*, and that if animals are *property* we cannot have legal (or moral) obligations to them, and they cannot have legal rights. Expressed with some vehemence, and apparently at the core of this set of views, was the assertion that to speak of animals as "property" is necessarily to analogize them to *inanimate objects*—and *therefore* to deny that they have value in themselves.

One of the most common claims made by proponents of the activist view of the legal status of animals is that if animals are property, they are no better than human slaves. In the following passage, one of whose authors is the activist philosopher and self-described "animal liberationist" Peter Singer, the legal status of animals as property is once again linked to human bondage. Singer describes the status of animals as property through history as follows:

> What, then did it mean, for a human being, to be a piece of property in classical Athens or Rome? . . . [W]e find one stable element: a condition of powerlessness. The treatment that the slave is accorded depends solely on the master. As a chattel, slaves have lost control over their own selves. And at the root of the master's power over the slave is the fact that the slave is not acknowledged by the community. Slave status is characterized by what the slave is not: slaves are not free, they cannot determine how to use their own labour, they cannot own property . . . Bought and sold as objects, liable to corporal punishment and sexual exploitation, they stand outside the protective moral realm of the classical community.
>
> . . . As with slaves in ancient times, so with animals today, treatment varies—from the affectionate care of the "pet-owner" to the naked exploitation of the factory farmer concerned only with maximising profits. The common thread, again, is that animals have suffered a total loss of control over their own lives. The difference in interests and capacities of human and animal slaves does not affect their fundamental identity of status. Like chattel slaves, nonhuman animals stand outside the protective moral realm of the modern community.[79]

Assume for the purposes of discussion that the historical claims in this passage are correct, and people who raise animals for food or fiber can be characterized as "factory farmers" interested only in maximizing profits. Can it seriously be maintained that pets, at least, suffer a total loss of control over their own lives? Is it not just inaccurate—but outrageously so—to assert that animals stand entirely outside the protective realm of the modern community? Do not cruelty and other laws seek, and sometimes succeed, in helping their lot? And even if such laws do not always succeed in protecting them as much as they ought to be protected and in providing them the kinds of control over their lives some ought to be provided, it is surely *obviously* not the case that animals stand *entirely* outside the protection of the moral realm of the community. In whole spheres of activity that are not even covered by laws, *clearly* many millions of animals are protected, respected, and loved.

It is probably fruitless to attempt to convince people who equate animal ownership with human slavery that the legal status of animals as property is consistent with high regard and respect for animals. Singer's assertions seem to show that it is really not the legal status of animals as property that he and other animal liberationists find most objectionable. They believe that animals and people deserve equal consideration.[80] Thus, the longest list of examples in which people treat their animals with regard, respect, and love will still evoke the same response: "By treating animals as property you are still not giving them equal consideration and they are still essentially slaves."

In fact, the legal status of animals as property is consistent with treatment of animals that most people find admirable. Nowhere is this fact more apparent than from the myriad of relations people have with companion animals. That dogs, cats, and other pets are property means that their owners have enormous authority over the lives of these animals. One consequence of this authority is that a pet owner need not provide an animal with the best veterinary care possible. Indeed, it is lawful for an owner to have an animal euthanized at any time, even if it is healthy and well-behaved, and certainly if providing it medical care or a good quality of life is an unwanted financial or emotional burden. (It is also lawful for veterinarians to refuse to kill healthy animals.) As I have argued, too many companion animals are not given appropriate care, and far too many are euthanized at the behest of owners who do not fulfill their moral obligations to these animals.[81] At the same time, many pet owners love their animals, and willingly provide medical care for them at great financial and psychological cost. Serious ethical issues can arise when owners seek to continue care when it would be humane and in the best interests of an animal to end a life of suffering or illness. These dilemmas, and the high regard for animals from which they arise, are not in any way precluded by the status of animals as property.

PROPERTY, CRUELTY, AND THE LIKELY LEGAL STATUS OF ANIMALS

I want to close by discussing some of the implications of the legal status of animals as property and how the law is likely to respond to the activist view.

1. Animals, and certainly the great majority of animal species with which people come in regular contact, will continue to be classified as property. Otherwise, it would be impossible to buy or sell animals, to pass their ownership on through inheritance, to tax their value, or to use them in a myriad of ways (such as sources of food and fiber) that will continue to be regarded as acceptable by the great majority of people.

2. The law can accommodate special status for certain kinds of animals within the general framework of property. As has been noted, cruelty and other laws can impose stringent prohibitions or conditions on certain uses of animals. The status of animals as property is consistent with affording animals certain kinds of protection that some activists seek for them. For example, the Great Ape Project argues that orangutans, gorillas, and chimpanzees ought never to be used for human purposes and should be left alone in natural areas free from human interference. The Project proposes to effect this goal by redefining the moral and legal "community of equals to include all great apes: human beings, chimpanzees, gorillas, and orang-utans."[82] However, under current principles of

Anglo-American jurisprudence, nonhuman primates are wild animals. These species are already the property of the government, which can therefore afford them protection from human interference or contact.

3. The legal status of property is consistent not just with animal rights, but with significant legal rights for animals that can severely constrain how people use or treat them. This follows from the fact that property status does not preclude the enactment of laws that impose legal duties upon people for the purpose of protecting animals.

4. The legal status of animals as property is consistent with a more realistic recognition by the law not just of their intrinsic value but of their importance and value to people. There are many ways the law can treat animals which would reflect more accurately their value and worth to people. For example, at present animal owners are generally unable to sue for compensation for their mental distress caused by someone's negligent injury to or destruction of their animal. Rather, they are limited to recovering the economic value of the animal, or their actual economic loss. Thus, if an animal dies because of the negligence of a veterinarian, the owner would be entitled to the market value of the animal, which in most cases is trivial compared with its value to its owner. It is far from obvious that the law should allow recovery for emotional distress in such situations, because this could require veterinarians, and ultimately their clients, to pay much more for malpractice insurance.[83] However, the law could allow recovery for emotional distress without negating the status of animals as property of their owners. Some courts have allowed some recovery for emotional distress caused by the negligent injury or destruction of an animal on the grounds that the true value of these beings can include the sentimental or emotional value to their owners.[84]

5. Categorizing animals as property does embody a judgment by the legal system that animals and their owners are not of equal value or worth. The concept of property does imply that an owner has substantial (but not absolute) control over his or her property. Moreover, property is usually capable of being bought and sold. Clearly, something that can be bought or sold is not of equal value in the eyes of the law to its owner. Many proponents of the activist view of the legal status of animals believe that the ability to sell property—and its implied judgment that property cannot be the moral equivalent of its owner—is a devastating argument *against* categorizing animals as property. However, this would be persuasive only to people who view animals as equals of people, and who therefore believe that owing an animal is no different from owning a human slave. As compelling as this comparison seems to some activists, it is not even remotely plausible to the vast majority of people, and the legal system always reflects in its most fundamental concepts and principles the views of the vast majority of people. In fact, animals (and certainly farm animals or pets) will *never* be viewed by most people as the moral equals of human beings. The law will continue to categorize them as property. Therefore, those who wish to protect animals and improve their lives would more usefully accept their status as property and work within and with this extremely flexible concept.

6. *Our legal system will continue to afford animals certain legal protections, but like the cruelty laws will probably leave most development of standards relating to animals to private ethics, that is, to discussion among people regarding our moral duties to animals and to individual decisions regarding such duties.* It can be argued that this is generally how the law should operate, but regarding animals in particular there are particularly strong arguments in favor of this approach. The American legal system is already bloated, inefficient, overburdened, and overpopulated with laws and lawyers. The system cannot afford to allow animals independent legal standing even if it ever had the inclination to do so. The law therefore will demand a scheme consistent with substantial ethical deliberation regarding animals.

7. *The legal concepts of property and cruelty are enormously flexible*

*and are consistent with a balancing of human and animal interests
that would favor the latter more strongly than is done at present.*
These concepts are consistent with more rigorous reasoning
about the justification of certain kinds of treatment of animals.
These concepts are consistent with greater specificity in cruelty
laws, with passing and enforcing additional cruelty laws in
order to protect animals, and with more vigorous enforcement
of existing cruelty laws. If cruelty laws reflect inconsistent or
unclear ethical reasoning it is because society's general
attitudes toward animals may be unclear or unpersuasive. The
way to change this is not to revamp the legal system's approach
to animals, which in any event will always in its actual
application reflect society's general ethical views.

*8. The legal concepts of property and cruelty allow animal protection
laws better characterized as noncruelty laws,* for example, laws or
regulations affording certain positive benefits to animals or to
some animals. Neither the legal status of animals as property
nor the strong interest the law has traditionally had in
protecting animals against negative mental states such as pain,
suffering, and distress preclude laws or regulations that
attempt to provide some animals positive mental experiences
and lives more in accordance with their inborn natures and
dispositions. An example of the latter approach is the
provision of the 1985 Amendments to the U.S. Animal
Welfare Act requiring "a physical environment adequate to
promote the psychological well-being of primates."[85]

*9. There is much more the law can do to improve the treatment of
animals.* Too many animals are treated inappropriately, in
laboratories, on farms, in the animal racing industry, and by
pet owners. However, the legal status of animals as property
does not of itself preclude major improvements in the use and
treatment of animals. To be sure, in the kind of world
envisioned by activists who believe that human beings must
never possess or use animals, animals would not be property.
Whether such a world will come to pass is doubtful, and
whether it should come to pass is certainly open to debate.

However, the great majority of people who believe that the humane use of animals is morally acceptable need not reject the legal status of animals as property. The history of this concept demonstrates its power and flexibility, and its ability to encompass vigorous deliberation regarding the moral status of animals and how they should be treated.

CONCLUSION: THE CONCEPT OF PROPERTY AND MORAL SPACE

I have argued that the legal concepts of property and cruelty are extremely flexible and serviceable. Unfortunately, some people interpret the kind of argument I have been making—and the insistence that animals are and will continue to be categorized as property—as intended to defend the *status quo*. Nothing could be farther from the truth. The point of my argument is not just that animals will remain property, and that this legal status is consistent with a variety of approaches to them, but that we need to work within this status if we are to improve their lives.

As some states have already done, cruelty laws should be amended to allow for felony convictions for the most serious kinds of animal abuse and neglect. Prosecutors can receive more adequate funding for cruelty investigations, they can choose cruelty cases carefully for maximum effect, and can prosecute forcefully and persuasively. Judges can be sensitized to the importance of imposing meaningful punishment in cruelty cases. Humane societies can educate the public, and its children, about the evils of animal cruelty. All government agencies entrusted with the protection of the welfare of animals must take their missions seriously.

Much, clearly, needs to be done to protect animals we human beings use for our own purposes, and there are doubtless many animals that ought not to be used for human purposes at all. However, it is folly to suppose that treating animals appropriately will require jettisoning fundamental legal concepts that have taken a millennium to develop. The American legal system does not quickly or lightly discard principles that have taken a millennium to develop, especially when such principles embody deeply ingrained ethical attitudes. The categorization of animals as property, and the fundamental approach of the cruelty laws, reflect society's deeply held view that while animals have some interests worthy of legal protection they are not our equals, and never will be. The protestations of some people to the contrary notwithstanding, animals will remain property. They should also be treated fairly.

Notes

[1] I am indebted to Daniel Dennett for this helpful way of characterizing the matter.

[2] Blackstone W: *Commentaries on the Laws of England* (1765), Sharswood edition, Philadelphia: J.B. Lippincott & Co., 1882, (hereinafter referred to as Blackstone: *Commentaries*), Bk. II, Chap. 1, *2 p. 393. Numbers preceded by asterisks refer to page numbers in Blackstone's original edition, a convention retained by subsquent renderings of the *Commentaries*.

[3] Blackstone: *Commentaries*, Bk. II, Chap. 24, *387, p. 698, emphases in original.

[4] Blackstone: *Commentaries*, Bk II, Chap. 25, *388, p. 698, emphases in original.

[5] Francione G: Animals, property and legal welfarism: "Unnecessary" suffering and the "humane treatment of animals." *Rugters Law Rev* 46: 737, 1994, hereinafter referred to as Francione: Legal welfarism. Francione has published an expanded version of the arguments offered in his paper. Francione G: *Animals, Property, and the Law*. Philadelphia: Temple University Press, 1995. The present discussion was in press at the time of the publication of Francione's book, the arguments of which could not be considered here.

[6] Francione and other proponents of the activist view do not, it must be made clear, deny that the law has traditionally viewed animals as property. Indeed, recognition of this fact forms the basis of much of their condemnation of the law's approach to animals.

Nevertheless, proponents of the activist view still view inanimate objects—things—as central to the concept of property in at least two respects. First, they believe that insofar as one would not be treating an animal as one would a lamp or a chair one would not be treating it merely as property, but at least to some extent as something else. Second, they believe that the *concept of property itself* cannot in some sense do justice to animals because (they assert) the concept must call for like treatment of all kinds of property, most of which are inanimate objects.

[7] Locke J: *First Treatise of Government*, Chap. IX, § 92. In Laslett P (ed): *Locke's Two Treatises of Government*. Cambridge: Cambridge University Press, 1960, p. 209.

[8] Francione: Legal welfarism, 734.

[9] Locke's views regarding natural law, and natural law theories in general, did not play a major role in the development of the common law. When Parliament began asserting its power against the monarchy, it did not turn to natural law for ammunition, but to the common law. It was not Locke but the great Chief Judge of the Common Pleas Edward Coke, whose views stimulated and infused the 16th and 17th century consolidation and supremacy of the common law. Plucknett TFT: *A Concise History of the Common Law*. Boston: Little, Brown, 1956, pp. 242–244, 280–284. Locke's theory of separation of powers and natural law as a limitation of State authority became a foundation of the Whig party's advocacy of a limited monarchy. As Plucknett observes, Locke's views provided a philosophical justification of the compromise struck between Parliament and William III after the expulsion of the Stuarts. *Id.*, pp. 63–64. Locke's views on property and natural rights, popularized by Blackstone, were widely read in the American colonies and greatly influenced the founders of the United States. But the common law, which was retained in the laws of the original 13 states, had already developed most of its central features from foundations much older than and quite different from Locke's natural law philosophy.

[10] Blackstone is not always a reliable source regarding the history of English (and related developments in American) law. The eminent legal historian Theodore Plucknett writes of Blackstone that "his great skill consisted in affording a reasonable explanation for the state of English law as it then existed. Some portions he explained on logical grounds, others from history. His history was not very profound, for like so many practicing lawyers of that time (and later), he expected little more in history than a plausibility at first sight. . . .

Blackstone was therefore in harmony with the thought of his age when he regarded our legal history as an object of 'temperate curiosity' rather than of exact scholarship. His equipment in jurisprudence was also somewhat slender, but his freedom from excessive learning was an actual merit; he found explanations which seemed adequate, clear, and above all interesting . . ." Plucknett TFT: *A Concise History of the Common Law*. Fifth edition. Boston: Little, Brown, 1956, p. 286.

[11] Blackstone: *Commentaries*, Bk. II, Chapt. 25, *388–390, pp. 698–699. Blackstone's account of why wild and domestic animals are treated differently provides an example of his tendency to try to explain logically a historical development that had less to do with logic than with the unfolding of events. Wild animals were treated as possessions of the Sovereign not so much because domestic animals were tractable as because the King owned the land of the nation and wild animals (like wild plants and flowers, which the King also "owned") were viewed as a feature or accompaniment of the land itself. Also, selling licenses to hunt and fish provided the King a welcome source of revenue.

[12] For a readable introductory account of this history see Plucknett TFT: *A Concise History of the Common Law*. Boston: Little, Brown, 1956. The classic treatment remains Pollack F and Maitland FW: *The History of English Law*, Vols I and II (SFC Milsom ed), hereinafter referred to as Pollack and Maitland: *History*. The lectures of Professor Maitland, the founder and greatest figure in modern historical studies of the common law are reprinted in Maitland FW (W.J. Whittaker ed): *The Forms of Action at Common Law*. Cambridge: Cambridge University Press, 1971.

[13] See Ganshof FL: *Feudalism*. (P. Grierson trans) New York, Harper and Row, 1964.

[14] Pollack and Maitland: *History*, Vol. 2, p. 153.

[15] For example, by 1200 the law fastened on primogeniture (the passing of a man's estate to his eldest son) as a way of preventing a man's right to possess land from reverting back to his grantor upon his death. This left the younger sons (and daughters) with nothing. The practice then arose in which a father took an oath of homage of service from his younger sons in return for a grant of part of his land. Fathers also obtained consent of their eldest son in order to transfer rights in land to other sons, and consent from any older son to transfer rights to a younger. The fiction was subsequently introduced that when the donee (son) gave the donor (father)

homage, this homage would somehow descend upon the donor's heirs thus binding them to warrant the land to the donee and his heirs. It was not until the early 13th century that a father could give to heirs other than his eldest son without that son's consent, but even here he did so by making a *statement* that he "and his heirs" warranted the gift. This monumental development elevated someone to ownership because the owner of the land could dispose of it as he wanted. But even here, remnants of the old feudal practices remained, because the donor still stated that he and his heirs accept the transfer of rights. These are but a few of the steps English law took on its tortuous route to allowing the free alienability of land.

[16] Pollack and Maitland: *History*, Vol. II, p. 153.

[17] Pollack and Maitland: *History*, Vol. II, p. 151.

[18] Pollack and Maitland: *History*, Vol, II, pp. 32, 151.

[19] Pollack and Maitland: *History*, Vol, II, p. 151.

[20] Pollack and Maitland: *History*, Vol, II, pp. 151–152.

[21] Pollack and Maitland: *History*, Vol, II, p. 154. In 1854 the English Common Law Procedure Act removed the option from the defendant of deciding between paying the value of the chattel or restoring it to a successful plaintiff, giving this determination to the court. Maitland FW: *The Forms of Action at Common Law*. Cambridge: Cambridge University Press, 1971, p. 58.

[22] Pollack and Maitland: *History*, Vol, II, pp. 178–179.

[23] Maitland FW: *The Forms of Action at Common Law*. Cambridge: Cambridge University Press, 1971, pp. 60–61.

[24] At the present time, the pendulum in American courts and legislatures appears to be swinging toward greater restrictions on the ability of government to restrict people's use of their real property. See, Greenhouse L: High court, in 5–4 split, limits public power over private property. *New York Times*, June 25, 1994, A1.

[25] The legal philosopher A.M. Honoré distinguished eleven "leading incidents" of ownership: "the right to possess, the right to use, the right to manage, the right to the income of the thing, the right to the capital, the right to security, the rights or incidences of transmissibility and absence of term, the prohibition of harmful use, liability to execution, and the incidence of residuarity." Honore AM: Ownership. In Guest AG (ed): *Oxford Essays in Jurisprudence*. Oxford: Oxford University Press, 1961, p. 113.

[26] Ackerman B: *Private Property and the Constitution*. New Haven: Yale University Press, 1977, pp. 26–27, emphases in original. Ackerman characterizes this way of describing property as part of the

"scientific" approach to law, which he distinguishes from what he calls the approach of the "ordinary observer" who seeks to understand legal concepts by placing their foundations in the "ordinary talk of nonlawyers." (p. 10) This "scientific" characterization accords with the concept of property as it developed historically.

[27] *Martin* v. *Waddell*, 41 U.S. (16 Peters) 367 (1842); *Geer* v. *Connecticut*, 161 U.S. 519, 528 (1896) (holding that the "attribute of government to control the taking of animals *ferae naturae*, which was thus recognized and enforced by the common law of England, was vested in colonial governments, where not denied by their charters, or in conflict with grants of the royal prerogative" and that this "power which the colonies thus possessed passed to the states with the separation from the mother country, and remains in them at the present day, in so far as its exercise may not be incompatible with, or restrained by, the rights conveyed to the Federal Government by the Constitution").

[28] Blackstone: *Commentaries*, Bk. II, Chapt. 25, *394–395, p. 702.

[29] Blackstone: *Commentaries*, Bk. II, Chapt. 25, *391, p. 699.

[30] Blackstone: *Commentaries*, Bk. II, Chapt. 25, *394, p. 702.

[31] *Sentell* v. *New Orleans and Carrollton Railroad Co.*, 166 U.S. 698, 701 (1897).

[32] See, *Wallace* v. *State*, 30 Tex. 758 (1868); *Brown* v. *State*, 26 Ohio St. 176, 178 (1875); *Chappell* v. *State*, 35 Ark. 345 (1880).

[33] Quoted in Leavitt ES and Halverson D: The evolution of anti-cruelty laws in the United States. In *Animals and Their Legal Rights*. Fourth edition. Washington, DC: Animal Welfare Institute, 1990 (hereinafter referred to as Animal Welfare Institute: *Animals and Their Legal Rights*), p. 1.

[34] Morison SE: *Builders of the Bay Colony*. Boston: Houghton Mifflin, 1930, p. 232 (relating a prosecution for cruelty to an ox).

[35] Quoted in Leavitt ES: Introduction to the first edition. In Animal Welfare Institute: *Animals and Their Legal Rights*, pp. xiii-xiv.

[36] French RD: *Antivivisection and Medical Science in Victorian Society*. Princeton: Princeton University Press, 1975, p. 29. French's book is the most complete historical account of the passage of early English cruelty laws in general and the Act of 1876 in particular, in the context of controversies about the use of animals in experimentation.

[37] New York Statutes, Pt. IV, Chapt I, Tit. VI. (1828), quoted in quoted in Leavitt ES and Halverson D: The evolution of anti-cruelty laws in the United States. In Animal Welfare Institute: *Animals and Their Legal Rights*, p. 2.

[38] Sec. 26. quoted in Leavitt ES and Halverson D: The evolution of anti-cruelty laws in the United States. In Animal Welfare Institute: *Animals and Their Legal Rights*, p. 5.

[39] Sec. 2, quoted in Leavitt ES and Halverson D: The evolution of anti-cruelty laws in the United States. In Animal Welfare Institute: *Animals and Their Legal Rights*, p. 5.

[40] Sec. 2, quoted in Leavitt ES and Halverson D: The evolution of anti-cruelty laws in the United States. In Animal Welfare Institute: *Animals and Their Legal Rights*, p. 6.

[41] Sections 3 and 4, quoted in Leavitt ES and Halverson D: The evolution of anti-cruelty laws in the United States. In Animal Welfare Institute: *Animals and Their Legal Rights*, p. 6.

[42] Section 5, quoted in Leavitt ES and Halverson D: The evolution of anti-cruelty laws in the United States. In Animal Welfare Institute: *Animals and Their Legal Rights*, p. 6.

[43] Sec. 10, quoted in Leavitt ES and Halverson D: The evolution of anti-cruelty laws in the United States. In Animal Welfare Institute: *Animals and Their Legal Rights*, p. 5.

[44] Sec 1, quoted in Leavitt ES and Halverson D: The evolution of anti-cruelty laws in the United States. In Animal Welfare Institute: *Animals and Their Legal Rights*, pp. 5–6.

[45] N.Y. Agric. & Mkts. Law § 353 (McKinney 1991). Compare the current Massachussets statute: "Whoever overdrives, overloads, drives when overloaded, overworks, tortures, torments, deprives of necessary sustenance, cruelly beats, mutilates or kills and animal, or causes or procures an animal to be so overdriven, overloaded, driven when overloaded, overworked, tortured, tormented, deprived of necessary sustenance, cruelly beaten, mutilated or killed; and whoever uses, in a cruel or inhuman manner in a race, game, or contest, or in training therefor, as lure or bait a live animal, except an animal if used as lure or bait in fishing; and whoever, having the charge or custody of an animal, either as owner or otherwise, inflicts unnecessary cruelty upon it, or unnecessarily fails to provide it with proper food, drink, shelter, or protection from the weather, and whoever, as owner, possessor, or person having the charge or custody of an animal, cruelly drives or works it when unfit for labor, or willfully abandons it, or carries it or causes it to be carried in or upon a vehicle, or otherwise, in an unnecessarily cruel or inhuman manner or in a way and manner which might endanger the animal carried thereon, or knowingly and willfully permits it to be subjected to unnecessary torture, suffering or cruelty of any kind shall be

punished by a fine of not more than one thousand dollars or by imprisonment for not more than one year, or both. . . ." Mass. Gen. Laws Ann. ch. 272, § 77 (West 1990).

[46] For a comprehensive and detailed review of state cruelty laws and differences among them see Animal Welfare Institute: *Animals and Their Legal Rights*. The first edition of this classic survey appeared in 1968. In light of the argument of this paper, the title of the book is noteworthy.

[47] However, this does not mean another law cannot address the situation. For example, in states that exempt animal researchers from cruelty prosecutions, researchers may still be subject to punishment by state or federal authorities that have jurisdiction over the research. Veterinarians practicing in states in which they are exempted from prosecution for cruelty can still have their license to practice suspended or revoked by their state licensing board for causing an animal unjusifiable pain or discomfort.

[48] See, e.g., N.H. Rev. Stat. Ann § 644:8(III) (1994) (making a second or subsequent conviction of cruelty a felony); § 644:8(III-a) (making a person guilty of a felony "who purposely beats, cruelly whips, tortures, or mutilates any animal or causes any animal to be beaten, cruelly whipped, tortured, or mutilated").

[49] See, Mo. Ann. Stat. § 578.009 (Vernon supp. 1995), *Animal neglect*, and § 578.012, *Animal abuse*.

[50] Tannenbaum J: *Veterinary Ethics*. Second edition. St. Louis: Mosby, 1995 (hereinafter referred to as Tannenbaum: *Veterinary Ethics*), pp. 120–123.

[51] *People ex rel. Freel* v. *Downs*, 136 N.Y.S. 442, 445–446 (N.Y.C. Mag. Ct. 1911).

[52] Tannenbaum: *Veterinary Ethics*, pp. 425–427.

[53] For a study of bonding between horses and people, see Lawrence E: *Hoofbeats and Society: Studies of Human-Horse Interactions*. Bloomington, IN: Indiana University Press, 1985.

[54] A good example of society's present disinclination to apply cruelty laws to animal "pests" is evidenced by the recent case of a New Jersey man who found a live rat in a trap he had placed in his garden. The rat was partially outside of the cage, and afraid that the animal would escape or bite someone, he killed it by beating it over the head with a broom handle. The local prosecutor instituted a criminal action charging the gardener with cruelty. After the case was subjected to withering ridicule in the mass media, it was dismissed—but over the objection of the defendant who wanted his

day in court to argue that the prosecution was misguided. Hanley R: Rat's tale, the sequel: Prosecutor backs off. *New York Times*, August 10, 1994, B4; Nieves E: Much ado in death of a rodent. *New York Times*, August 16, 1994, B4; Rat killer summons is piped out of court. *New York Times*, August 25, 1994, B6.

[55] Francione: Legal welfarism, 753.

[56] Francione: Legal welfarism, 756, emphasis supplied.

[57] *Commonwealth* v. *Higgins*, 277 N.E. 536 (Mass. 1931).

[58] *Knox* v. *Massachusetts Society for the Prevention of Cruelty to Animals*, 425 N.E.2d 393 (Mass. App. Ct. 1981).

[59] See, e.g., Newman JH: *Sermons Preached on Various Occasions*. Second edition. quoted in Passmore J: The treatment of animals. *J Hist Ideas* 36:203, 1979 ("(w)e have no duties toward the brute creation; there is no relation of justice between them and us . . . they can claim nothing at our hands; into our hands they are absolutely delivered"); Kant I: *Lectures on Ethics*. (L. Infield trans) New York: Harper and Row, 1963, p. 239.

[60] Lawrence J: "A Philosophical Treatise on Horses and the Moral Duties of Man towards the Brute Creation." quoted in Leavitt Introduction to the first edition. In Animal Welfare Institute: *Animals and Their Legal Rights*, p. xiv.

[61] Angell GT. Five questions answered, pp. 4–5, emphases supplied. In Angell GT: *Autobiographical Sketches and Personal Recollections*. Boston: Franklin Press, 1884. Angell was also keenly aware that, for some people, "when a man strikes his fellow-man, he expects to be arrested; but when he strikes *his property*, and an officer intervenes, he regards it as an impertinent interference with his personal rights and would be glad to do the officer an injury." (*id.*, emphases in original) Angell's solution to this attitude was not to reject the notion that animals are property, but to argue that animal protection and enforcement of cruelty laws could most effectively be accomplished through the creation of animal protection societies, officers of which would not be regarded as "ordinary police." (*id.*) Angell's recommendation has been widely adopted.

[62] Angell GT: Ten lessons on kindness to animals. Second lesson, p. 5. In: Angell GT: *Autobiographical Sketches and Personal Recollections*. Boston: Franklin Press, 1884.

[63] Galfunt, Judge: In *People* v. *O'Rourke*, 369 N.Y.S.2d 335, 341–342 (N.Y.C. Crim. Ct. 1975).

[64] *Stephens* v. *State*, 3 So. 458, 459 (Miss. 1898). This decision also justifies cruelty laws on the ground that "human beings should be

kind and just to dumb brutes; if for no other reason than to learn how to be kind and just to each other."

[65] *State* v. *Karstendiek*, 22 So. 845 (La. 1897).

[66] Feinberg J: The rights of animals and unborn generations. In Feinberg J: *Rights, Justice, and the Bounds of Liberty*. Princeton: Princeton University Press, pp. 159–160, emphases in orginal.

[67] Feinberg J: Human duties and animal rights. In Morris RK and Fox MW (eds): *The Fifth Day: Animal Rights and Human Ethics*. Washington, DC: Acropolis Books, 1978, p. 47.

[68] See, Francione: Legal welfarism; Regan T: Animals and the law: The need for reform. In Regan T: *All that Dwell Therein*. Berkeley: University of California Press, 1982, pp. 148–164; Tischler J: Rights for non-human animals. A guardianship model for dogs and cats. *San Diego Law Rev*, 14:484, 1987.

[69] Prosser W: *Handbook of the Law of Torts*. Fourth edition. St. Paul, MN: West Publishing Co., 1971, p. 586.

[70] E.g., Are laboratory animals treated humanely? Associated Press, October 28, 1985 (NEXIS, Current library), cited in Garvin LT: Constitutional limits on the regulation of laboratory animal research. *Yale Law Journal* 98:388, 1988 (poll showing that 76% of Americans believe that animals have rights, 81% think it is necessary to use animals in some applied medical research, and 42% of those who believe that animals have rights think that their use in research violates these rights).

[71] An interesting issue concerns whether giving animals standing to sue would render them something other than property. I would argue that by itself standing would not be inconsistent with their status as property. Suppose that under certain circumstances courts appointed legal guardians to sue an animal's owner on its behalf to order its owner to treat it in certain ways. (One could imagine, for example, a court directing an owner to provide and pay for veterinary care for a serious malady.) It would not follow that this person ceased to be an "owner" and the animal his property—if he retained ordinary rights associated with ownership, such as the right to sell the animal to someone else.

[72] Francione: Legal welfarism, 756.

[73] Francione: Legal welfarism, 723.

[74] For example, in 1990 Scotland and Wales enacted legislation, designed to effect "prevention of cruelty," that prohibits such practices. In Animal Welfare Institute: *Animals and Their Legal Rights*, p. 304.

[75] In Animal Welfare Institute: *Animals and Their Legal Rights*, pp. 305–307.

[76] Tannenbaum: *Veterinary Ethics*, pp. 457–463.

[77] See, Friend CF: Animal cruelty laws: The case for reform. *Univ of Richmond Law Rev* 8:201–231, 1974.

[78] See, Tischler J: Rights for non-human animals: A guardianship model for dogs and cats. *San Diego Law Rev*, 14:484, 1987; Francione GL: Personhood, property, and legal competence. In Singer P and Cavalieri P (eds): *The Great Ape Project* (hereinafter referred to as Cavalieri and Singer: *The Great Ape Project*), pp. 254–255.

[79] Singer P and Cavalieri P: The great ape project—and beyond. In Cavalieri and Singer: *The Great Ape Project*, pp. 305–6.

[80] See, Singer P: *Animal Liberation*. Second edition. New York: Avon Books, 1990, pp. 1–22.

[81] Tannenbaum: *Veterinary Ethics*, pp. 345–349.

[82] A declaration on great apes. In Singer and Cavalieri: *The Great Ape Project*, p. 4.

[83] Tannenbaum: *Veterinary Ethics*, p. 190.

[84] See, e.g., *Jankoski* v. *Preiser Animal Hospital, Ltd.*, 510 N.E.2d 1084 (Ill. App. Ct. 1987); *Brousseau* v. *Rosenthal*, 443 N.Y.S.2d 285 (N.Y. Civ. Ct. 1980).

[85] 7 U.S.C. § 2143(a)(2).

Keynote Address

Stephen Jay Gould

This text is an edited transcript of a keynote address given at the conference "In the Company of Animals" in New York City on April 6, 1995. Written and spoken English are entirely different languages, and this address, prepared only to be spoken, must be fairly awkward in this printed version. In addition, the talk depended upon a large number of slides, which could not be included here. I either deleted the sections based entirely on pictures or kept the text when I felt that the points could be understood without the visual material.

Stephen Jay Gould

W E all know the conclusion to that most famous of all poems about invertebrates—namely, Robert Burns's "To A Louse." The louse speaks from its position in a hairpiece of an upper-class lady, if I remember correctly. "Oh, would some power the giftie gi'e us, to see ourselves as others see us. It would from many a blunder free us, and foolish notion." Very familiar lines that you all know. Apparently, unfortunately, no such power exists, and so everything we know about animals we see in our terms.

I want to illustrate the general theme of how we're always seeing not only animals, but everything else, in our terms, by giving you four quick examples of what, in a way, is the most egregious kind of misinterpretation we can make—namely, when we try to identify the attributes of animals as a result of nothing more than the arbitrary name that we happen to have given to them. It's bad enough that we backread our *features* into organisms, but when we backread an arbitrary name that we happen to give to an organism, and then assume that its characteristics flow from this arbitrary name, then that's the ultimate example of the backreading fallacy.

Steve Glickman this afternoon talked about T.H. White's

bestiary translation. If you look at the attributes in medieval bestiaries, they always discuss where the names of animals come from—why is the goat Capra, for example. White's bestiary tells us that you only have to invert the syllables. It's aspera captet: "he seeks the rough places." And you reverse it, and then it's capra. Now let's go forward, to the age of Newton and Sir Thomas Browne, and we come to various myths, such as the famous myth of the beaver—that to elude the hunter, the beaver bites off his own testicles. It's a very old myth. And Sir Thomas Browne debunks it in his *Pseudodoxia Epidemica*— that is, his *Epidemic of Falsities*, the first of the great exposés of foolish wisdom, so to speak, written in the 1640s. He first goes through the various reasons why people ever would have believed such nonsense and points out (I am quoting): "Some have been so bad grammarians as to be deceived by the name." The name for the beaver is Castor, and many people thought the name came from castrate.

Browne is marvelous—his use of language—so, if you don't mind we'll go on just a little bit. He says—after pointing out that the name castor does not share the same root as "castration," but ultimately derives from a Sanskrit word for musk; that comes later. He then cites the factual evidence of intact males, and the reasoned argument that a beaver couldn't even *reach* his own testicles if he wanted to bite them off. Now quoting from Browne: "The testicles properly so called are of a lesser magnitude, and seated inwardly upon the loins. And, therefore, it were not only a fruitless attempt but impossible act to eunuchate, or castrate themselves, and might be an hazardous practice of art if at all attempted by others."

And then we move forward another 100 years to Linnaeus. Linnaeus, among his many inventions—we'll come back to mammalia later in this talk—did coin, as a name for the order that includes us among the mammals, *Primates*, or first. Now, that leads to all sorts of trouble—this time in the opposite direction, because this is one example where the animal meaning almost takes over. In this particular case, the original use of "primate" in the English language is not for monkeys and apes, but for heads of the Anglican church. This slide shows Mr. Ussher, who gave the famous date of 4004 B.C. to the earth; Jacobus Usurius, totius Hibernia primas (primate of all Ireland.

Linnaeus then used the name primate for monkeys and apes, and everybody thinks that is the original meaning. This example is just an excuse, because I wanted to read one of the most marvelous letters that I ever got. This is a letter from the Reverend Michael Ingham, who is the principal secretary to the Primate of the Anglican Church of Canada, and he wrote it to a John Hearn, the Director of the Wisconsin Regional Primate Research Center—who, clearly, thought that this guy represented the other kind of primate. In any event, it is a terrific letter. Ecclesiastics must live for this once-in-a-lifetime opportunity. "Dear Dr. Hearn: Thank you for your letter of December the fourth addressed to Dr. George Cram of the Primate's World Relief and Development Fund, in which you seek information for your international directory of primatology. I should perhaps inform you that the term 'primate' in our context refers to the senior archbishop and chief pastor of the Anglican Church of Canada. The Relief and Development Fund over which he presides is an agency for the alleviation of global poverty and hunger on behalf of Anglican Christians. I think the primates in your study are perhaps of a different species. While it is true that our primate occasionally enjoys bananas, I have never seen him walk with his knuckles on the ground or scratch himself publicly under the armpits. He does have three children, but this is a far cry from 'breeding colonies of primates,' as your research project mentions. Like you, we do not import our primates from the wild, however. They are elected from among the bishops of our church. This is occasionally a cause of similar, though arcane, comment. The subject of primate biology might be of great importance in your field, but, alas, not so in ours. There are a mere 28 Anglican primates in the whole world. They are all males, of course, and so far we have had no problems of reproduction."

But lest you think this type of error has disappeared, I point out a scene from "Jurassic Park," right in the beginning, when the paleontologist, Mr. Grant is trying to persuade folks on his field trip out West that dinosaurs are related to birds. He gives a whole bunch of perfectly good arguments based on anatomy, which is how the argument should be made. Then he turns around—because the fossil he is talking about is *Velociraptor*—and says, "Even the *word* 'raptor' means bird of prey—and that is his closing argument. That is the crowning glory: Even the word "raptor" means bird of prey—so, of course, dinosaurs must be birds—because this dinosaur is named *Velociraptor*. It is a wonderful example, because "raptor," in English, was used for humans centuries before. I think it was Linnaeus, again, who first used it as a term for birds of prey. It comes from the Latin *rapere*, to seize by force. And I hope you recognize that "The Rape of Europa," the next time you see Titian's painting at the Gardner Museum in Boston, refers to the abduction of Europa—not to whatever happened afterwards. It is the seizure by force, or rape in the original sense. Easy to see how it acquired the other meaning, but it was a human word for centuries before it became a bird word.

Okay, that is the introduction to the theme of this talk, which now proceeds in two parts. First, that backreading—that is, the placement of human characteristics into animals—is really the only way that we have ever proceeded. There are a few exceptions among honorable scholars, I suppose, and others, and fishermen and hunters, I am sure—but let us just say by far the overwhelmingly predominant way of understanding animals is by backreading human characteristics into them. And then, of course, people also fall into the further funny fallacy responsible for so much biodeterminist nonsense. You then, having identified human phenomena in animals, and called them that—like cuckoldry, or adultery, or whatever— you then rederive them as natural for humans. Because if rape exists in mallard ducks, as has been claimed, then, clearly, it is a biologically conditioned feature in human beings—which is nonsense on many criteria.

So this will be part one of the talk, and there's really nothing particularly original here. Part two—not only do we backread characteristics of ourselves when we are talking about animals, but Protagoras was right when he said that man is the measure of all things—whether he meant just male human beings or was using the Greek term for all of us, I leave aside for now. And that, therefore, even the most abstract and universal issues of science and philosophy are often really, at root, inquiries about humans—particularly validations of human hegemony in the face of fear that we are not quite so powerful as we think we are.

And this leads me, at the very end, to a particular argument, which, in a way, is the key to this talk, and its only possible point of mild originality—that, since human beings are a contingent product of history, and not a predictable outcome of the laws of evolution or other natural laws, therefore, these abstract universals, which we have always seen as transcendently general, are really tales from historical science after all—since they are, fundamentally, ways in which we justify our own status, and since our status is historically contingent, rather than conditioned by laws of nature, some of these very deepest and most abstract issues are really discussions about historical particulars, and not the transcending generalities they have always been assumed to be.

Let me then go to the first point—backreading as the only way we have looked at or tried to understand animals. First let me point out that there is limited legitimacy to this tactic on occasion, because genuine homology exists between humans and other creatures. That is, we *are* animals, and we are evolved from other animals, and we do have varying degrees of kinship with animals. And, as you know, evolutionary biologists make a key distinction when judging similarities that exist among different organisms, into homology and analogy.

Analogies are similarities held because evolution has produced, independently, pretty much the same form over and over again: the wing of a bat; the wing of a bird; the wing of an insect; the wing of a pterosaur. These are analogous features, because the common ancestor for any pair of these creatures had no wings—and the wings evolved separately in each lineage. Homologies are traits shared by common descent. The bones in the arm of the whale, the horse, the bat, and me are effectively the same topologically, but the whale swims, the bat flies, the horse runs and I gesticulate. Clearly, this is not a result of separate evolution for common function, but is a tie to history. We all have the same bones because we have common ancestors in mammals, which have this configuration.

For evolutionary biologists, homology has primacy. Homological similarity—similarity by history and descent—is overwhelmingly powerful. An analogous similarity—a convergence, as we call it—can never be anything other than superficial. You cannot get independent evolution of hundreds of similar features—it is just a mathematical probability argument. If you have complex similarity, it is homologous; and homology is, therefore, deep, and fundamental, and important. Now, since we know ourselves best, if you can make a genuine argument from homology, it may not be invalid to analyze a feature in us that we understand better, because we know it from personal experience, and use that analysis to interpret animal features if they are truly homologous. This procedure is not invalid.

Let me give you my absolutely favorite example of a brilliant argument by homology—in this case more to explain humans from animal models—from the third of Darwin's evolutionary books. All his books were evolutionary in one sense, but I like to see his great evolutionary books as a trilogy: *The Origin of Species* (1859); *The Descent of Man* (1871); and the one that deserves to be much more read, the third member of the trilogy, *The Expression of the Emotions in Man and Animals* (1872). An absolutely brilliant book because Darwin does what he always does so well. Intellectuals are supposed to be people with the deepest knowledge about the broadest questions. Put an intellectual on television, and he will tell you what the course of the future is going to be. But this is not right. Often the most important thing is to recognize the *limits* and to stick to what *can* be done.

The brilliance of Darwin's book about the emotions is that he's not trying to interpret in evolutionary terms what cannot be done—namely, the deep meaning of the emotions, their moral value, and so on. He is talking about the expression of the emotions in man and animals—that is, the physical appearance, the gestural capacities. The whole book is one brilliant argument—namely, that if you look at the universal *form* of those emotional expressions that *are* universal across human cultures—and Darwin took great pains trying to establish that universality—and you interpret these gestures by seeing their homological similarity with expressions in other mammals, then you can understand that their origin must be evolutionary, and cannot be by divine creation—because although the expressions have no functional meaning in humans—they seem arbitrary, and they do not have to mean what they universally do in other animals. When you see what the other animals are doing, those in which this expression of emotion first arose, the gesture is very functional there. And, therefore, homology is retained in human expression, even though the original purpose is no longer followed in humans.

Consider my favorite example (and I am sure Darwin is right here) rage. What we do? We snarl; we raise the side of our lips, thereby exposing our canine teeth—which are no bigger in humans than any other teeth, and therefore not a threat at all. But in other animals the raising of the lips and the exposure of the canines shows the large dangerous teeth, as in dogs—and, therefore, the gesture makes sense in its original form. But we still do it, and this must be a homological retention.

I do not want to leave the subject of homology without

saying something about the most exciting development in evolutionary theory during the last decade—namely, the beginnings of understandings through genetic tracings of the actual course of embryological development as genetically mediated. We have found that the extent of homology is vastly greater than anyone would have thought. In 1963 Ernst Mayr, the greatest evolutionist of this generation said: It would be vain to look for any homology—that is genetic identity, similarity based on common ancestry—in the genetic sequences of genes from different phyla—such as insects and humans, to choose the obvious example—that have been separate for at least 550 million years. The attempt would be vain because we know that natural selection is so powerful and has so completely altered every aspect of the genome, that whatever similarities might have once existed have clearly been wiped out by independent evolution over all that time and in such different directions.

Yet it is not so. The homologies between distant phyla are quite stunning, and often produce very similar features. If any of you saw the cover of *Science* recently, you will have read about remarkable experiments showing that a single gene—*eyeless* in Drosophila—can, when expressed in a part of the body that does not normally form eyes, produce fully functional eyes. You can make them on the wing; you can make them on the tips of the antennae; you can make them on the legs—you can make them practically anywhere. Even more fascinating, the homologue of that gene in humans—called *aniridia*—and in mice works just as well on the insects to produce eyes. The development of eyes in insects, in humans and in squid is *the* classic textbook example of convergence—that is, of independent evolution. And, clearly, in an anatomical sense, this old view is correct. But it turns out that there is homological underpinning of the developmental pathway. They all carry the same PAX-6 gene that controls the developmental pathway for making eyes.

Let me discuss the example that has been most in the

press—body segmentation in the famous homeobox story of arthropods and chordates—that is, the groups that include insects and vertebrates. It is an old argument, dating to Etienne Geoffroy Saint-Hilaire in 1830, who made an argument, which turned out to be wrong in detail, and for which he was ridiculed, that insects and vertebrates share a common structured plan based on the archetype of the vertebra—just as Goethe, his friend, had argued that the common form of plant structures is the leaf—the *Urpflanze*. And Geoffroy did not shy away from the strange implications of his expanded homology, which is, let's face it, that an insect—who has an external skeleton—lives inside its own vertebrae and walks on its ribs.

Now, that comparison is wrong, and people therefore threw aside the entire notion that there could be similarity in a genetic, homological sense in segmentation—until all the new discoveries—based, initially, on the so-called homeotic mutants of *Drosophila*.

The ordinary antenna of drosophila, the fruit fly, consists of two parts—the antenna and the so-called arista at the end. There are a set of odd mutations, which have been known for a long time, called the homeotic mutations, which place a body structure in a "wrong" place, so to speak. An example is *antennapedia*, in which a leg appears where an antenna ought to be. That is not as weird as it sounds, because, evolutionarily, legs and antennae are based on the same ancestral structures.

Now, about 20 years ago, Ed Lewis at Cal Tech made a brilliant suggestion and confirmed a genetic model that explained how these homeotic mutations work. This slide shows an ordinary insect as it develops: the larva is on the left, the adult fly on the right. We note a series of segments, H atop his head. T1, T2 and T3 are the three segments of the thorax—that is all we have to be concerned with here. In insects, each thoracic segment bears a leg. That is why insects have six legs, because there are three thoracic segments and each bears a leg. The second thoracic segment bears a pair of wings. Most insects have two pairs of wings, and the third thoracic segment also bears wings. In flies, which have only two wings, unlike most insects, the third thoracic segment bears a vestigial set of wings, called halteres. And then you get a bunch of abdominal segments behind.

Ed Lewis figured out that there is a wonderfully simple model for how differentiation proceeds in the right order. There are a series of genes—*Drosophila* has only four chromosomes; these genes are one arm of the third chromosome, and they are lined up in a row. They are products of a single ancestral gene that duplicated and put its duplicate copies right next to each other along a line. What happens is the following. The genes that produce the correct differentiating of the segments, turn on in sequence. The first gene turns on in the second thoracic segment and then is expressed all the way back. The second gene turns on further back—it turns on in A1, and then is expressed all the way further back. The third gene, number two, starts in A2 and is expressed all the way back, and so on.

The whole point of this, the only thing you have to grasp, is that this process yields a gradient, where the maximum gene product is at the back of the organism—because all of the genes are expressed in A8 (the last abdominal segment) and a *minimum* amount of the gene product occurs up front. Each segment differentiates according to how much of the gene product it has. Here is a simple prediction, the affirmation of which broke the dike in understanding this system: mutations that intensify the gradient—that give you more gene product than you ought to have—make posterior structures. But most mutations are so-called deletion, or loss-of-function mutations—you get less of the gradient. This means that structures that ought to be up front appear further back, because there is less gene product further back than there ought to be, and less gene product means that a segment is differentiating as though it was further forward on the body.

Now, that simple theme explains all the really weird

homeotic mutations at the back end of *Drosophila*. The most famous of all is *bithorax*, the four-winged *Drosophila*. It looks as though it recovered an evolutionary past and has four wings again. But this is not true—for *bithorax* is just a loss-of-function mutation in which, because there is less gene product, the third thoracic segment—which ought to develop vestigial halteres—thinks it ought to be another second thoracic segment. And so it grows its third thoracic as though it were another second, so you have two seconds. And since seconds grow wings, you have a four-winged fly. It's not really recreating its ancestry.

And then we have an even weirder one, called *bithoraxoid*, the eight-legged fly. Insects have six legs. Here's one that seems to violate the definition of its class. But it is the same thing—a loss-of-function mutation. There is not enough gene product in the first abdominal segment. Therefore, it thinks—you see, we use backreading of intent-language all the time!—that is, the segment thinks it ought to differentiate as a supernumerary thoracic segment, and so it does. Instead of being a first abdominal, it differentiates as another third thoracic. The third thoracic has legs, so now we have eight legs instead of six.

So far this is just an insect story. But here is the great discovery of the last 10 years. The same genes exist in vertebrates. In fact, they exist in fourfold repetition. The whole sequence is repeated four times in vertebrates. This is probably why we do not have weird homeotic mutations in vertebrates, because there are four copies of all these genes. So if one of them mutates, the other three are still presumably expressing the normal state and can overwhelm it—whereas in insects there is only one copy, so if it mutates it expresses. But is that not stunning? I mean, Geoffroy was right after all. There is a fourfold repetition of gene clearly homologous to those of insects—they are 90 to 95 percent similar after 550 million years of separation from the insects. You might say, "Yeah, but so what? If it is differentiating segments in insects, what is it doing in vertebrates? If it is doing something totally different, then so what." Well, it is doing something similar in vertebrates—and that is the fascinating thing.

It turns out that vertebrate backbone segments are not the same thing as insect body segments—that is where Geoffroy was wrong. But what modern scientists had forgotten is something that all the great nineteenth-century embryologists knew—that the brain, the mid and hind brain, as it differentiates in embryology, develops as a set of segments, called rhombomeres. And you might say, "But it is all erased in the adult brain." But it is not, because the tongue structures and the cranial-nerve divisions are largely reflective of this old segmentation. In this slide, you can see four of the HOX genes—that is, the mammalian homologues of the invertebrate genes—and their anterior expression boundaries are not in the spinal column but in the rhombomeres. So, clearly, they are determining the rhombomeres, which are the homologues of the insect segments. It is just fascinating. And this slide is a mouse embryo showing that most of these genes—you can see them along the top—are expressing in the rhombomeres.

The initial vertebrates, by the way, had very small backbones—that is, the part that is not homologous—and very large gill baskets—which are the homologues, which differentiate from the rhombomeres. So the initial vertebrates in the fossil record are mostly expressing the system of segmentation that is the homologue of the insect case. It is just a fantastic story.

One last point about homology—we'll just look at the next slide—this is one of the most poignant pictures I have ever seen. This is the gravestone of Baby Fae. You may remember her story at Loma Linda University. A baboon's heart was engrafted into her, and she died. Something so touching—call it "vernacular art"—but the two hearts on a tombstone—her own, that failed, and the baboon heart that failed. Or is it her mother and father who love her?—I do not know. I do not want to make a big point about this, and I do not know whether she could have been saved any circumstances. Let us just say it was foolish in the extreme and not respectful of evolutionary principles, if the experiment was to be done at all, to use a baboon heart and not a chimpanzee heart. Baboons are 30 million years evolutionarily distant from humans. Immunological acceptance or rejection is a question of overall genetic similarity, which is homology. Chimpanzees are six to eight million years different. If you look at Dr. Bailey's justification for why he did the procedure, he justified it only in functional terms. Well, a baboon heart is about the right size; chimpanzee hearts are hard to get. But then we come to the key point: Dr. Bailey is a Seventh Day Adventist—he does not believe in evolution. Sometimes, if you do not acknowledge what it is all about, you can make some tragic errors. I will leave it at that.

Although homology is a legitimate theme, there are many fallacies based on false usages of evolution. The one I have written about most in my own career is gradualism, progressionism, and continuity theories in general. Not everything is homology. Consider the chimp-language debate—I do not want to insert myself in something that I do not know a great deal about, but I think everyone would agree that many errors were made in assuming that there could be a kind of strict continuity between basically gestural systems of organisms that are close to our ancestry and our own language faculty, which is uniquely human. Many people even misread Chomsky as a quasicreationist, because he says there is no continuity. He does not mean that God put it in there. He means that what we call the language organ may have been co-opted from some other mental function. Certainly, evolution does not always work by adaptive gradualistic continuity, though this is one of its modes.

And then we have other fallacies, the main one being the supremacist, or progressionist fallacy—the paradox of seeing animals as both lesser than us, but also defined by us—or even by our arbitrary words in the examples I gave at the beginning.

You would have thought, perhaps, that evolution might

have made it better, by showing kinship with animals and partnership with the earth. The argument of evolution *could* be used that way, but, historically, this has not been its primary weight. The notion of backreading, of lesser-than-but-defined-by, goes right through. Evolution's not the watershed that one might hope for. It ought to have been, by Freud's famous observation that all great revolutions in the history of science kick human arrogance off one pedestal after another of our claims for cosmic self-importance. First the Copernican revolution that made our place in the universe peripheral, then the Darwinian revolution that relegated us to descent from an animal world. And then, in what I like to call the least modest statement of intellectual history, his own, that taught us we did not have rational minds, by discovering the unconscious. But it does not work, because we can spin-doctor the story—that is, we can accept the Darwinian revolution—relegation to descent from an animal world, but we spin-doctor the result. The revolution is not complete, in Freud's very prescient sense, until we accept the pedestal-smashing consequences. And that is what we are not willing to do. We want to still read animals in our terms as lesser-than-although-defined-by. We want to see ourselves as the top of the heap. We want to see evolution as progressive, complexifying and sensibly leading towards us in a predictable manner. And then we can spin-doctor the Darwinian revolution to avoid the Freudian implications.

Let me very quickly talk about two preevolutionary versions—I call them "pinnacle theories" and "embodiment theories," both of which point to the same direction. First pinnacle theories, in which we see ourselves at the top of a progressive sequence. You do not need evolutionary theory. Progressive sequence can be constructed as a static chain of being. Charles White, *Regular Gradation in Man* (1799), was mentioned by a previous speaker. This slide shows his main chart—in which you see a motley collection arranged in a so-called progressive sequence, from birds at the lower left, to dogs, to primates. And then up the conventional racist ladder of human groups: from African blacks, to American Indians, to Greek statuary on the right. So this scheme also clearly has social implications.

And then we have embodiment theories—not the claim that we are at the pinnacle, but that all the lower creatures are imperfect embodiments of us. Now, the most wonderful example of that theory occurs among the German Naturphilosophen of the early nineteenth century. I did a study of Lorenz Oken in my book *Ontogeny and Phylogeny*, who published his *Lehrbuch der Naturphilosophie* between 1809 and 1811, and his theory is based upon a this marvelous notion. The whole book, this very thick book, is nothing but a series of 4,000 oracular pronouncements. And the basic notion is that all development begins with a primal zero and progresses to complexity by the successive addition of organs in a determined sequence. The sequence of additions follows Oken's ordering of the four Greek elements— earth processes, or nutritive organs, first; water processes, or digestion, second; air processes, respiration, third; and ether, or fire processes— motion—fourth. "Man"—his word—contains all the organs within himself, thus, he represents the entire world—because all the organs are in humans. Quoting now: "In the profoundest, truest sense, a microcosm. Man is the summit, the crown of nature's development, and must comprehend everything that had preceded him. In a word, man must represent the whole world in miniature." All lower animals, as imperfect or incomplete humans, contain *fewer* than the total set of organs. "The animal kingdom"—this is the most famous pronouncement in the *Lehrbuch*— "is only a dismemberment of the highest animal —that is, of man."

And then, just as White invokes the racial implications, let me read you the closing oracular pronouncements. Now, poor Oken was just a romantic liberal, but you can see how notions like this can be used for other German social philosophies that came after him. He talks about the sequential ordering of human skills. "The first science is the science of language, the architecture of science, the earth." (I am not saying that any of this makes sense, I am just quoting.) "The second science is the art of rhetoric—the sculpture of science, the river"—in other words, water. Remember the sequence—earth, water, air, fire. "The third science is philosophy, the painting of science—the breath"—air. "The fourth science is the art of war [*Krieg-skunst*]— the art of motion, dance, music, the poetry of science, the light"—fire. "As all arts are united in poetry, so are all arts and sciences united in the art of war. The art of war is the highest, the most exalted, the most godly [*göttliche*] art. The hero [*der Held*] is the highest man. The hero is the god of mankind. Through the hero is mankind free. The hero is the prince—the hero is God." That is how the book ends.

And then, of course, we also encounter the myth of meliorism in our interpretations—namely, that once you get to the evolutionary interpretation, or at least get towards it, things ought to be better, right?—because now we are getting closer to a truthful biological theory. So we come to Linnaeus, who is not yet an evolutionist but at least is trying to place humans into nature.

Consider the wonderful story of Linnaeus's coining of the term "mammal." This is not my argument—it comes from Londa Schiebinger, whose book on the subject impressed me greatly. I knew before I read her work, that Linnaeus had invented the term "Mammalia" for our vertebrate class in the *Systemae Naturae* of 1758, but I thought that he had simply promoted an old vernacular word to a new technical meaning. However, Londa showed that Linnaeus truly invented the word—that no language had ever before referred to the group of warm-blooded, hair-sporting, live-bearing vertebrates as mammals. All previous systems had treated and named our relatives differently. Aristotle had established a vertebrate group called Quadrupedia—four legs—with a primary subdivision into Oviparia, scaly and egg-laying—including reptiles and some amphibians—and Viviparia—that is, hair and live-bearing—thus including most mammals but, please remember, excluding such creatures as bats, whales and, most importantly, humans—who thus could remain separate. By Linnaeus' time, our group had a better definition but no recognized name. John Ray, for example, the greatest of Linnaeus' predecessors, had suggested Pilosa—meaning hairy—as a way of annexing obviously related animals that did *not* exhibit Aristotle's defining feature of four legs.

So why did Linnaeus choose a new name? And why, particularly, did he choose such a peculiar term as Mammalia—referring, obviously to the female breast? We must grasp the extreme unconventionality of Linnaeus' decision. Most generally, and for the usual sexist reasons, we tend to personify active phenomena as male, and organisms judged most complex should ordinarily fall under this sad convention. By the way, in contemporary English we still invariably refer to an unsexed animal as he—as in, "Isn't he cute?" or "Look at him go!" If Linnaeus had been an explicit egalitarian—out to sink a bad habit by example—he might have chosen Mammalia for this overt political reason. But Linnaeus was a social conservative and a conventional sexist. More particularly, zoologists have long translated this general cultural convention into technical practice. Do you realize that in formal taxonomy it is still stated that the so-called type specimen—that is, the defining name-bearer for a species—has to be male? That is still a rule of biological naming.

Why, then, did Linnaeus choose a female trait to define the highest group—apparently adding insult to male injury by selecting a feature that males also possess, but in a rudimentary and useless state. Schiebinger argues cogently that Linnaeus made this decision for an ideological reason, one very distant from any notion of sexual equality. Linnaeus had been deeply engaged in a different and equally important battle—this time, or so most of us would judge today, on the right side: namely, his campaign to classify humans *into* nature with other animals at a time when many naturalists still insisted on a separate human *kingdom* for beings with a soul, and created in God's image. Our propagandists have always recognized that an adroit choice of name can convey great power of persuasion. Nature has almost invariably been personified as female, in a cultural and linguistic tradition that dates at least to Chaucer. If one wishes to gain some rhetorical advantage in a struggle to place humans *within* nature, then choose a female feature to define our larger group—thereby emphasizing our closeness to Mother Earth and her other animate productions. Interestingly, in the same work that defined our larger group as *within* nature, Linnaeus sought to separate us as a species for our mental prowess, and here he chose a male designation: *Homo sapiens*—although the Latin "Homo," I realize, may be taken more generally in the old sense of humankind, while *vir* is more specifically a male person—from which we obtain, by the way, and for sexist reasons, the notions of virtue—although the word is feminine in most European languages—and virility.

So Linnaeus is a step in the right direction in the meliorist tradition, and Darwin is the next step in melioration because he finally got us to evolutionary theory. Everything should now be right and factual; our biased way of treating animals in human terms should now cease. But it does not work. It does not work in the small, and it does not work in the grand. I want to give two examples—a small example of the errors we still make for a particular case, and then a large example in how we look at the whole history of life.

This slide shows a study done by a colleague of mine in the sociobiological research tradition. I want to criticize his backreading of human traits into animals, not his sociobiology *per se*. Let me just tell you what he did. This is meant to be a study of the adaptive meaning of a behavior in mountain bluebirds. He took two nests. The females tend to sit at the nest, the males go out foraging for food. While the males were out foraging, he took a stuffed male and placed it by the nest, and saw what the returning male would do. Would the returning male be aggressive to this presumed potential imposter, and would he be aggressive to the female?

Now, on the vertical axis of the graph we see the number of aggressive encounters towards the male stuffed bird (the model) and toward the female. My colleague did this procedure at various times. He did it for the first time—that is, exposed the stuffed bird—after the nest was begun but before any eggs had been laid, and there were a lot of aggressive encounters. In fact, in one case the female was thrown out of the nest. Then, after the eggs were laid, he did it again, and he found fewer aggressive approaches towards the stuffed bird, and none towards the female. And then he did it one last time, after hatching of the eggs, and found even fewer aggressive encounters towards the supposed intruder.

I want to read you his interpretation of this, because it is such an amazing example of how we read human traits into our language, and then rederive them: "The results are consistent with the expectations of evolutionary theory. Thus, aggression toward an intruding male, the model, would clearly be especially advantageous early in the breeding season, when territories and nests are normally defended. The initial aggressive response to the mated female is also adaptive, in that given a situation suggesting a high probability of adultery the presence of the model near the female—and assuming that replacement females are available—obtaining a new mate would enhance the fitness of males." You see the point. As a good Darwinian male bluebird, you do not want to help this female raise someone else's genes. However, after the eggs are laid, the pressure of other males ceases to matter, because you know your genes are in there, if you have been watching carefully before.

So he goes on, and his explanation could be right. It is just the language I am talking about. "The decline in male-female aggressiveness during incubation and fledgling stages could be attributed to the impossibility of being cuckolded . . . " (Now, is this not wonderful? Here is a word that, of course, comes from animals—cuckoos—is then used in a human sense and is now being reimposed on other birds.) " . . . the impossibility of being cuckolded after the eggs have been laid."

This is a particularly good example, because it is so obviously subject to the following different interpretation: The model— that is, the stuffed bird—is exposed for the first time. The male comes back, pecks at the stuffed bird, and gets mad at the female, too. Then, a few days later, the same returning male encounters the stuffed bird, pecks at the stuffed bird a few times and says to himself: "It's that goddamned stuffed bird again"—I am backreading, too, you see—and he does not bother the female. It does not have to have anything to do with genetic adaptive behaviors.

Now for my example in the large. We still view the entire history of life as a grand backreading. We see this whole history of life as predictably preparatory to us. We see all previous creatures as precursors, or avatars, of the eventual appearance of humans.

[At this point in the talk, I presented a long series of slides showing our conventional iconographies of evolution as progressive sequences leading to *Homo sapiens* and therefore defined by the predictability of our eventual and inevitable appearance. I showed several different versions from a wide variety of times and cultures, from pre-evolutionary Biblical natural histories of the early eighteenth century (the progressive sequence of Genesis leading from initial chaos to Adam and Eve) to modern advertising (the evolution of computers from a hairy ape bent over from the weight of holding an old vacuum tube computer to a white male in a business suit [thereby encoding other iconographic biases of our culture] standing straight and tall because he only has to hold a light power book).]

Now for the second part of the talk. We, as Protagoras said, regard ourselves as the measure of all things, so that even the most abstract and universal issues of science and philosophy are often backread from desires to assert our primacy. Thus, many of the great timeless abstractions of philosophy really arise from the contingent history of us. A great dichotomy pervades the sciences, as popularly understood. One kind of science represents the stereotype we all learned—science is experimental, predictive, quantitative. You simplify and bring material into a controlled laboratory. You make predictive statements. You find out the laws of nature. This is good, or hard, science. But other scientists are, in a sense, relegated to explaining those uniquenesses that can occur but once in all the detailed glory of history—much of cosmology, evolution, geology, paleontology—and these are the lesser, or the soft sciences. The hierarchy goes from adamantine physics at top to squishy subjects like psychology at the bottom. I feel some affinity to this bottom, because paleontology is regarded as pretty squishy, too.

At Harvard, we actually set up our science curriculum in the general-education program, the core curriculum, in a somewhat innovative way. We did not just make the standard division into natural and physical sciences, or physical and social sciences. We actually recognized these two styles, of the ahistorical-predictive and the historical-explanatory. But we called them A and B, and guess which one was A?

Well, as a member, and proud of it, of one of the B sciences, paleontology, I've tried to institute something of a campaign to get people to recognize the virtues, the excitements, the power, the equal explanatory role of the sciences of contingency—that is, the sciences that are not trying to explain things by subsumption and prediction from nature's laws, but by the actual sequence of the antecedent states that happened to occur, but could have unfolded in a totally different manner, thus leading to an entirely different outcome. This is a very different mode of explanation, but when you have enough evidence about antecedent states, it is just as powerful, just as good.

The point I want to make—and this is a quick summary of the arguments in my book, *Wonderful Life*—is that humans, though we have tried to interpret our origin under the predictive models of the A sciences—hence, all our iconography of depicting evolution as something that was, if not bound to occur in exactly this form, is at least expectable and understandable under nature's law of evolution as complexification. But evolution does not work in this way at all. We are products of a contingent history. Rewind the tape of life to the early history of multicellular forms, and you get a whole different set of solutions every time—most of which, although equally explainable, do not include the origin of any self-conscious creature to have conferences like this.

We are quite comfortable with contingentist explanations for human history. We know that they apply to our affairs. This slide shows the Angle of Gettysburg, the clump of trees towards which Robert E. Lee directed his men in Pickett's Charge. The power of Gettysburg for us lies in our knowledge that the war could have gone the other way. It was not foreordained by the strength of their army that the Union would win. July 4th, 1863 was a very crucial time—though Vicksburg fell to Grant on the same day. Draft riots were about to break out in New York. And in my city of Boston, the 54th regiment of black volunteers was being armed—not for any abstract sense of racial justice, but for a desperate need for bodies. Northern victory was not assured. Had the Civil War been, as MacPherson argues, a war of conquest, the South could not have conquered the North, but it was not this kind of war. The South's aim was simply to hold on long enough to induce sufficient warweariness to get the North to recognize its boundaries—and this almost happened. MacPherson argues powerfully that at least up until the reelection of Lincoln in 1864, the war could easily have gone the other way. And we know why the South lost at Gettysburg as a result of a whole set of events and errors: Not taking the high ground in the beginning; Chamberlain and his Maine division holding Little Round Top; Lee thinking that the Northern battery had gone silent because his guns had knocked it out. Lee knew he had made a total mistake the minute he heard the Union canons firing on his men. We know why the South lost at Gettysburg, but the explanations are not laws of nature—they are particulars of history. The battle could have gone the other way, and all of American history might have been different.

We are comfortable with this mode of explanation for human history. We should be just as comfortable for life's history, and for our own origin as human beings. But, unfortunately, the other viewpoint is encoded into our iconography. We see the history of life as a cone of increasing diversity. From a common point of origin, a correct view under evolutionary theory, lineages move up and out. The cone is very narrow at the base, so you can only have a few lineages at the start, and these must be predictably preparatory to the ones that come later. Up is only supposed to mean geologically younger, but it is so easy to conflate up with higher on the ladder of being, and so things move up and out towards necessary progress and diversification. This slide shows the first great tree of life—from Ernst Haeckel, 1866. Let me show you why it is a biased iconography, in case you never recognized this. The problem is that you're forced to put at the top, where there's most space, the group that you think is maximally advanced. But what if that group is not very diverse, as in this case? Haeckel took the conventional view that mammals are most advanced. But there are only 4,000 species of mammals—not very many. There are a million described species of insects. Haeckel spreads mammals across the entire top of the tree. But all of insects, all of those million species, lie on one little branch down here. One little branch, because that is all the room he has lower down on the cone, in the region of low zoological status, where insects must share a limited space with other groups. This is a biased iconography that promotes the predictabilist, meliorative view. That is why its replacement by another iconography is so vital.

I suggested an alternate iconography in *Wonderful Life* as shown on the next slide. Note that you still have a common point of ancestry, but then a maximal spread of lineages occurs very early in history. And only a few of the initial possibilities survive, though these few may be enormously successful, and you may end up with more species than you ever had but restricted to far fewer anatomical groups. Now, this pattern is subject to a conventional predictabilist reading—namely, there was a grand struggle for Darwinian reasons during this early period and the good guys won. But this new picture is also subject, as the conventional iconography is not, to a radically different contingency-based explanation, in which each initial group only gets a lottery ticket in the greatest lottery ever held on the history of this planet. The survivors are effectively those who were fortunate. And I think that the evidence of the Burgess Shale, the great soft-bodied fauna from the early days of life's history, supports this point of view. I do not have time to give you the rationale but only to show you some of the organisms.

In this one fauna, from the early history of multicellular life, 530 million years ago in Western Canada, we have greater diversity than in all modern oceans put together. Arthropods today represent 80 percent of animal species. There are three major groups of arthropods today: the insect group, the spider-scorpion group, and the marine group—crustaceans: lobsters, shrimp, barnacles, and so on. Moreover, we do not know why the creatures died out. I recognize the limits of negative evidence, but we have no argument that the ones who survived did so for cause in the conventional sense of predictable superiority.

The Burgess Shale also includes a very insignificant creature named *Pikaia*, the one name you might want to remember because *Pikaia* is the first chordate, the first known member of our phylum. I do not say it's the only chordate living, and I do not say *Pikaia* is our direct ancestor. But if you had been an impartial observer 530 million years ago, I do not think you would have identified the chordates as a group with enormous probability of success. If, like most groups in this lottery, they had died, all of vertebrate history would have been wiped out of the fossil record. All of us—from trout to hippopotamuses to all humans.

The theme of contingency is fractal. It does not only work at this grandest level—it works down to every sequence of scales. An asteroid wiped out the dinosaurs 65 million years ago. If it did not hit, we would still be in a world of dinosaurs, and mammals would probably still be little creatures in the interstices of their world. Why not? That situation had prevailed for 100 million years before. It's only been 65 million years since.

Last point. The most abstract questions of science and philosophy are really about human beings as historical particulars. That is why we have struggled with these issues, and that is why we still grapple with them today. Since we are historical particulars, some of these grand questions—though we frame them as universals—really turn out to be inquiries about an individual pathway of history. Descartes starts with a whole series of pictures illuminating bodies as *la machine*—that is, humans shown as mechanical devices for the eyes, the hands, and so on. But as soon as he gets to the brain, every picture shows the pineal organ—the seat of the soul and its dualistic mediation of the material aspects of reality outside.

Is there any deeper philosophical tradition than dualism? This is John Milton, *Il Penseroso*: O, let my lamp at midnight hour / Be seen in some high lonely tower / Where I may oft out watch The Bear [that is, the plow or the Big Dipper] or unsphere the spirit of Plato / to unfold what worlds / or what vast regions hold / the immortal mind that hath forsook / her mansion in this fleshly nook. Is this not a great image? The immortal mind's spirit which has forsaken its mansion in this fleshly nook, and can soar up into and with the stars. Dualism is really an inquiry about us and why we are superior.

Theology—where it's clearer, perhaps, that all these great issues are about us. The two versions of creation in Genesis—the sequential story of Genesis 1; and the second story of Genesis 2, where the animals are brought to Adam, and he is given the power to name them. Genesis 1, the conventional story of the days of creation, is subject to two quite different interpretations, but both united by putting humans on top. First, the traditional interpretation—which I think is wrong—an additive sequence: First, God makes the earth, then the firmament, then the plants, then the animals. Then consider the alternative model, which I think is right but is hard for us to see, because we are living with different presuppositions—that Genesis 1 really talks about successive divisions and differentiation from an initial, or primary, chaos. If you look at the mosaics in the south dome of the narthex of San Marco's Cathedral, you will see this alternative iconography. First there is chaos filling the whole space, then a division of light and darkness, then the firmament from the earth. Here we encounter an old paradox: How can there be light and darkness before there's a sun and a moon? Of course there can, because the sun versus the moon represents a later division within the realm of light—one for the night and one for the day. And then the earth brings things forth—that is, there is a differentiation of the earth into its waters and its plants, and so it goes. And finally, the ultimate sexist differentiation—Eve comes from Adam.

Conclusion: How do we transcend all this? How do we get to the possibility of partnership and respect as better models for our relationship with animals? I do not know, but, for one thing, we need a new iconography. That is one humble suggestion I can make. Go to the newly revised mammal hall at the Museum of Natural History, because they have done a brilliant thing. Virtually every evolution hall in the world shows organisms in a conventional linear sequence. If there is a hall of mammals, primates are at the end. And nobody ever questions that—because there is a linear order through the hall. You go to the new hall and primates are in the middle. And you are bound to wonder and ask: "Now, why are primates in the middle, next to bats and other creatures like that." Ah! They are ordering the hall by branching sequence, not supposed progress. Primates are part of a lineage that branched off from the main line of mammals early in our history. If you arrange creatures by the order of branching sequence, primates come early. This new and unconventional arrangement really gets you thinking.

But why do we do usually proceed by backreading? I think that we act primarily from fear. We are so afraid that maybe we *are* insignificant that we have to spin-doctor evolution in our form. The greatest document for this issue, by the way, is Psalm 8, not Genesis 1. Psalm 8—the expression of fear, first of all: "When I consider thy heavens, the work of thy fingers, the moon and the stars which thou hast ordained, what is man, that thou art mindful of him?" And how could human life mean anything in the face of the vast heavens? The answer in Psalm 8: "For thou hast made him a little lower than the angels, and hast crowned him with glory and honor. Thou madest him to have dominion over the works of thy hands. Thou hast put all things under his feet. All sheep and oxen, yea, and the beasts of the field, the fowl of the air and the fish of the sea, and whatsoever passeth through the paths of the sea. O Lord, our Lord, how excellent is thy name in all the earth?"

Now, contrast this conventional hubris with Darwin's answer to the same question: What is man? Consider a wonderful letter that he wrote to Asa Gray in 1860, after Gray had written to him and said: Look, Darwin, I can accept the principle of natural selection—it makes sense—but I cannot avoid the conclusion that there must be some deep God-given meaning to the totality. Darwin writes back: You may be right. Science cannot adjudicate big issues like this. But, he says, even if that is so for the unknowable totality, "the details, whether good or bad" must be "left to the workings out of what we may call chance"—by which he does not mean chance in the technical sense of randomness. He means contingency, as I use the term. That is clear from the rest of the argument.

He then goes on to make a brilliant set of points, trying to lead Gray towards the position that we are an accident, a contingent part of history. He says: Look, Gray, if a man is caught in a thunderstorm on top of a mountain, is hit by lightning and dies, he died for a reason, based on the physics of lightning. But no one would say that his death was meant to be in some cosmic sense. It was an accident that he was on the mountain. So we have contingency for the death of an individual. How about the birth of an individual? A child is born with terrible mental retardation. This undoubtedly occurred for some reason not yet understood in the mechanics of development and embryology. But no one, if God be just, would claim that this situation was meant to be. It happened; it was an accident. Contingency then for individual life and death. How about the death of a species? Death of a species is also by accident; species become extinct. If the death of species is an accident—then the birth of a species should be viewed as an accident, too. And humans are species, like all others. So, from the death of a man by lightning on a mountain, which is clearly due to contingency, Darwin has subtly led Gray towards an acceptance of human origin as one of life's contingent details. Remember what he said: "With the details, whether good or bad, being left to the workings out of what we may call chance." The realm of details is enormous, and it includes the human species. I suggest to you that it also includes most of the grand questions of philosophy, politics, and so forth.

I want to end with my favorite sonnet from Frost, because it is such a brilliant statement about the problem of design or intent in nature. So many horrible things occur in nature. If we have to say they are a result of design, how can we honor anything? Frost answers that they are not—they are details in the realm of contingency.

The poet is taking a walk, and he notices a remarkable scene. He sees a heal-all, a flower, which is usually blue. But this one is white, so that is rare. And on the flower he sees the wings of a moth that has been eaten, and they are white, too. The moth has been eaten by a white spider—a rarity as well—and the spider is still there. So the poet notes three white objects, each with a different geometry: the starburst of the flower; the solidity of the spider; the two-dimensional structure of the moth wings. This must have some general meaning—three white things of different geometries, all together? And yet, if it has meaning, what meaning could be expressed? The moth has been eaten—it is a horrible scene. So Frost writes:

I found a dimpled spider fat and white
On a white heal-all, holding up a moth
Like a white piece of rigid satin cloth.
Assorted characters of death and blight

Mixed ready to begin the morning right
Like the ingredients of a witch's broth:
A snow-drop spider, a flower like a froth
And dead wings carried like a paper kite.

What had that flower to do with being white?
The wayside blue and innocent heal-all.
What brought the kindred spider to that height,
Then steered the white moth thither in the night?

What but design of darkness to appal
If design governed in a thing so small.

You see, the point is that *Homo sapiens*—and this goes back to Darwin on Gray—is also a "thing so small" in a vast universe—a wildly improbable evolutionary event well within the realm of contingency. Make of such a conclusion what you will. Some find the prospect depressing. I have always regarded it as exhilarating and a source of both freedom and consequent moral responsibility toward other animals as well as fellow humans. Thank you.

REPRESENTATIONS

Introduction

The essays in this section represent explorations of variously fictive ways in which the representations of animals have been shaped and employed in human discourse. Humans have deified animals or given them ancillary roles in their pantheons; they have imagined them as the objects of mythological metamorphosis and given them quasi-human status; they have given them language, reason, character, and personality in both oral and written fictions; they have continued to observe animals, both domestic and wild, and tried in different ways to give what could be considered variously authentic accounts of them. Imaginative constructions of the names, natures, and dispositions of animals throughout human history, thus, have been based upon knowledge and observation and the very fabric of metaphor out of which those constructions are built. The ways in which other creatures are seen to be identified with, or similar to, the humans who construct those identifications are themselves varied, and the matters considered in the papers which follow are samples of a whole range of others.

Nicholas Howe's exploration of the realm of animal fable remarks on the long history of a literary mode starting in our literature with Aesop and continuing with undiminished fecundity to the present day. Immune to the pressures of realism that gradually shaped the development of the novel during the last two centuries, the exemplary tale of animals possessed of human speech and displaying human moral and social traits seems to have an imaginative life of its own. In the first of these essays, Howe looks at some extremely sophisticated medieval revisions of Aesopian models.

The treatment of animals in the history of prose fiction since the eighteenth century, on the other hand, has attempted primarily to engage the actualities of—primarily domestic—

animals in their interaction with humans in society. The particular interest of Gerald Vizenor in the second essay in this section is to examine some of the more poetic resources of later twentieth-century American fiction, particularly that of Native American writers, in making animals, both wild and domestic, present in non-naturalistic ways.

John Hollander

Fabling Beasts:
Traces in Memory

Nicholas Howe

> We will make a Zion out of this Sinai and
> we will build there three tabernacles, one
> for the Psalms, one for the Prophets, and
> one for Aesop.
>
> (Martin Luther)[1]

> "Animals don't find it hard to die," he
> says gently. "Perhaps we should take our
> lesson from them. Perhaps that is why they
> are with us here on earth—to show us that
> living and dying are not as hard as we
> think."
>
> (J. M. Coetzee, 1994, p. 208)

Sour grapes, tortoise and hare, country mouse and city mouse, *Stuart Little*, mouse and lion, Detroit Lions, *The Jungle Book*, William Wegman dog photographs, Saturday morning cartoons, Big Bird and Miss Piggy, Babar, a certain purple dinosaur, *Animal Farm*, Mickey Mouse, Art Spiegelman's *Maus*, wolf in sheep's clothing, sly as a fox, Reynard the Fox, Tony the Kellogg's Tiger, the Esso Tiger in Your Tank, cars called Jaguar or Cougar or Lynx, Puma athletic shoes, *The Lion King*, *Old Possum's Book of Practical Cats*, *Cats* (the musical), the boy who cried wolf, the goose that laid the golden eggs—these are but a few of the traces left in our memories, both personal and cultural, by the human act of fabling beasts. The term "fabling beasts" denotes something quite simple and, therefore, hard to define precisely: the activity of telling stories in words about animals, sometimes with and sometimes without people, in order to point a moral that humans believe should hold throughout the natural world. As H. J. Blackham emphasizes, the fable is "about what can be thought of without time or place" (1985, p. 210). In this act of fabling, animals can imitate the behaviors of humans, humans can imitate the behaviors of animals, but most of all the genre calls into question all categorical distinctions between animals and humans. If animals can speak in human language, and they must if there are to be beast fables, then the most cherished of our modern distinctions between human and animal—that based on language as a creative, recursive faculty—seems pointless, even evasive.

The relation between beast fables and Saturday morning cartoons for children is, of course, more obvious than that between beast fables and many of the traces listed above. If advertisements can persuade us that driving a Jaguar will give us a dark aura of feline grace, or if calling teams Tigers, Sharks, Bulls, or Eagles makes us believe they will be more likely to win, then distinctions between humans and animals blur in a different way: for these names allow humans to acquire some of the force we attribute talismanically to animals and also, we like to think, some of their elegance of purpose. Our notions about animals do seem most childish when they become the nomenclature of big-time sports and mass advertising. For who but a child can believe that a Jaguar gives its driver feline grace, or that calling a hockey team from San José the Sharks insures that it will defeat a team from Anaheim (and the Disney Empire) called the Mighty Ducks? Yet, if in the natural world sharks are more ferocious or aggressive than ducks, so by a discernible logic it must be in the world of a child's desiring. After a certain age, however, believing such things about Jaguars or Sharks suggests that one has never quite left behind the world of children's books and beast fables.

For we as adults rarely think much about beast fables, and if we do, we like them least when they are most preachy. Our tolerant skepticism toward the genre is finely registered by the title of Lloyd W. Daly's collection, *Aesop without Morals* (1961). If we must have collections of fables, his title suggests, let them be free of tedious moralizings tacked on at the end to convince us that stories about animals have didactic value. As two more recent translators of Aesop, Patrick and Justina Gregory, remark: "It seems to us that these moral tags not only jar with the fables' sophistication of form, but also deprive them of one of their prime functions: to make the reader think" (*The Fables* . . . , 1975, p. 2). That we burden animals by asking them to teach us how to behave like human beings seems no more than yet another way of exploiting them. We force animals to do physical labor, we raise them under cruel conditions, we mistreat them in all sorts of ways, and then we domesticate them most fully by moralizing them. Far better, it would seem, to read accounts by naturalists who observe animals in their own environments to learn about the natural world, who resist treating animals as figures to be written into beast fables to confirm our moral categories.

If we use animals to write about our concerns, let it be with a gentle reserve, with the sense that we are crossing lines or confusing categories that they themselves do not cross or confuse. Consider, as such an example, "Apartment Cats" (1971), a poem by our contemporary Thom Gunn. The title registers the poet's awareness that the animals he describes exist in a space separate from even the outside world of domesticated pets. These are cats who live entirely amid human surroundings. The poem's speaker describes coming home and being greeted by "The Girls," as he calls the cats, who go through a routine of awakening, stretching, sniffing his shoe "rich with an outside smell," and rolling on the floor to be petted. Gunn continues: "Now, more awake, they re-enact Ben Hur / Along the corridor, / Wheel, gallop . . . ." The final stanza moves toward its quietly observed truth:

And then they wrestle: parry, lock of paws,
    Blind hug of close defence,
    Tail-thump, and smothered mew.
    If either, though, feels claws,
  She abruptly rises, knowing well
How to stalk off in wise indifference (Gunn, 1994, p. 194).

Only the allusion to cats racing in the corridor like charioteers in *Ben Hur* seems to impose human knowledge on feline behavior; the rest is simply observation about two cats in an apartment. And yet the phrase "Ben Hur" is enough to remind us that all of the poem is attributed human knowledge, that it is our reading of cat behavior to draw an unobtrusive but fully moralized point: that when they play, cats follow rules that we humans would do well to emulate. For like them we should, when the rules of the moment are transgressed, know "to stalk off in wise indifference."

There are at least three reasons this poem does not seem a beast fable: it speaks of two specific cats ("The Girls" in the poet's affectionate phrasing); it makes no claim that all cats behave in a predictable way as fables require their animals to do (a dull-witted fox would be absurd by the rules of the genre); and it requires interpretation to understand its moral (and thus allows one's interpretation to be rejected as overly determined, as a misreading). By contrast, the fables we remember always portray typical beasts who remain true to their species (sly foxes always want grapes); they maintain quite strict conventions of behavior for animals and people (foxes cannot climb trees to get at grapes); and they state their morals explicitly ("So it is with men, too. Some who can't do what they want because of their own inability blame it on circumstances" [Daly, 1961, p. 268]). Or so it seems as we hold one or two fables in memory: the fox that consoles itself by calling the grapes it cannot reach sour; the goose that lays golden eggs but is not itself made of gold; the boy who relieves the tedium of watching sheep by raising false alarms about wolves. In a curious way, though, as we remember individual fables we distort the genre because we mistake traces of memory for the form itself. Knowing a moralized fable or two rather than living with many of them is like studying a particular animal in its natural environment without refering to anything else in that same environment. It is convenient, perhaps necessary, but it leads to overly certain findings.

That beast fables have left even these limited traces in our memory suggests that they were once a more vital form of representation, that they were once not reserved for children. For the questions they raised about the mutual company of humans and animals were not matters simply for children. Consider, to offer a surprising example of the fable's lost prestige, Plato's account of Socrates' last days in the *Phaedo* (60D–61C). Socrates, as he calmly awaits death, has become in the Western tradition the exemplar of philosophical self-possession. Plato tells us that Socrates spent his time while imprisoned writing poetry so that he could satisfy what he believed to be the injunction of a prophetic dream he had one night. Lacking any capacity for invention, Socrates says, he decided to versify some of Aesop's fables because he had them at hand (which means most likely that he had them in memory). And then Socrates tells Cebes, another character in the dialogue, to report his versifying of Aesop to Evenus: "So tell Evenus that, Cebes, and bid him farewell, and tell him, if he is wise, to come after me as quickly as he can. I, it seems, am going to-day; for that is the order of the Athenians" (*Plato: Phaedo*, 1990, pp. 210–13).[2]

That Socrates consoled himself with Aesop is itself a kind of fable, all the more haunting for its unlikeliness, about our reasons for reading fables. Socrates versifying Aesop became a fable for later fabulists because it suggested that such stories could retain value when other ways of thinking or other forms of philosophical discourse faded in the face of elemental circumstances. In the Preface to his great collection, La Fontaine retells this story of Socrates to prove that his own act of fabling beasts is a work of serious literary merit and concludes: "Socrates was not alone in judging poetry and our fables to be sisters" (Spector, 1988, p. xxix).[3] In an oblique way we recognize the value of beast fables—their simplicity of form—when we put them in the children's bookcase along with other troubling works that cross adult distinctions between the comforting and the frightening, between the human and the animal. As Annabel Patterson observes shrewdly, "Everybody has been, since childhood, familiar with Aesop's fables, and almost everyone, consequently, believes them to be children's literature" (Patterson, 1991, p. 1). And here we might remember such other "children's books" as *Gulliver's Travels* and *Alice's Adventures in Wonderland*, works which feature talking horses and long-legged birds that serve in a game of croquet.

Socrates was not alone in valuing beast fables as moral illustrations. Aristotle in the *Rhetoric* (II.20) recommends them to public speakers addressing popular assemblies, and adds "they have this advantage that, while it is difficult to find similar things that have really happened in the past, it is easier to invent fables" to serve as analogies or illustrations (Aristotle, 1991, pp. 276–77). Aristotle's claim that beast fables are easy to invent is striking because it suggests not only that they are valuable in rhetorical persuasion but also that they come out of a shared body of experiential knowledge. They are easy to make up, in other words, because everyone knows about animals, whether they be of the household, barnyard, or wild variety. (So, too, their currency helps to explain the persuasive force of fables.) As we think about the animals we know—or, at least, that most of us know well who do not study or work with them daily—we see our pets, our companion animals, our cats and dogs and birds and mice. It seems much harder to invent fables from them because they do not display as wide a range of behavior as do animals in the older fable collections. Perhaps that is why an anthology such as John Hollander's *Animal Poems* (1994) should contain so few examples that we can read as fables. For people in the industrialized West over the last several centuries, when most of the poems in this anthology were written, have grown apart from a diverse company of animals. We live in cities and suburbs with our pets; we visit zoos and watch television documentaries to see exotic animals; we use animal metaphors and read animal books. But we have never encountered anything as routinely necessary for the workings of a beast fable as a fox raiding our chicken coop. The ways we live with animals, or do not live with them, affect the stories we can tell about them. So, too, the variety of animals we keep daily company with affects the variety of stories we can tell and understand about them.[4]

Aristotle's claim that the beast fable functions well as illustration was repeated by Isidore of Seville, the seventh-century Christian bishop and encyclopedist. His testimony is historically vital because his great work, the *Etymologies*, was an essential reference throughout the Middle Ages and into the early Renaissance. Anyone wanting to know how to define *fabula* (fable) in the period from the seventh through the fourteenth centuries would have been likely to consult Isidore's statement that fables are intended "to produce a recognizable picture of human life through the conversations of imaginary dumb animals" (1911, I. 40).[5]

The idea that fables have their origin in what seems to us a paradox—the human conversations of mute animals—must also have been a puzzle to the ancient Greeks. As he does with most such puzzles, Aesop explains this one with a fable (number 240): At the direction of Zeus, Prometheus fashioned men and beasts. But when Zeus saw that there were more of the dumb animals, he ordered him to destroy some of the beasts and make them over into men. When he did as he was told, it turned out that the ones who had not been fashioned as men from the start had human form but were bestial in spirit (Daly, 1961, p. 193). This is not a frequently anthologized fable, nor has it left its trace with a memorable phrase like 'sour grapes'. It is, in fact, unsettling as a story about why there can be beast fables, stories which record the human conversations of mute animals, because it posits that animals and humans were made by Prometheus from the same material. Had it been otherwise, it would hardly have been possible for him to remake animals into humans, at least in their outward form. Now, if some of these humans retained their animal nature, that also explains why in turn some animals could possess human characteristics, at least to the extent of speech. The absolute distinctions between humans and animals that today make us increasingly uncomfortable were confused, one might suggest, from the very start. Or so this fable teaches us.

The fable as a form explores those regions where human and animal overlap, where it becomes not only hard but also counterintuitive to separate them. The fable is not simply the metaphorizing of human behavior in animal guises, that is, in guises created from human-made or anthropomorphic conventions about animals. When modern writers such as George Orwell in *Animal Farm* and Art Spiegelman in *Maus* represent human evil as being literally bestial, their mistake lies not in imagining animals to be capable of such behavior. Rather, it lies in treating the beast fable as an allegory or literal equivalence in which, as Speigelman draws it, mice are Jews, cats are Nazis, pigs are Poles, dogs are Americans, and reindeer are Swedes. The difficulty is that if one finds it hard to imagine cats as Nazis, or, more crucially, if one finds representing Nazis as cats to be inadequate (for lack of a better word) to that historical experience, then *Maus* as a fable collapses. One can argue this point while also understanding and even admiring Spiegelman for depicting that most inescapable, if over-represented event, the Holocaust, through the moral obliqueness of a beast fable.[6]

That the fable requires a more elemental blurring of human and animal not at the extremes of history (such as the Holocaust) but as a fact of daily life is evident from a wonderful, if spurious *Life of Aesop* written in the first century A.D., long after its subject would have died in the fifth century B.C.—that is, if he were ever alive to die.[7] This *Life of Aesop* seems to have been written less out of biographical curiosity than out of a felt need to invent an origin story for the genre of beast fable. It is literary theory masquerading as life story. According to the *Life*, Aesop was a slave of "loathsome aspect," of "portentous monstrosity" (Daly, 1961, p. 31). He was also born mute. He is described by characters in the *Life* as being one or another ugly beast or object, as when the Samians cry out: "What a monstrosity he is to look at! Is he a frog, or a hedgehog, or a pot-bellied jar, or a captain of monkeys, or a moulded jug, or a cook's gear, or a dog in a basket?" (1961, p. 74). At another moment, when Aesop's master orders his steward to sell him in the slave-market, the steward responds: "Who will want to buy him and have a baboon instead of a man?" (1961, p. 35). The point is clear: Aesop in his bestial ugliness and muteness is for all intents and purposes an animal, most likely a baboon. But then one day he shows great kindness to a priestess of Isis, and the goddess herself rewards him with the gift of speech. So, too, each of the Muses gives him some gift of her own art (1961, pp. 33–4).

Only after this transformation does Aesop's master succeed in selling him to a slave-dealer who just this once is searching the area to buy not slaves but pack animals (1961, p. 36). Again, the implication is clear: Aesop is at least as much animal as human, even if he possesses speech. After the dealer succeeds in selling Aesop to the philosopher Xanthus on the island of Samos, a maid of the house greets him in this way:

> 'Are you the new slave?'
> Aesop said, 'I'm the one.'
> The maid: 'And where's your tail?'
> Aesop took a look at the girl and, realizing that she was making fun of his dog's head, said, 'My tail doesn't grow behind in the way you think, but here in front' (1961, p. 46).

Aesop's obscene response to the maid's taunt of bestiality is the first sign in the *Life* that his wit comes from a sense of being somewhere between the categories of animal and human. If she accuses him of having a tail, he will in turn allude to his penis, which is, as we later learn, enormous (1961, p. 67).

In time, the Samians come to respect Aesop for his wisdom and even prefer his fables to the teachings of his master, Xanthus. When the citizens ask Aesop for advice, he responds with fables which allude to a past time "when animals talked

the same language as men" (1961, pp. 77, 88), that is, to a time
when fables were not necessary because there could be direct
communication among species. The *Life* explains that Aesop
wrote the majority of his fables after he was freed and
"deposited them in the library," a fact that explains how it is
that they survived to circulate as written texts. After wandering
through many lands, the *Life* tells us, Aesop met a grisly death
in Delphi where, by a decree of the citizens, he was pushed
over a cliff for honoring the Muses rather than Apollo. For
remaining loyal to the nine sisters who had each given him
some of her gift after he gained the power of speech, Aesop
fell victim to "the ritual sacrifice of the scapegoat (*pharmakos*),"
a tellingly ironic fate for the creator of the beast fable (Perry,
1965, p. xlii). In turn, the Delphians are punished with a
plague, the *Life* tells us, and an injunction by Zeus that they
expiate the killing of Aesop.

Without ever slipping into explicit statement, the *Life*
records the birth of the beast fable as a genre by tracing the
stages of Aesop's career. As grotesque and inarticulate as an
animal, as rational and economically valuable as a man
(because a slave and, thus, a saleable commodity), Aesop seems
a liminal creature. By his act of human kindness to a divine
stranger, he gains speech so that he can utter his wisdom. But
his wisdom is not that of academic philosophers, like his master
Xanthus. It is instead practical, moral instruction that can be
understood by anyone who hears or reads his fables. It is a
wisdom that derives from a time before animals lost their
ability to use speech. Only when Aesop is freed and no longer
a slave, a category neither quite animal nor quite human, does
he compose and inscribe his fables. And only then does he fall
victim to the kind of human behavior that seems so
incomprehensibly evil that we distance it by labelling it bestial.

The *Life of Aesop* is by my reading an attempt to explain the
riddle posed by the fable as a genre: How is it that anyone can
know enough to narrate all of the various creatures it depicts?
How can anyone know what words to give the fox or what

behavior to ascribe to the tortoise? And more, how can any being that knows all this also know how to speak to human beings? In other words, as you trace out the *Life*, you comprehend the genre. In this way, the name of Aesop came in antiquity and afterwards to designate the author of all fables (Ziolkowski, 1993, p. 16). By an irony he might have predicted, Aesop has become as stock a figure as any of the animals or people who appear in the genre: the fabulist of beasts.

This *Life* of Aesop survives in a tenth-century A.D. manuscript held by the Pierpont Morgan Library (MS M.397).[8] In this Greek manuscript from the south of Italy (tenth or eleventh century), Aesop is depicted with a misshapen head that gives him something of a canine appearance. That collections of fables, whether by Aesop or such later writers as Marie de France and William Caxton, are frequently illustrated with drawings reminds us that the fable is itself an illustration, a sketch in words that creates a picture.[9] Fables typically present a moment or moralized scene that can be represented visually to good effect. There are recent translations of, to cite only the most famous names in the genre, Aesop and La Fontaine which feature new illustrations.[10] Tellingly, the title page of the Aesop volume names the illustrator, David Levine, before the translators, Patrick and Justina Gregory. We might also consider that such illustrations, especially when drawn by a gifted cartoonist such as David Levine, function much like those moralizations we modern readers claim to dislike. For these illustrations make explicit to the eye what the text presents implicitly and, thus, do the work of moralizing.

Medieval men and women who met Aesop in a manuscript had, in a direct and palpable way, a richer and yet more ambiguous experience as readers than we can have. As they read fables written on the prepared skin of an animal, most likely of a sheep or a calf, they saw that the words spoken by animals were inscribed quite literally on the back of a beast so that they could be read by human beings. This circumstance at the very least would have reminded medieval readers that fables merited being written on the same rare and expensive material as did the most sacred of texts, the most learned of commentaries, the most jealously guarded of legal privileges. To translate my point into our economic terms, medieval fables were not relegated to the children's hour of Saturday morning television.

Nowhere is the medieval reverence for the beast fable more evident and yet also more perplexing to the modern reader than in Chaucer's *Canterbury Tales*. The most elusive of the *Tales*, the one most stubbornly resistant to any of our sophisticated theories of interpretation, is a beast fable called *The Nun's Priest's Tale*. The title refers to the pilgrimage character who is supposed to tell it, and gives no hint that it is a fable about a widow, a rooster, and a fox. At some 625 lines, *The Nun's Priest's Tale* is far longer than any of Aesop's fables, but, like some of them, it turns on how one animal tricks another into opening its mouth and thus dropping its prize, as in the well-known example of the fox persuading the crow to drop its morsel of meat (number 124).

Chaucer's tale opens by describing the widow's barnyard and its various inhabitants: the rooster Chauntecleer; his harem of seven hens, including his favorite Pertelote; three sows; three cows; a sheep called Malle; and three dogs called Colle, Talbot, and Gerland. One morning, Chauntecleer awakens from dreaming about an unknown beast that looks like a hound, is between yellow and red in color, has a tipped tail and ears, and two glowing eyes. Chauntecleer has no name for this beast—obviously a fox—perhaps because the naming of species is a human concern, perhaps because this wild beast inhabits a different sphere and is thus unfamiliar to Chauntecleer. He does tell Pertelote that the beast wanted to seize and eat him. She dismisses his dream and its value as oracular warning with an extended lecture on medieval dream physiology and psychology. Simply put, she tells Chauntecleer that his bad dream is a result of too much "red choler" and advises him to take a laxative and purge himself, much as we might take an Alka-Seltzer after awakening from a nightmare caused by eating too much pizza too late at night.

At this moment, *The Nun's Priest's Tale* moves into its most crucial revision of the beast fable, one that raises very high claims for the genre, by presenting a long and learned discussion between Chauntecleer and Pertelote on the ways in which dreams are to be understood as a form of knowledge. Readers, especially those familiar with this discussion elsewhere in medieval philosophy, are likely to forget that Chaucer's text is in fact spoken by chickens. They display all the features of learned speakers in medieval texts: they bolster their arguments by alluding to honored authorities; they score debaters' points; they talk to hear the sound of their own voices. Their resemblance to academics may explain why I once heard an eminent Chaucerian open a talk about *The Nun's Priest's Tale* by asking: "Do we really think this is a tale about chickens?" The audience, polite to a fault, murmured appreciatively at this suave opening gambit, but now I wish I had said "Yes, it is about chickens and how they look at the world. And if it is not about chickens, then it is about nothing at all." The precise substance of the discussion in Chaucer's tale is not of immediate concern, but I can translate it into contemporary terms by asking you to imagine my cats, Mimi and Lola, interrogating Wittgenstein's dictum that "the limits of my language mean the limits of my world." (An aside: when I said this to my wife, she asked, "What else do you think they talk about all day?")

After a long discussion about dreams, complete with learned allusions to Cato, Macrobius, and other revered authorities, the substance of Chauntecleer's dream comes to pass: Don Russell the fox enters the barnyard, flatters Chauntecleer into singing for him, then seizes him by his outstretched neck, and makes off for the woods. A cry of alarm is raised, the old widow and her animals chase after Chauntecleer in a scene that the narrator compares to such moments from epic history

as the Greeks sacking Troy and Nero burning Rome. When
Russell with Chauntecleer firmly in his mouth enters the
woods, the barnyard rooster urges the woodland fox to turn
on his pursuers and tell them that they must return home
because they are violating their nature by chasing him into his
territory. The fox agrees, opens his mouth to speak, and, thus,
Chauntecleer flies away to make good his escape to the safety
of a tree. Part of Chaucer's fable is to revise the fable's generic
convention by showing that the fox is indeed too sly for his
own good.

That a rooster and a hen can talk intelligently about dream
psychology and epistemology might seem at first no more than
Chaucer's joke about the folly of human concerns. By one
reading, *The Nun's Priest's Tale* is simply brilliant satire. But
such a reading neglects many issues: that throughout his work
Chaucer takes dream psychology and epistemology very
seriously; that the chickens have shrewd points to make about
these philosophical concerns; that Chaucer elsewhere casts
animals (often birds) as authority figures; and, most crucially,
that at the end of *The Nun's Priest Tale* Chauntecleer uses what
he has learned from interpreting his dream to persuade the
fox to open his mouth. That is, this one fable illustrates the
value of fables as a genre; it proves that knowing them may
help you save your own neck or, conversely, your dinner.

Anticipating that we might find *The Nun's Priest Tale* to be a
piece of foolishness, Chaucer ends with a brief speech of
authorial instruction:

> But ye that holden this tale a folye,
> As of a fox, or of a cok and hen,
> Taketh the moralite, goode men.
> For Saint Paul seith that al that writen is,
> To oure doctrine it is ywrite, ywis;
> Taketh the fruyt, and lat the chaf be stille.
> (Benson, 1987, p. 261, 11.3438–43)

The Pauline injunction to take the fruit and let the chaff be seems no more than an injunction here to take the moral of the fable and let its narrative be. Yet it is not that simple. For if there can be no fruit without the chaff, so there can be no moral without the narrative. Or, in the terms of the genre, without the words of chickens there can be no truth for humans to ponder. And in one way, at least, it is the fact that chickens can and do speak about troubling issues of existence which may be the first truth we humans need to ponder. This reading may also reassure us when we stop to notice that properly speaking *The Nun's Priest's Tale* has no pithy moral at its end.

The Nun's Priest Tale can be read as providing a key to the genre of the beast fable and also as suggesting an explanation for its decline. Central to both readings is the fable's setting. It takes place chiefly in a barnyard bound on one side by the human (the house of the old widow who owns the rooster and the rest of the domestic animals) and on the other side by the wild (the woods where the fox makes his home). This kind of symbolic setting features two clearly demarcated zones (the house and the woods) as well as an intermediate zone that links the two and, thus, gives each its meaning (the barnyard).[12] It is the talking rooster, the figure at once animal and human, who makes this symbolic topography explicit when he urges the wild fox to tell the human widow and her barnyard animals that they cannot go into the woods because it is for them forbidden territory. This sense of place does not govern each and every fable ever told, but it does underlie the fable's sense that the human and animal exist together as something other than a rigid binary opposition. Put another way, the setting of *The Nun's Priest's Tale* and the sequence of events in the *Life of Aesop* are alike in explaining how it is that humans and animals can do what is so palpably obvious they do: speak to each other.

Just as barnyards have disappeared from the common experience of people living in cities, so fables have slipped out of our normal range of reference and discourse. A few traces remain, perhaps more than are listed at the start, but what does seem gone is the impulse as well as the ability to create *new* fables. We are in that sense surprised by William Hazlitt's avowal that "I would rather have been the author of *Aesop's Fables*, than of *Euclid's Elements!*" (Blackham, 1985, p. xvi). In making this claim, I must admit at least one exception: contemporary writers like Gerald Vizenor have created tricksters or shapeshifters that move as characters between the human and the animal. But these writers are novelists, not fabulists, and they would be uncomfortable if I attached morals to their tales. And there lies another reason for the slow disappearance of the beast fable. For much as we claim that fables can exist apart from their morals, we know that this really is not true. We know that the very act of telling a story in which an animal does something it cannot do in life—speak Latin or English or French—becomes a moral critique of the order by which we interpret our experience. That fables tell stories about another order of experience, one in which humans and animals each speak to one another, suggests a longstanding unhappiness with the way things are.

When we lived and worked in barnyards and knew animals as something other than pets or exotica, we understood that the beast fable could have a moral function. This knowledge came with belonging to a group defined at least in part by its knowledge of animals, knowledge not strictly speaking scientific but rather traditional or experiential. Within the group that tells them, fables must depend on reasonably fixed associations between a particular species and a certain form of animal behavior as it can be used to gloss human behavior. In its allusiveness, the fable demands a community of knowledge. Once we stopped knowing animals as a direct matter of survival—as partners in work, as quarry to hunt, as predators to evade—fables could be read as stories about cute animals that could safely be given to children. Similarly, we find the infantilization of animals in cartoons, such as the gradual softening of Mickey Mouse's sharp rodent features that

Stephen Jay Gould noted in his essay on that figure's fiftieth birthday (1980). Cartoon animals must be cute and never die because we want our children to be cute and never die. Under these conditions, the beast fable cannot be a vital, or even a possible, form through which adult human beings can explore their place in the larger scheme of the natural world. But let me end by suggesting that as we continue to reimagine our relations with animals as one of continuum rather than difference, then we can also reimagine our representations of them. Perhaps as adults we can listen to them and learn from them. Aesop would, I think, like that.[13]

Notes

[1] Quoted in Ziolkowski, 1993, p. 24.

[2] For a reading of this episode in the *Phaedo* that illuminates the political value of the fable as a genre, see Patterson, 1991, especially Chapter 1.

[3] "Socrate n'est pas le seul qui ait considéré comme soeurs la poesie et nos fables" (Spector, 1988, p. xxviii).

[4] See Salisbury, 1994, pp. 13–41.

[5] "Quae ideo sunt inductae, ut fictorum mutorum animalium inter se conloquio imago quaedam vitae hominum nosceretur." For animals and fables in pre-modern Europe, see Beagon, 1992, especially Chapter 4; French, 1994, especially Chapter 6; and Dronke, 1974.

[6] That the beast fable can be an exercise in liminality between species may explain why Spiegelman should quote this statement by Adolf Hitler as the epigraph for *Maus*: "The Jews are undoubtedly a race, but they are not human." For Orwell on the beast fable, see his "Preface to the Ukrainian Edition of *Animal Farm*" in Orwell, 1970, pp. 455–59.

[7] For the Greek text of the *Life*, see Perry, 1952; for an English translation, see Daly, 1961; for commentary, see Perry, 1936.

[8] Through the courtesy of William M. Voelkle, this manuscript was included in the show "Animals as Symbol in Medieval Illuminated Manuscripts" held at the Morgan Library from April 6-September 7, 1995.

[9] See, for easily accessible examples, the illustrations reproduced in Marie de France, *Fables* (Spiegel, 1987) and Lenaghan, 1967. See also Salisbury, 1994, p. 107.

[10] For Aesop, see *The Fables* . . . , 1975; for La Fontaine, see Shapiro, 1988.

[11] To translate lightly into modern English: "But you that hold this tale a folly, About a fox, or a cock and hen, Take the morality, good men. For Saint Paul says that all that is written, Is written for our doctrine indeed; Take the fruit, and let the chaff be still."

[12] For another medieval example of this three-fold distinction, see Ladurie, 1979, pp. 293–96.

[13] I dedicate this essay to the memory of my mother, Thalia Phillies Feldman, among whose books I found *Aesop Without Morals*.

References

Aristotle, *Art of Rhetoric*, John Henry Freese, trans., Loeb Classical Library (Cambridge, MA: Harvard University Press, 1991).

Beagon, Mary, *Roman Nature: The Thought of Pliny the Elder* (Oxford: Clarendon Press, 1992).

Benson, Larry D., ed., *The Riverside Chaucer* (Boston: Houghton Mifflin, 1987).

Blackham, H.J., *The Fable as Literature* (London: Athlone Press, 1985).

Coetzee, J.M., *The Master of Petersburg* (New York: Viking, 1994).

Daly, Lloyd W., ed. and trans., *Aesop without Morals* (New York: Thomas Yoseloff, 1961).

Dronke, Peter, *Fabula: Explorations into the Uses of Myth in Medieval Platonism* (Leiden: E.J. Brill, 1974).

French, Roger, *Ancient Natural History* (New York: Routledge, 1994).

Gould, Stephen Jay, "A Biological Homage to Mickey Mouse," in *The Panda's Thumb* (New York: Norton, 1980).

Gunn, Thom, *Collected Poems* (New York: Farrar, Straus and Giroux, 1994).

Hollander, John, ed., *Animal Poems* (New York: Knopf, 1994).

Isidore of Seville, *Etymologiae sive Origines*, W.M. Lindsay, ed. (Oxford: Clarendon Press, 1911).

Ladurie, Emmanuel LeRoy, *Montaillou: The Promised Land of Error*, Barbara Bray, trans. (New York: Vintage, 1979).

Lenaghan, R.T., ed., *Caxton's Aesop* (Cambridge, MA: Harvard University Press, 1967).

Orwell, George, *The Collected Essays, Journalism and Letters of George Orwell, Vol. 3: As I Please* (New York: Penguin, 1970).

Patterson, Annabel, *Fables of Power: Aesopian Writing and Political History* (Durham, NC: Duke University Press, 1991).

Perry, Ben Edwin, *Studies in the Text History of the Life and Fables of Aesop* (Haverford, PA: American Philological Association, 1936).

Perry, Ben Edwin, ed., *Aesopica*, Vol. I (Urbana: University of Illinois Press, 1952).

Perry, Ben Edwin, ed., *Babrius and Phaedrus*, Loeb Classical Library (Cambridge, MA: Harvard University Press, 1965).

Plato: Phaedo, Harold North Fowler, trans., Loeb Classical Library (Cambridge, MA: Harvard University Press, 1990).

Salisbury, Joyce E., *The Beast Within: Animals in the Middle Ages* (New York: Routledge, 1994).

Shapiro, Norman R., trans., *Fifty Fables of La Fontaine* (Urbana, IL: University of Illinois Press, 1988).

Spector, Norman B., ed. and trans., *The Complete Fables of Jean de la Fontaine* (Evanston, IL: Northwestern University Press, 1988).

Spiegel, Harriet, ed. and trans., Marie de France: *Fables* (Toronto: University of Toronto Press, 1987).

The Fables of Aesop, selected and illustrated by David Levine, Patrick and Justina Gregory, trans. (Boston: Gambit, 1975).

Ziolkowski, Jan M., *Talking Animals: Medieval Latin Beast Poetry, 750–1150* (Philadelphia: University of Pennsylvania Press, 1993).

Authored Animals: Creature Tropes in Native American Fiction

Gerald Vizenor

The anthropomorphist ascribes and traces human emotion and motivations to animals and nature; these modes of narration cause misconceptions in both science and literature.

John Stodart Kennedy named *feelings*, *motivations*, and *thought* the three sources of mental experiences, sources that are subjective and independent of motion or human action. Granting animals the same introspection as humans, without pretense or intentional tropes, would be "unwarranted anthropomorphism" (Kennedy, 1992, p. 9).[1]

Arguably there are *warranted* anthropomorphic ascriptions in narratives; literary ascriptions that are figurative and create a creature *presence* rather than a causal representation of animal consciousness.

William James considered consciousness, the assumptions of introspection, and wrote that "everyone assumes that we have direct introspective acquintance with our thinking activity as such, with our consciousness as something inward and contrasted with the outer objects which it knows. Yet I must confess that for my part I cannot feel sure of this conclusion" (James, 1992, p. 432).[2]

Kennedy pointed out in *The New Anthropomorphism* that the "most widely held scientific reason for assuming that there must be some measure of consciousness in animals is the Darwinian principle that evolution has been a continuous process" (Kennedy, 1992, p. 15). Novelists and scientists, with tropes and theories, create a sense of nature and a mode of evolution, the narratives of our creation and presence. Kennedy observed,

> Altogether, then, it seems likely that consciousness, feelings, thoughts, purposes ... are unique to our species and unlikely that animals are conscious. If we were entirely logical about it these probabilities would be enough to make us try to avoid anthropomorphic descriptions of animal behavior. But we are not entirely logical about it, and we have to ask why scientists as well as laymen should be so addicted to anthropomorphic expression (Kennedy, 1992, p. 24).

Jeffrey Moussaieff Masson, for instance, asserted in the introduction to *When Elephants Weep: The Emotional Lives of Animals* that "animals cry. At least they vocalize pain or distress, and in many cases seem to call for help. Most people believe, therefore, that animals can be unhappy and also that they have such primal feelings as happiness, anger, and fear. . . . I try to show that animals of all kinds lead complex emotional lives" (Masson and McCarthy, 1995, pp. Xii, xxii, 219).[3]

The novelist creates a *presence* of animals and nature with tropes and descriptions that are not bound to the modes of scientific causation or objective representations. "The fundamental difficulty, however, when we wish to avoid anthropomorphism, lies in the nature of our ordinary language," argued Kennedy. "Our everyday language would be crippled without its constant use of metaphors and analogies," and anthropomorphic analogies "readily generate misunderstanding." On the other hand, "wholly objective language is almost impossible to achieve completely and attempts at it are usually clumsy and prolix because they are inevitably strained compared with our everyday speech" (Kennedy, 1992, p. 158, 159).

Nature is a narrative creation, and nature is a trope; pristine nature is untamed, unnamable, elusive, and precarious. At the same time, nature, *natura,* is *our* creation in a lazy loan word.

We trace our *presence* in animals, a *warranted* narrative creation. The memories of oral performances are silenced and creation deferred as cultural evidence in the causal discoveries and translations of the social sciences.

The nature of *authored creation* is silence, a written narrative of tropes, discoveries, observations, representations, comparisons, and transitive closures. Likewise, the *authored animals* in literature are wild tropes, fantastic creatures, and others are mundane similes of domestication. The most elusive animals are in the heart of the native hunter, to be sure, and in the mind of the novelist.

The animals created in literature are no more distinct than their animal authors, distinctions, to be sure, and simulations of the abstruse *other* in the similes of descent and evolution. The author, as the animal and unaccustomed hunter, overcomes *wild* animals in the *authored familiarities* of literature, the episteme of authored animals.

Roland Barthes observed that no one without formalities can "pretend to insert his freedom as a writer into the resistant medium of language because, behind the later, the whole of History stands unified and complete in the manner of a Natural Order. Hence, for the writer, a language is nothing but a human horizon which provides a distant setting of *familiarity*, the value of which, incidentally, is entirely negative" (Barthes, 1968, p. 9).[4]

Common sense, the outcome of causal reason, representation, speciesism, and theories of evolution are conversions of chance and pristine nature. The most inscrutable animals are tamed in the *authored familiarities* of human nature, the inescapable consequences of reason, iconic *silence*, and the philosophies of grammar; at the same time, certain animals are memorable characters with their own manners, consciousness, and points of view in literature.

Louis Owens, the novelist, created an animal with an enormous head and a thyroid condition, a giveaway dog named Custer. The presenter said, "Indians and dogs go together. . . . It's an ancient, honorable alliance. A good dog warms the lodge during those hard winters and warns when the stealthy enemy approaches. And Custer's a sweet dog; look at that face. He's just a little nervous right now" (Owens, 1994, p. 147; Vizenor, 1995, p. 190).

Jack London created a clever, heroic, proletarian animal character in *The Call of the Wild*. "Buck did not read the newspapers, or he would have known that trouble was brewing, not for himself, but for every other tide-water dog." Buck, with domestic deference, "accepted the rope with quiet dignity. To be sure, it was an unwonted performance: but he had learned to trust in men he knew, and to give them credit for a wisdom that outreached his own."

London was an evolutionist and an advocate of social justice; moreover, he celebrated selective variations of heroic individualism in his stories and novels. Sometimes his animals and humans were matched to the same temperament and seasons. "It was beautiful spring weather, but neither dogs nor humans were aware of it" (London, 1982a, pp. 1, 7, 55). Nature was a superior character in one of his short stories, and the point of view was much wider than individualism. "Nature was not kindly to the flesh. She had no concern for that concrete thing called the individual. Her interest lay in the species, the race" (London, 1982b, p. 367).

Mary Allen observed in *Animals in American Literature* that London created "romantically realistic heroes in his dogs," an inclination that dismissed naturalism. "His dogs not only survive but they triumph. Within the realm of actual behavior, the exceptional dog is capable of deeds that humankind finds noble. Because adaptability is more important than sheer savagery, the triumphant animal is much more than the most powerful predatory beast" (Allen, 1983, pp. 78, 79).[5]

Such heroic and ironic animals are the encore of more than mere realism and evolutionism; rather, these are distinct creations of *mutant omniscience*, a marvelous punctuated equilibrium in literature. Surely some authors are punctuationalists at heart, but the contradictions of their animal characters are denatured on the "human horizon" of existential reason. Animals that are *seen* more in iconic silence than *heard* in nature are *poselocked* in evolution, the mutant caricatures of anthropomorphism.

The forms of reason "are not independent of our animal nature; rather, they depend crucially on that animal nature," argued George Lakoff in *Women, Fire, and Dangerous Things*. "Imagination is not mere fancy, for it is imagination, especially metaphors and metonymy, that transforms the general schemas defined by our animal experience into forms of reason . . . " (Lakoff, 1987, pp. 368, 586).[6]

My rifle was cold that autumn. Colder now in this remembrance of the death of a common red squirrel. Oak leaves rattled on the wind. The squirrels hunched in the trees out of reach but not sight, their natural escape distance. They sensed my distance, a hunter without the earned honor and chance of the chase. The squirrels were in cold sight; their thighs the amusement of my aim. My weapon, not the bond of my imagination, was the instrument of their death, but not now in the silence and closure of these words.

"Men with only hand weapons do not need to invent stern codes to insure that hunting is a challenge rather than an amusement," wrote Paul Shepard in *The Tender Carnivore*. "The hunter's confrontation of the enigmas of death and animal life inspire attitudes of honor and awe expressed in ceremonial address" (Shepard, 1973, p. 150, 154).[7]

The sun shimmered on the leaves. I waited and then fired my rifle at a red squirrel. The bullet shattered the bones in his shoulder. He tumbled to the ground near the trunk of an oak, bounced once, and then reached out with one paw to climb back into the tree. The other paw was bloodied, loose, turned under, dead. He reached with his other paw to the tree, to climb out of my sight; again and again he fell back. Blood flowed down his body. He watched me in the distance of one eye. His escape from me would be eternal. The hunter and the author were the entire cause of his miseries, silence, and death.

The Boy Scouts of America and the Izaak Walton League taught me and other hunters of my generation the monomercies of the coup de grâce. We learned as hunters, and later as authors, never to let a wounded animal suffer. Wounded animals were put out of their miseries, at heart *our* miseries of the animal other in literature.

The first mercy bullet tore the fur and flesh from his skull. The second bullet smashed his jaw and exposed his teeth, a hideous death mask. I fired twice more to end his miseries; the bullets shattered his forehead and burst through one eye. He held onto the rough trunk of the oak tree with one paw; at last, the blood bubbled slower and slower from his nostrils. I touched his back; my hands were warmed with his blood. At last his death was mine that autumn.

Erich Fromm in *The Anatomy of Human Destructiveness* wrote that the hunter

> returns to the natural state, becomes one with the animal, and is freed from the burden of the existential split. For modern man, with his cerebral orientation, this experience of oneness with nature is difficult to verbalize and to be aware of, but it is still alive in many human beings. . . . It is amazing how many modern authors neglect this element of skill in hunting, and focus their attention on the act of killing. After all, hunting requires a combination of many skills and wide knowledge beyond that of handling a weapon (Fromm, 1973, pp. 132, 133, 136).[8]

The hunter, then, is a natural presence; the contrarious author is a secular *creationist*, the eternal animal, the animal other of animism and the untramundane. The creation of animals, the literal and simile versions, is traced and measured in nature, authors, and literature. Otherwise, the animals in literature would be mere impostures of nature in the minds of their authors.

Paul Shepard observed in *Thinking Animals*:

> When totemic thought, with its instinct for animal imagery, is carried forward into totemic culture, the wild animal groups are regarded as sets containing necessary secrets for human conduct, translated by myth and applied to actual human situations by speculative thought. When it is carried forward into caste or class thought, on the other hand, the wild animals are likely to be seen as atavisms ruled by mysterious powers or mindless passions, while the domesticated animals represent puerile expressions of our civilized egos (Shepard, 1978, pp. 256, 260).[9]

Mary Allen noted that metaphorical animals "far outnumber the literal animals in literature," and authored animals are bestial, carnal, utilitarian, pastoral, exotic, stoical, heroic, and more, but "man's creations did not outdo the uncanny subjects in nature."

God, for instance, "created his whale easily in a day," observed Allen. Herman Melville "shows us the wrenching difficulty of the task." Gradually he "builds his literal whale alongside the myths of whales, later showing the view from the outside working back in. . . . Not only does he inspire a wealth of metaphysical possibilities, but as a literal animal the sperm whale is so composed as to make an extraordinary but distinctly American character (Allen, 1983, pp. 5, 6, 7, 12, 19, 33, 34).[10]

Native American Indians are commonly perceived as being in close association with nature and the natural *presence* of animals; these associations are sources of native omniscience and consciousness. The literary interpretations of this presence have presumed the doctrines of nativism, animism, naturalism, realism, and other theories.

N. Scott Momaday created a sacred landscape of bears and eagles in the myths, metaphors, and traces of native ceremonies. He imagined an environment that is twice real, in the same sense that an oral performance is both real and imagined in the sovereignty of motion; authored landscapes are twice natural as the stories and the seasons change. He wrote in his memoir *The Names*: "The names at first are those of animals and of birds, of objects that have one definition in the eye, another in the hand, of forms and features on the rim of the world, or of sounds that carry on the bright wind and in the void. They are old and original in the mind, like the beat of rain on the river, and intrinsic in the native tongue, failing even as those who bear them turn once in the memory, go on, and are gone forever" (Momaday, 1976, p. 3).

Joseph and Barrie Klaits observed in the introduction to *Animals and Man in Historical Perspective* that people "love and pity animals; we also use, abuse, and fear them. Animals are our companions, our amusements, and our sustenance. We pursue animals to satisfy our tastes in food and fashion or to enjoy the raw pleasure of destruction. The pursuit can also be creative, as when we seek their images through art and literature" (Klaits and Klaits, 1974, p. 1).

Animals are imagined in nature and literature, translated, and compared in memories, narratives, and cultural contexts. The animals of literature are twice their nature: the real in visions and environments and the authored animals of a narrative creation. The authored animals are as diverse as the *real*, the species of imagination: captured, domestic, wild, transmuted, and fantastic.

Language, then, is one of the *real* environments of the authored animals, the names, memories, and manners of the real as a narrative. Authored animals are real in the nature of tropes, the figurations that trace and redouble both the imagination of the author and the baited reader in what becomes a marvelous conception, an arcane animal of the shared pleasures of creation.

Authored animals, however, are burdened, as they would be in any environment, by the literary styles of their authors, the

very sustenance of creation. The animals of literature must
depend upon the turns of tropes, the practices of simile,
metaphor, and metonymy, as the necessities of authored
nature.

"In the language of prose, besides the regular and proper
terms for things, metaphorical terms only can be used with
advantage," observed Aristotle in *Rhetoric*. "Metaphor, moreover, gives style clearness, charm, and distinction as nothing
else can: and it is not a thing whose use can be taught by one
man to another. Metaphors, like epithets, must be fitting,
which means that they must fairly correspond to the thing
signified: failing this, their inappropriateness will be conspicu-
ous; the want of harmony between two things is emphasized by
their being placed side by side" (1984, 1404b32–34, 1405a8–
13). The difference between metaphor and simile, he noted,
"is but slight."

Thomas McLaughlin, in his essay "Figurative Language,"
pointed out more than a "slight" divergence in the measure of
tropes. "A *simile* is a comparison of terms. Unlike metaphor
which requires the reader to do the work of constructing a
logic of categories and analogies, a simile states explicitly that
two terms are comparable and often presents the bases for the
comparison. . . . *Metonymy* accomplishes its transfer of meaning
on the basis of associations that develop out of specific contexts
rather than from participation in a structure of meaning."
Moreover, metonymy "places us in the historical world of
events and situations, whereas metaphor asserts connections
on the basis of a deep logic that underlies any use of words"
(Lentricchia and McLaughlin, 1990, pp. 83, 84).

John Searle, in his essay "Metaphor," wrote that the meaning
of "metaphorical utterances" is systematic. The "knowledge
that enables people to use and understand metaphorical
utterances goes beyond their knowledge of the literal meaning
of words and sentences." He argued that a "literal simile" is a
"literal statement of similarity," and that "literal simile requires

no special extralinguistic knowledge for its comprehension. . . .
" (Searle, 1979, pp. 93, 105, 123).[11]

Philip Wheelwright declared that metaphor is an "element
of tensive language" (1962, pp. 70, 148).[12] Donald Davidson
explained "metaphor as a kind of ambiguity: in the context of
metaphor, certain words have either a new or an original
meaning, and the force of the metaphor depends on our
uncertainty as we waver between the two meanings" (Davidson,
1978, pp. 32, 33). Robert Rogers, in a psychoanalytic
exploration of figurative language, noted that one "puzzling
aspect of the expressive capaciousness of metaphor takes the
form of an image's potential for focusing both thought and
emotion in a particularly intense, economical way" (Rogers,
1978, p. 7).

Janet Martin Soskice, in *Metaphor and Religious Language*,
defined metaphor as that "figure of speech whereby we speak
about one thing in terms which are seen to be suggestive or
another." She noted that the "greatest rival of metaphor,
simile, in its most powerful instances does compel possibilities.
Simile is usually regarded as the trope of comparison and
identifiable within speech by the presence of a 'like,' or an 'as,'
or the occasional 'not unlike.'" Still,

> to regard simile as necessarily mere 'same-saying' of the trivial
> sort is greatly to misrepresent that trope. Simile may be the
> means of making comparisons of two kinds, the comparison of
> similars and dissimilars, and in the latter case, simile shares
> much of the imaginative life and cognitive function of its
> metaphorical counterparts. For this reason, we can say that
> metaphor and simile share the same function and differ
> primarily in their grammatical form (Soskice, 1985, pp. 15, 58,
> 59, 60).[13]

Soskice, it seems, would share the more classical view that
the difference between metaphor and simile "is but slight."
However, the author noted, a simile cannot "be used in
catachresis." Simile cannot create the lexicon, as does "dead
end" or the "leaf of a book."

George Lakoff and Mark Johnson provide one of the most accessible descriptions of metaphor in *Metaphors We Live By*.

> Metaphor is for most people a device of the poetic imagination and the rhetorical flourish—a matter of extraordinary rather than ordinary language. Moreover, metaphor is typically viewed as characteristic of language alone, a matter of words rather than thought or action. For this reason, most people think they can get along perfectly well without metaphor. We have found, on the contrary, that metaphor is pervasive in everyday life, not just in language but in thought and action. Our ordinary conceptual system, in terms of which we both think and act, is fundamentally metaphorical in nature (Lakoff and Johnson, 1980, pp. 3, 4).[14]

Authored animals are the tropes of human severance, an environment of iconic silence. Metaphor and simile are the traces of creations, memories, and narrative conversions, the *nature* and *presence* of animals in literature. Tropes are the associations of naturalism and sources of animal consciousness. Likewise, there are many epistemic tropes, literary styles, and critical distinctions, the essence of authored animals in literature.

Bestialities, brute consciousness, and other memorable tropes show that the monotheistic separation of animals has never been sincere. That human horizon of authored animals must reveal the diversities of creation in native literature.

N. Scott Momaday, Leslie Silko, Louise Erdrich, Louis Owens, and Gordon Henry, Jr. are native novelists with diverse cultural experiences and distinctive literary styles. The authored animals in their novels are both mythic and mundane; the metaphors and similes of animal creations are as diverse as the authors.

Grey, the main character in *The Ancient Child* by Momaday, is a metaphor of native identities. "Her father was Kiowa and her mother Navajo, and the two cultures came together in her easily, more or less." She imagined the death of Billy the Kid, and she "dreamed of sleeping with a bear. The bear drew her into his massive arms and licked her body and her hair. It hunched over her, curving its spine like a cat, until its huge body seemed to have absorbed her own. Its breath which bore a deep, guttural rhythm like language, touched her skin with low, persistent heat" (Momaday, 1989, p. 17).[15]

The authored bear is a metaphor, a dream, a mythic character in the narrative. The simile of the bear is the comparison and description of sound and motion. The presence of the bear is a metaphor of transcendence. The heat of the authored animal is natural, human, and animalism. Formost, the bear is the mythic healer of human separation in a narrative. That separation is never closed, but metaphor is a sense of presence, the source of shared imagination in a novel.

"Living beyond civilized life, sexually and boldly aggressive, the bear gives vent to a massive and uncontrolled appetite, upsetting rule and restriction," wrote Paul Shepard and Barry Sanders in *The Sacred Paw*. "But in its display of maternal care and concern, the bear is the very essence of civility and order. Standing for both male and female characteristics, the bear would appear to have no gender" (Shepard and Sanders, 1985, p. 130).[16]

Set is an artist in *The Ancient Child*, separated from a sense of native presence; he is an orphan and returns with the spiritual power of his ancestors. The return is a shared occurrence in the memories of the bear. His return is a metaphor of motion, survivance, and native sovereignty.

> Set took the medicine bundle in his hands and opened it. The smell of it permeated the whole interior. When he drew on the great paw, there grew up in him a terrible restlessness, wholly urgent, and his heart began to race. He felt the power of the bear pervade his being, and the awful compulsion to release it. Grey, sitting away in the invisible dark, heard the grandmother's voice in her mouth. When Set raised the paw, as if to bring it down like a club, she saw it against the window, huge and phallic on the stars, each great yellow claw like the horn of the moon (Momaday, 1989, pp. 303, 304).

The metaphor of the bear and the medicine bundle are

obscure and unnamable powers, but the simile of the "horn of the moon" is direct and abates the uncertainties and shadows of the medicine.

Momaday told Charles Woodard in *Ancestral Voices* that he was "serious about the bear" and "identified with the bear" because he is "intimately connected with that story. And so I have this bear power. I turn into a bear every so often. I feel myself becoming a bear, and that's a struggle I have to face now and then" (Woodard, 1989).

Momaday turns the metaphors of authored animals into an unrevealed presence, the tensive myths of creation and native solace in *House Made of Dawn*. The mere mention of the bear is traced in sound, motion, and the memories of the characters.

Abel came to cut wood for three dollars. Angela "watched, full of wonder, taking his motion apart." The sound of the axe was incessant. "Once she had seen an animal slap at the water, a badger or a bear. She would have liked to touch the soft muzzle of a bear, the thin black lips, the great flat head. She would have liked to cup her hand to the wet black snout, to hold for a moment the hot blowing of the bear's life. She went out of the house and sat down on the stone steps of the porch. He was there, rearing above the wood." Later, they came together. "He was dark and massive above her, poised and tinged with pale blue light. And in that split second she thought again of the badger at the water, and the great bear, blue-black and blowing" (Momaday, 1968, pp. 31, 32, 33, 64).

Susan Scarberry-Garcia observed, in her study of *House Made of Dawn*, that this "sensual image which depicts Abel's bear-like physique and presence has been dismissed in the criticism as a forlorn white woman's fantasy about having a dark elemental man as her ideal lover" (Scarberry-Garcia, 1990, p. 52).[17]

Momaday told Charles Woodard that many "things happen in *House Made of Dawn* that I can't explain in a logical way. They are based upon insights which I think are valid, but those insights are not fully conscious. That is, they weren't consciously developed. They exist beneath the level of everyday consciousness, but they are nonetheless real" (Woodard, 1989).

The metaphor of the bear is an ancient presence in the novel, the motivation of natural reason, and the style is an inscription of native realism. "The elements constituting a work obey an internal logic, not an external one," observed Oswald Ducrot and Tzvetan Todorov in the *Encyclopedic Dictionary of the Sciences of Language*. "Motivation is thus a variant of realism. It is not conformity to the genre, but a cloak that the text casts prudently over the rules of the genre" (Ducrot and Todorov, 1979, pp. 261, 263).[18]

Leslie Silko encircles the reader with witches, a hard metaphor that turns the creation stories in *Ceremony*. Alas, the hardhearted witches *invented* white people, a distinctive trope that overcomes the temptations of simile and the mere comparison of opposition and wicked extremes.

> The old man shook his head. 'That is the trickery of the witchcraft,' he said. 'They want us to believe all evil resides with white people. Then we will look no further to see what is really happening. They want us to separate ourselves from white people, to be ignorant and helpless as we watch our own destruction. But white people are only tools that the witchery manipulates; and I tell you, we can deal with white people, with their machines and their beliefs. We can because we invented white people; it was Indian witchery that made white people in the first place.

> Long time ago
> in the beginning
> there were no white people in this world
> there was nothing European
> And this world might have gone on like that
> except for one thing:
> witchery.
> This world was already complete
> even without white people.
> There was everything
> including witchery
> (Silko, 1977, pp. 132, 133).

Louis Owens pointed out that "Betonie's words and the story of witchery underscore an element central to Native American oral tradition and worldview: responsibility." This sense of responsibility, of course, is a metaphor that denies closure; the actions, connections, and intentions are not causal but obscure ceremonies. "To shirk that responsibility and blame whites, or any external phenomenon, is to buy into the role of helpless victim" (Owens, 1992a, p. 93).

The authored animals are connected to the environment, not to the similes of human consciousness. Silko creates animals with a *natural* character; the metaphors are their presence and motivation. Simile is used in the novel to describe motion and to compare animals to the environment, not to assay human characteristics. For instance, the "mountain lion came out from a grove of oak trees in the middle of the clearing. He did not walk or leap or run; his motions were like the shimmering of tall grass in the wind. . . . Relentless motion was the lion's greatest beauty, moving like mountain clouds with the wind, changing substance and color in rhythm with the contours of the mountain peaks: dark as lava rock, and suddenly as bright as a field of snow" (Silko, 1977, pp. 195, 196).

The authored animals in novels by Momaday, Silko, Owens, and others are metaphors that are motivations of character; some of their animals are introspective and with consciousness. These authors use simile as motion, comparisons to be sure, but not as mere attributions of animal and human characteristics. The simile, of course, is more than the signature of "like" and "as" or "mere 'same-saying,'" as Janet Martin Soskice has pointed out. However, the style of simile is a common comparative in narratives; the most limited or direct is the "literal simile" described by John Searle.

Louise Erdrich has used a style of tropes in *Tracks* that is closer to the literal or *prosaic simile* than to the obscure metaphors of motivation; the other authors mentioned here seldom used the literal style of simile. For instance, "she shivered all over like a dog," and he's "hiding from you like a dog," and "I'd go off in the bush like a sick dog first, alone," and his "head shaggy and low as a bison bull," and she "leaned over the water, sucking it like a heifer," and the "bear followed, heeling her like a puppy," and the "man was bearded, hug, clumsy-looking like a weak-sighted bear," and more (Erdrich, 1988, pp. 10, 37, 54, 60, 89, 168).

Erdrich names moose, pigs, bears, cats, and other animals, but the most common authored animal in *Tracks* is the dog. The animal and dog are generic creations more often than not, and few animals are characters in a natural environment. The generic animal is a generic and literal simile.

"I think like animals, have perfect understanding for where they hide," said Nanapush. The generic animal is the binary beast in a prosaic simile. "Moses, who had defeated the sickness by turning half animal and living in a den," is an authored human as the beast. "He said the animals understood what was happening, how they were dwindling," is an instance in the novel of generic animal consciousness (Erdrich, 1988, pp. 35, 40, 139).

Tracks has more dogs on the page than other animals. One unnamed authored animal has character and is more memorable than the other generic dogs in the novel. "Lily had a dog, a stumpy mean little bull of a thing with a belly drum-tight from eating pork rinds. The dog was as fond of the cards as Lily, and straddled his barrel thighs through games of stud, rum poker, *vingt-un*. The dog snapped at Fleur's arm that first night, but cringed back, its snarl frozen, when she took her place" (Erdrich, 1988, p. 18).

Custer, the authored animal in *Bone Game* by Louis Owens, is a memorable character in the tensive metaphor of his name. The names of some animals are ironic, and other authored animals have a sense of presence and character without a name. In another scene, a character in the novel recommends a guard dog: "I mean it. You should get a dog. . . . A big, mean one. And remember, dogs don't like ghosts or witches. We keep them around the hogans at home just for that" (Owens, 1994, p. 96).

Owens, in *The Sharpest Sight*, creates a tricky bond between a rabbit and hunter. "Cherokee rabbits were smart. They lived by tricks in a world of words and had a good time doing it." Cole "raised the rifle and aimed the notched sight at a spot just below the rabbit's ear. 'Time for a trick,' he whispered to the rabbit. He pulled back the hammer and shounted" (Owens, 1992b, p. 12). The presence of the rabbit is not lost in a prosaic simile.

Gordon Henry, Jr., for instance, has created a wild and comic scene of two dogs stuck together in a natural sexual dance, a white dog, a brown dog, and, in the end, one dead dog, in his novel *The Light People*.

Boozhoo tried to separate the dogs, and then he turned to his own magic, "my sideline vocation, my years of training. First I tried mental magic: I attempted to project the image of a piece of meat into the mind of the dancing white dog. For a minute I thought it worked, since I heard the white dog give off a low growl and I thought I saw his mouth water. But the dogs remained stuck together. . . . "

LaVerve was drunk; he arrived with his rifle and shot the white dog. "But the only thing that changed was the dog dance; the white dog whimpered as he remained fast to the brown female, following behind her in a more frantic dance that still failed to separate them." LaVerve tried again to pull the dogs apart, and bloodied his hands before he took another drink and drove away drunk in his pickup.

Boozhoo carried the dogs

back to Seed's place. . . .Seed studied the dogs. He spoke without looking up at me. 'We've got to get this dead one off before he stiffens up.' Seed slid out of his chair then and leaned over the dead white dog. He whispered into the dog's ear. I didn't hear the words or the language, but when he settled back into his chair he told me to take the white dog out and bury it. I did as he told me, and I expected the white dog to remain fast to the brown, but when I lifted the dead dog the two dogs separated and the brown dog ran off into the darkness (Henry, 1994, pp. 170, 171).

The presence of authored animals in these selected novels are *real* as tropes of imagination. The creatures in native literature are seldom mere representations of animals in nature or culture, wild, domestic, generic, or otherwise; however, there is an unnamable presence, traces of a familiar nature, comic motivation, native reason, and author introspection. Owens, Momaday, Silko, Erdrich, and Henry use metaphor to turn the pleasures of the unnamable into animal characters. Momaday transmutes both humans and animals on a landscape of lyrical metaphors and descriptions. Erdrich uses literal and prosaic simile in her novel more than the other authors. Owens and Henry create a comic and ironic presence of native mongrels.

Monotheistic creation is a separation of animals and humans in literature and nature; the common unions since then have been both domestic and aesthetic. Literal simile is a familiar disseverance of authored animals on a human horizon. The more obscure tropes in literature must be closer to nature and animal consciousness than a literal simile. The authors are animals, the readers are animals, the animals are humans, and the authored hunters bond with animals in their own novels.

Meanwhile, those dogs who move behind the wheel as drivers when their masters park the car are very close to the *real* tropes of consciousness; those dogs waiting behind the wheel are certain to be seen as human. The dogs are either ready to be seen as humans or to drive automobiles into our consciousness and prove that punctuated equilibrium is not hindered by prosaic similes, domestic loyalties, or learned critiques of anthropomorphism.

Notes

[1] Kennedy wrote that "we are directly aware of these things only in ourselves, through introspection." The "intentional stance" is what

he calls "mock anthropomorphism" that "can be valuable for the hypotheses it generates about the functions of the animal's behavior . . ." (Kennedy, 1992).

[2] "Whenever I try to become sensible of my thinking activity as such, what I catch is some bodily fact, an impression coming from my brow, or head, or throat, or nose," wrote James. "It seems as if consciousness as an inner activity were rather a *postulate* than a sensibly given fact, the postulate, namely, of a *knower* as correlative to all this known; and as if '*sciousness*' might be a better word to describe it" (James, 1992).

[3] "Some animals have senses humans do not possess, capacities only recently discovered," wrote Masson. "Other animal senses may remain to be discovered. By extension, could there be feelings animals have that humans do not, and if so, how would we know? It will take scientific humility and philosophical creativity to provide even the beginning of an answer" (Masson and McCarthy, 1995).

[4] Barthes wrote that language "is not so much a stock of materials as a horizon, which implies both a boundary and a perspective; in short, it is the comforting area of an ordered space. The writer literally takes nothing from it; a language is for him rather a frontier, to overstep which alone might lead to the linguistically supernatural; it is a field of action, the definition of, and hope for, a possibility" (Barthes, 1968).

[5] "What most dramatically sets London apart from the mainstream of naturalism in his Klondike stories," wrote Allen, "is that he makes Darwinism literal and presents the *animal characters themselves*" (Allen, 1983).

[6] Lakoff wrote that many scholars "take it for granted. . . that conceptual categories are defined solely by the shared essential properties of their members, that thought is the disembodied manipulation of abstract symbols, and that those symbols get their meaning solely by virtue of correspondences to things in the world. The view of reason as abstract, disembodied, and literal is well-established" (Lakoff, 1987).

[7] Shepard (1973) wrote that the "ways of the hunters are beginning to show us how we are failing as human beings and as organisms in a world beset by a 'success' that hunters never wanted." The failures of nature show in the success of commercial literature, failures that some authors and hunters anticipated but never wanted.

[8] "Fortunately, our knowledge of hunting behavior is not restricted to speculations; there is a considerable body of information

about still existing primitive hunters and food gatherers to demonstrate that hunting is not conducive to destructiveness and cruelty, and that primitive hunters are relatively unaggressive when compared to their civilized brothers," wrote Fromm (1973).

[9] "The language of everyday life has thousands of figures of speech, phrases, colloquialisms, and neologisms that not only employ animal imagery but which assume a profound and fundamental habit of thought, relating consciousness of language" (Shepard, 1978).

[10] "The realism that superseded romanticism focused for the most part on social man, an increasingly urban man, a context in which animals play little part. While realism follows no particular style, the tendency is away from symbolism, in some cases away from metaphor altogether; thus the figurative animal occurs less frequently. . . . " The "realistic animals are wild, terrestrial beings. Many are big and violent, though their spirit is more important than their force. They are usually disciplined, clean, and utilitarian. They are celibate males, free or fighting to be free. And they are markedly independent" (Allen, 1983).

[11] "The question, 'How do metaphors work?' is a bit like the question, 'How does one thing remind us of another thing?' There is no single answer to either question, though similarity obviously plays a major role in answering both. Two important differences between them are that metaphors are both restricted and systematic; restricted in the sense that not every way that one thing can remind us of something else will provide a basis for metaphor, and systematic in the sense that metaphors must be communicable from speaker to hearer in virtue of a shared system of principles" (Searle, 1979).

[12] "The primitive thinker, unlike ourselves, did not start from a known world of inanimate things and then pretend they were otherwise. He started with a world that was not clearly animate or inanimate but hovered between the two conditions, sometimes partaking more of the one and sometimes more of the other" (Wheelwright, 1962). The author is the animal who hovers between imagination and the other environment of animals twice real; once, unnamed in nature, and the second *real* is in the tropes of language.

[13] "In the light of similarities between metaphor and simile, should we not dispense with the distinctions between the two and, for instance, call them both metaphor if the difference is only one of superficial grammar? Apart from a natural reluctance to abandon a distinction commonly made in our speech, there is a good reason why the distinction should be kept; it is that while metaphor and simile

differ primarily in grammatical form, there is one important role which, by virtue of grammatical form, metaphor performs and simile cannot. . . . Simile cannot, for reasons of syntactic form, be used in catachresis. A gap in a lexicon is filled by a term, such as 'leaf' of a book, or by a phrase such as 'dead end,' but not by an 'is like' clause" (Soskice, 1985).

[14] "Primarily on the basis of linguistic evidence, we have found that most of our ordinary conceptual system is metaphorical in nature" (Lakoff and Johnson, 1980).

[15] "Identity is acquired through an act of self-imagination, Momaday has explained: 'We are what we imagine. Our very existence consists in our imagination of ourselves. Our best destiny is to imagine, at least, completely, who and what, and that we are. The greatest tragedy that can befall us is to go unimagined.' Finally, Momaday has told critical biographer Mathias Schubnell, 'I believe that I fashion my own life out of words and images, and that's how I get by,' " wrote Louis Owens (1992a).

[16] In the bestiaries "bears are perceived as particularly sexual, for 'they do not make love like other quadrupeds, but, being joined in mutual embraces, they copulate in the human way" (Shepard and Sanders, 1985).

[17] "Peter Beidler in his article 'Animals and Human Development in the Contemporary American Indian Novel' comments about these scenes: 'Her identification of Abel with bear is, of course, part sexual fantasy. . . . her association of Abel with bear triggers in Abel much later an awareness of his own bear nature" (Scarberry-Garcia, 1990). Beidler is reductive, and his representations of myth and metaphor are more anthropological than literary.

[18] "The desire to provide a narrative with full motivation is not unrelated to the arbitrariness of the sign. Signs are arbitrary; names are not inscribed in things. But any used of a sign system tends to naturalize it, to present it as though it were self-evident. the tension resulting from the opposition gives rise to one of the dominant currents in literary history" (Ducrot and Todorov, 1979).

References

Allen, Mary, *Animals in American Literature* (Urbana, IL: University of Illinois Press, 1983).

Aristole, *The Complete Works of Aristotle*, in *Rhetoric* (Princeton, NJ: Princeton University Press, 1984).

Barthes, Roland, *Writing Degree Zero* (New York: Hill and Wang, 1968).

Davidson, Donald, "What Metaphors Mean," in Sheldon Sacks, ed., *On Metaphor* (Chicago: University of Chicago Press, 1978).

Ducrot, Oswald and Todorov, Tzvetan, *Encyclopedic Dictionary of the Sciences of Language* (Baltimore, MD: The Johns Hopkins University Press, 1979)

Erdrich, Louise, *Tracks* (New York: Harper and Row, 1988).

Fromm, Erich, *The Anatomy of Human Destructiveness* (New York: Holt, Rinehart and Winston, 1973).

Henry, Jr., Gordon, *The Light People* (Norman, OK: University of Oklahoma Press, 1994).

James, William, "Psychology: Briefer Course" in *William James* (New York: The Library of America, 1992).

Kennedy, John Stodart, *The New Anthropomorphism* (Cambridge: Cambridge University Press, 1992).

Klaits, Joseph and Klaits, Barrie, *Animals and Man in Historical Perspective* (New York: Harper and Row, 1974).

Lakoff, George, *Women, Fire, and Dangerous Things* (Chicago: University of Chicago Press, 1987).

Lakoff, George and Johnson, Mark, *Metaphors We Live By* (Chicago: University of Chicago Press, 1980).

Lentricchia, Frank and McLaughlin, Thomas, eds., *Critical Terms of Literary Study* (Chicago: University of Chicago Press, 1990).

London, Jack, *The Call of the Wild* in *Jack London* (New York: The Library of America, 1982a).

London, Jack, "The Law of Life" in *Jack London* (New York: The Library of America, 1982b).

Masson, Jeffrey Moussaieff and McCarthy, Susan, *When Elephans Weep: The Emotional Lives of Animals* (New York: Delacorte Press, 1995).

Momaday, N. Scott, *House Made of Dawn* (New York: Harper and Row, 1968).

Momaday, N. Scott, *The Names: A Memoir* (New York: Harper and Row, 1976).

Momaday, N. Scott, *The Ancient Child* (New York: Doubleday, 1989).

Owens, Louis, *Other Destinies* (Norman, OK: University of Oklahoma Press, 1992).

Owens, Louis, *The Sharpest Sight* (Norman, OK: University of Oklahoma Press, 1992b).

Owens, Louis, *Bone Game* (Norman, OK: University of Oklahoma Press, 1994).

Rogers, Robert, *Metaphor: A Psychoanalytic View* (Berkeley, CA: University of California Press, 1978).

Scarberry-Garcia, Susan, *Landmarks of Healing* (Albuquerque, NM: University of New Mexico Press, 1990).

Searle, John, "Metaphor" in Andrew Ortony, ed., *Metaphor and Thought* (Cambridge: Cambridge University Press, 1979).

Shepard, Paul, *The Tender Carnivore and the Sacred Game* (New York: Charles Scribner's Sons, 1973).

Shepard, Paul, *Thinking Animals: Animals and the Development of Human Intelligence* (New York: Viking Press, 1978).

Sherpard, Paul and Sanders, Barry, *The Sacred Paw: The Bear in Nature, Myth, and Literature* (New York: Viking Penquin, 1985).

Silko, Leslie, *Ceremony* (New York: Viking Penguin, 1977).

Soskice, Janet Martin, *Metaphor and Religious Language* (Oxford: Clarendon Press, 1985).

Vizenor, Gerald, ed., "The Last Stand" in *Native American Literature* (New York: HarperCollins, 1995).

Wheelwright, Phillip, *Metaphor and Reality* (Bloomington, IN: Indiana University Press, 1962).

Woodard, Charles, *Ancestral Voices: Conversations with N. Scott Momaday* (Lincoln, NE: University of Nebraska Press, 1989).

SAMENESS *and* DIFFERENCE

Introduction

Similarities and differences between us and animals have been a great focus of human interest for millennia, in cultures of the most various sorts. The title of this session is striking, though, in that it does not ask us to consider sameness*es* or difference*s* between us and animals but *sameness* and *difference*. The title itself, then, suggests an all-or-nothing issue: is there some *really big essential difference* between us and animals—or is there some *fundamental sameness*? The title alludes to attempts, past and present, to see the relation between us and them in these all-or-nothing terms. So, for example, Descartes asserted that our essence as rational beings marked a great gap between us and animals. *We* have immaterial minds which are, in this life, intimately connected with our bodies. Animals, in contrast, are *material* beings; they are intricate machines, utterly without the kind of awareness that distinguishes us. Hume, in contrast, argued for continuity, for a fundamental similarity between us and animals. Even our reason, in which we take such pride, is, he argued, continuous with that of animals. He rejects the Cartesian conception of reason as a distinctively human capacity and treats it instead as a kind of instinct, shared with animals, although ours is greatly superior in degree to that of animals. And later Darwin famously argued for continuity, giving many instances, including emotional and aesthetic responsiveness.

Duane Rumbaugh, one of the three participants in this session, notes that human beings have repeatedly "sought lines of argument and fact" which would mark us out as totally different from animals. When such attempts to find a fundamental difference are made, the capacity to use language, which we have and which animals seem to lack and which is so closely linked to rationality, has been a central issue. Stephen Clark has argued about attempts to distinguish human beings from animals on the basis of the use of language that, whatever exactly we define as linguistic capacity, when animals are shown to have *that*, we go ahead and redefine linguistic capacity. Language is being *made* to serve as the basis of difference; so whatever *they*—the animals—can do, we shall mean something *else* by language.[1] Clark made that point in 1977; in the years since then, the developing body of experimental work on the linguistic capacities of the great apes has been met by new accounts of why what the great apes do is, supposedly, not really language. Not only is it not really language, the argument goes, it never will be—no matter what else imaginative experimenters contrive to get the great apes to do in the future. In our own thought about this issue, we can profit greatly from seeing what the data really are—what the animals actually are able to do. Rumbaugh himself provides a marvelous close-up view of what some of our nearest relatives *can* do when they are brought up in circumstances which strongly encourage the development of linguistic capacities like our own.

Both Daniel Dennett and Colin McGinn are concerned with a twentieth-century version of a question raised by Descartes, whether the kind of awareness associated with a self or mind is a unique possession, ours alone, or whether it is shared by all animals—all, that is, who are able to see or hear or smell or use any sense, who are able to feel fear or hunger or cold or pain. The question is an important one in part because it has been taken to have such weighty implications for our *moral* relations to animals. The topic of this session, thus, has a pivotal significance within the conference as a whole. We may be concerned about whether there is some fundamental difference between us and animals because such a difference, it seems, would justify subordinating them entirely to our interests. But we may ask also whether the idea of such a difference is precisely something we construct to rationalize such subordination. Both McGinn and Dennett point out that our moral views may shape our answers to the question of whether animals do share such important properties of ours as sentience. McGinn emphasizes that "invasive" experimentation on animals can be squared with our conscience more easily if we persuade ourselves that animals are not genuinely subjects of experience, and he claims too that it is no accident that psychologists who have engaged in the most morally question-able sorts of experimentation have been behaviorists. And, while Dennett does note that Descartes himself has been falsely accused of engaging in callous experimentation justified by his view of animals as machines, it is certainly true that a Cartesian view of animals has been used to justify treatment of animals which would otherwise seem to raise serious moral questions. It is easy to agree with Dennett that much thinking about animal consciousness has been distorted by people's desires to support their moral and practical commitments, but we might note that the examples which he gives of one-sided treatment of the evidence are all examples of people who take animals to share fundamental human capacities. The one-sidedness is hardly all on that side.

The dispute between McGinn and Dennett has at its center the question whether animals which have sense experience— sense experience which is important in their capacity to find food, to mate, to escape predators, and so on—are in general actually aware of what is experienced. During the conference, much discussion focused on the possibility of what one might speak of as "unfelt experience," as, for example, the "unfelt pain" caused by the position of one's limbs during sleep, which can lead one to change the position of the limbs (see the *Postscript* section of Dennett's paper). Might all the experiences of animals be like such experiences, that is, capable of playing a role in the adaptability of the animal to its surroundings yet not the object of any awareness? If this were so, animals could indeed be said to have pain, but they would not genuinely *suffer*. (I might mention that the first time I ever heard it argued that although animals had pain they did not suffer, the argument was put by a man defending his experimental procedures.) Dennett himself does not argue that if the pains and other experiences of snakes, say, are not anything the snake is aware of, it follows that we may treat snakes as we treat tires, but McGinn would claim that that conclusion does follow. And opponents of the animal rights movement have used arguments like Dennett's to draw exactly the conclusion which Dennett himself does not draw.[2]

Here are two questions which are suggested by the extremely interesting papers in this session: (1) We are concerned especially with the linguistic capacities of animals and with animal awareness. These are Cartesian issues, and the conception of animal awareness in the discussion was itself a Cartesian conception, in a modern form exemplified for McGinn by Gottlob Frege and for Dennett by Thomas Nagel; the conception of what it would be for animals to lack Cartesian awareness was also Cartesian. It was a conception of animals as non-conscious machines; but is the only alternative to an updated Cartesian machine view the metaphysical selfhood argued for by McGinn? During other sessions, discussions of animal capacities, and of their similarities and differences from us, were not framed in Cartesian terms. Much of Vicki Hearne's account of her work with animals, for example, is framed in terms which come from the sharing of life and work with animals. Might our understanding of questions about animals and our relationship to them be improved if we were able to break free of the hold of the Cartesian alternatives?

(2) Do we not need to examine the idea that our moral relation to animals depends on whether they share some fundamental property? McGinn begins his paper by saying that it would be widely agreed (he presumably means agreed by contemporary moral philosophers) that moral concern can be appropriate for some entity only if it satisfied some psychological condition. But should not that widespread agreement itself lead us to wonder whether we are not taking for granted something which needs more questioning than it

gets? Many people, on the basis of that assumption, do conclude that a very seriously retarded infant is not a proper object of moral concern. But equally many would regard this conclusion with horror and would reject the assumption on which it rests. Our moral and imaginative vision may itself shape our understanding of likeness and difference, as came out in other sessions of the conference. "Am not I a fly like thee?"

Cora Diamond

Notes

[1] Clark, 1977, pp. 94–105, on the attempt to draw a sharp distinction between us and animals on the basis of rationality and language.

[2] See, for example, Carruthers, 1992, Chapter 8.

References

Carruthers, Peter, *The Animals Issue* (Cambridge: Cambridge University Press, 1992).

Clark, Stephen R.L., *The Moral Status of Animals* (Oxford: Clarendon Press, 1977).

Animal Consciousness: What Matters and Why

Daniel C. Dennett

Aʀᴇ animals conscious? The way we are? Which species, and why? *What is it like* to be a bat, a rat, a vulture, a whale?

But perhaps we really do not want to know the answers to these questions. We should not despise the desire to be kept in ignorance—are there not many facts about yourself and your loved ones that you would wisely choose not to know? Speaking for myself, I am sure that I would go to some lengths to prevent myself from learning all the secrets of those around me—whom they found disgusting, whom they secretly adored, what crimes and follies they had committed, or thought I had committed! Learning all these facts would destroy my composure, cripple my attitude towards those around me. Perhaps learning too much about our animal cousins would have a similarly poisonous effect on our relations with them. But if so, then let us make a frank declaration to that effect and drop the topic, instead of pursuing any further the pathetic course upon which many are now embarked.

For current thinking about animal consciousness is a mess. Hidden and not so hidden agendas distort discussion and impede research. A kind of comic relief can be found—if you go in for bitter irony—by turning to the "history of the history" of the controversies. I am not known for my spirited defenses of René Descartes, but I find I have to sympathize with an honest scientist who was apparently the first victim of the wild misrepresentations of the lunatic fringe of the animal rights movement. Animal rights activists such as Peter Singer and Mary Midgley have recently helped spread the myth that Descartes was a callous vivisector, completely indifferent to animal suffering *because of* his view that animals (unlike people) were mere automata. As Justin Leiber (1988) has pointed out, in an astringent re-examination of the supposed evidence for this, "There is simply not a line in Descartes to suggest that he thought we are free to smash animals at will or free to do so *because* their behavior can be explained mechanically." Moreover, the favorite authority of Descartes's accusors, Montaigne, on whom both Singer and Midgley also uncritically rely, was a gullible romantic of breathtaking ignorance, eager to take the most fanciful folktales of animal mentality at face value, and not at all interested in *finding out*, as Descartes himself was, how animals actually work!

Much the same attitude is common today. There is a curious tolerance of patent inconsistency and obscurantism and a bizarre one-sidedness in the treatment of evidence regarding animal minds. Elizabeth Marshall Thomas writes a book, *The Hidden Life of Dogs* (1993), which mixes acute observation and imaginative hypothesis-formulation with sheer fantasy, and in the generally favorable welcome the book receives, few if any point out that it is irresponsible, that she has polluted her potentially valuable evidence with well-meant romantic declarations that she could not have any defensible grounds for believing. If you want to *believe* in the consciousness of dogs, her poetry is just the ticket. If you want to *know* about the consciousness of dogs, you have to admit that although she raises many good questions, her answers are not to be trusted. That is not to say that she is wrong in all her claims, but that they *just will not do* as answers to the questions, not if we really want to know the answers.

A forlorn hope, some say. Certain questions, it is said, are quite beyond science at this point (and perhaps forever). The cloaks of mystery fall conveniently over the very issues that promise (or threaten) to shed light on the *grounds* for our moral attitudes toward different animals. Again, a curious asymmetry can be observed. We do not require absolute, Cartesian certainty that our fellow human beings are conscious—what we require is what is aptly called *moral* certainty. Can we not have the same moral certainty about the experiences of animals? I have not yet seen an argument by a philosopher to the effect that we cannot, with the aid of science, establish facts about animal minds with the same degree of moral certainty that satisfies us in the case of our own species. So whether or not a case has been made for the "in principle" mystery of consciousness (I myself am utterly unpersuaded by the arguments offered to date), it is a red herring. We can learn enough about animal consciousness to settle the questions we have about our responsibilities. The moral agenda about animals is important, and for that very reason it must not be permitted to continue to deflect the research, both empirical and conceptual, on which an informed ethics could be based.

A striking example of one-sided use of evidence is Thomas Nagel's famous paper "What is it Like to be a Bat?" (1991). One of the rhetorical peculiarities of Nagel's paper is that he chose bats and went to the trouble to relate a *few* of the fascinating facts about bats and their echolocation, because, presumably, those hard-won, third-person-perspective scientific facts tell us *something* about bat consciousness. What? First and least, they support our conviction that bats *are* conscious. (He did not write a paper called "What is it Like to be a Brick?") Second, and more important, they support his contention that bat consciousness is very unlike ours. The rhetorical peculiarity—if not outright inconsistency—of his treatment of the issue can be captured by an obvious question: if a few such facts can establish *something* about bat consciousness, would more such facts not establish more? He has already relied on "objective, third-person" scientific investigation to establish (or at least render rationally credible) the hypothesis that bats are conscious, but not in just the way we are. Why wouldn't further such facts be able to tell us in exactly what ways bats' consciousness isn't like ours, thereby telling us what it *is* like to be a bat? What kind of fact is it that only works for one side of an empirical question?

The fact is that we all do rely, without hesitation, on "third-person" behavioral evidence to support or reject hypotheses about the consciousness of animals. What else, after all, could be the source of our "pretheoretical intuitions"? But these intuitions in themselves are an untrustworthy lot, much in need of reflective evaluation. For instance, do you see "sentience" or "mere discriminatory reactivity" in the Venus Fly Trap, or in the amoeba, or in the jellyfish? What more than mere discriminatory reactivity—the sort of competence many robots exhibit—are you *seeing* when you *see* sentience in a creature? It is, in fact, ridiculously easy to induce powerful intuitions of not just sentience but full-blown consciousness (ripe with malevolence or curiosity or friendship) by exposing people to quite simple robots *made to move in familiar mammalian ways at mammalian speeds.*

Cog, a delightfully humanoid robot being built at MIT, has eyes, hands, and arms that move the way yours do—swiftly, re- laxedly, compliantly (Dennett, 1994). Even those of us working on the project, knowing full well that we have not even *begun* to program the high level processes that might arguably endow Cog with consciousness, get an almost overwhelming sense of being in the presence of another conscious observer when Cog's eyes still quite blindly and stupidly follow one's hand gestures. Once again, I plead for symmetry: when you acknowledge the power of such elegant, lifelike motions to charm you into an illusion, note that it ought to be an open question, still, whether you are also being charmed by your beloved dog or cat or the noble elephant. Feelings are too easy to provoke for them to count for much here.

If behavior, casually observed by the gullible or generous- hearted, is a treacherous benchmark, might composition—

material and structure—provide some important leverage? History offers a useful perspective on this question. It was not so long ago—Descartes's day—when the hypothesis that a material brain by itself could sustain consciousness was deemed preposterous. Only immaterial souls could *conceivably* be conscious. What was inconceivable then is readily conceivable now. Today, we can readily conceive that a brain, without benefit of immaterial accompanists, can be a sufficient seat of consciousness, even if we wonder just how this could be. This is surely a *possibility* in almost everybody's eyes, and many of us think the evidence for its truth mounts close to certainty. For instance, few if any today would think that the "discovery" that, say, lefthanders don't have immaterial minds but just brains would show unmistakably that they are just zombies.

Unimpressed by this retreat, some people today baulk at the *very idea* of silicon consciousness or artifactual consciousness, but the reasons offered for these general claims are unimpressive to say the least. It looks more and more as if we will simply have to look at what entities—animals in this case, but also robots and other things made of nonstandard materials—*actually can do*, and use that as our best guide to whether animals are conscious and, if so, why and of what.

I once watched with fascination and, I must admit, disgust while hundreds of vultures feasted on a rotting elephant carcass in the hot sun of a June day in Kenya. I found the stench so overpowering that I had to hold my nose and breath through a kerchief to keep from gagging, all the time keeping my distance, but there were the vultures eagerly shouldering each other aside and clambering inside the carcass for the tastiest morsels. (I will spare you the most mind-boggling details.) Now I am quite confident, and I expect you agree with me, that I was thereby given very good evidence that those vultures do not share my olfactory quality space. In fact, as I have subsequently learned, these Old World vultures, unlike their rather distant New World cousins, do not rely on olfaction at all; they use their keen eyesight to spot carrion.

The peculiar nauseating odors of rotting carrion, carried by such well-named amines as *cadaverine* and *putrescine*, are attractants to the New World turkey vultures (*Cathartes aura*), however, and the presumed explanation is that in the New World these birds evolved in an ecology in which they hunted for food hidden under a canopy of trees, which diminished the utility of vision and heightened the utility of olfaction. David Houston (1986) has conducted experiments using fresh, ripe, and very-ripe chicken carcasses, hidden from sight in the forests of a Panamanian island, to titrate the olfactory talents of turkey vultures. So we're making progress; we now know—to a moral certainty—something about the difference between what it is like to be an African vulture and what it is like to be a Central American turkey vulture.

So let's go on. What does a rotting chicken carcass smell like to a turkey vulture? At first blush it may seem obvious that we can confidently set aside the philosophers' problem of other minds in this instance and assume, uncontroversially, that these vultures rather go in for the smell of carrion. Or does anybody suppose that vultures might be heroic martyrs of the scavenger world, bravely fighting back their nausea while they perform their appointed duties?

Here, it seems, we correct one extrapolation from our own case by another: we dismiss our imputation to them of our own revulsion by noting their apparent *eagerness*—as revealed by their behavior. When we exhibit such eagerness, it is because we *like* something, so they must like what they are doing and feeling. Similarly, we do not worry about the poor seal pups on their ice floe, chilling their little flippers. We would be in agony, lying naked on the ice with the wind blowing over us, but they are designed for the cold. They are not shivering or whimpering, and indeed they exhibit the demeanor of beasts who could not be more content with their current circumstances—home sweet home.

"But wait!" says the philosopher. "You are being awfully sloppy in these everyday attributions. Let's consider what is *possible in principle.* Vulture revulsion is possible in principle, is it not? You would not make their observed behavior *criterial* of pleasure, would you? Are you some benighted *behaviorist?* The suggestion that it makes no sense for vultures to be disgusted by their designated diet is nothing but Panglossian optimism. Perhaps vultures have been misdesigned by evolution; perhaps vulture ancestors found themselves in a sort of evolutionary cul-de-sac, hating the taste and smell of the only food available in their niche, but having no choice but to overcome their distaste and gag it down; perhaps they have since developed a sort of stoic demeanor, and what you have interpreted as gusto is actually desperation!"

Fair enough, I reply. My rush to judgment was perhaps a bit rash, so let's explore further to see whether any supporting evidence can be found for your alternative hypothesis. Here is a relevant fact: turkey vultures are attracted by the smell of one-day-old or two-day-old carcasses, but they ignore older, still more pungent fare. It is conjectured that the toxin level in such flyblown remains eventually is too great even for the toxin-tolerant vultures, who leave them for the maggots. Insects, it is believed, use the onset of these later products of decomposition as their cue that a carcass is sufficiently decomposed to be a suitable site for egg-laying and, hence, maggot formation. This still leaves unanswered the residual question of whether turkey vultures actually *like* the smell of middle-aged carrion. At this point, my knowledge of actual or contemplated vulture research gives out, so I will have to consider some invented possibilities, for the time being. It would be fascinating to discover something along the lines of an incompletely suppressed gag-reflex as part of the normal vulture feeding behavior, or perhaps some traces of approach-avoidance opponent systems tugging away at each other in their brains, a sort of activity not to be found, we might imagine, in the brains of birds with more savory diets. Such discoveries would indeed add real support to your surprising hypothesis, but, of course, they would be just more "behavioral" or "functional" evidence. Once again, a superficially plausible but retrospectively naive or oversimple interpretation would be overthrown by more sophisticated use of behavioral considerations. And you can hardly accept the support of this imagined evidence without agreeing that *not* discovering it would count *against* your alternative and in *favor* of my initial interpretation.

This might be—indeed ought to be—just the beginning of a long and intricate examination of the possible functional interpretations of events in the vultures' nervous systems, but let us cut to the chase, for I imagine our dissenting philosopher to insist in the end, after one or another hypothesis regarding complexities of vulture *reactivity* to carrion had been effectively confirmed, that still no amount of such merely third-personal investigation could ever ("in principle") tell us what carrion *actually smelled like* to a vulture. This would be asserted not on the basis of any further argument, mind you, but just because eventually this is the "intuitive" card that is standardly played.

What I find insupportable in this familiar impasse is the coupling of blithe assertion of consciousness with the equally untroubled *lack of curiosity* about what this assertion might amount to, and how it might be investigated. Leiber (1988) provides a handy scorecard:

> Montaigne is ecumenical in this respect, claiming consciousness for spiders and ants, and even writing of our duties to trees and plants. Singer and Clarke agree in denying consciousness to sponges. Singer locates the distinction somewhere between the shrimp and the oyster. He, with rather considerable convenience for one who is thundering hard accusations at others, slides by the case of insects and spiders and bacteria; they, *pace* Montaigne, apparently and rather conveniently do not feel pain. The intrepid Midgley, on the other hand, seems willing to speculate about the subjective experience of tapeworms . . . Nagel . . . appears to draw the line at flounders and wasps, though more recently he speaks of the inner life of cockroaches.

The list could be extended. In a recent paper, Michael Lockwood (1993) supposes, as so many do, that Nagel's "what it is like to be" formula *fixes a sense of consciousness*. He then says: "Consciousness in this sense is presumably to be found in all mammals, and probably in all birds, reptiles and amphibians as well." It is the "presumably" and "probably" to which I want us to attend. Lockwood gives no hint as to how he would set out to replace these terms with something more definite. I am not asking for certainty. Birds aren't just *probably* warm-blooded, and amphibians aren't just *presumably* air-breathing. Nagel confessed at the outset not to know—or to have any recipe for discovering—where to draw the line as we descend the scale of complexity (or is it the cuddliness scale?). This embarrassment is standardly waved aside by those who find it just obvious that there is something it is like to be a bat or a dog, equally obvious that there is *not* something it is like to be a brick, and unhelpful *at this time* to dispute whether it is like anything to be a fish or a spider. What does it mean to say that it is or it isn't?

It has passed for good philosophical form to invoke mutual agreement here that we know what we're talking about even if we can't explain it yet. I want to challenge this. I claim that this standard methodological assumption has no *clear* pre-theoretical meaning—in spite of its undeniable "intuitive" appeal—and that since this is so, it is ideally suited to play the deadly role of the "shared" intuition that conceals the solution from us. *Maybe* there really is a huge difference between us and all other species in this regard; *maybe* we should consider "radical" hypotheses. Lockwood says "probably" all birds are conscious, but *maybe* some of them—or even all of them—are rather like sleepwalkers! Or what about the idea that there could be unconscious pains (and that animal pain, though real, and—yes—morally important, was unconscious pain)? *Maybe* there is a certain amount of generous-minded delusion (which I once called the Beatrix Potter syndrome) in our bland mutual assurance that as Lockwood puts it, "*Pace* Descartes, consciousness, thus construed, isn't remotely, on this planet, the monopoly of human beings."

How, though, could we ever explore these "maybes"? We could do so in a constructive, anchored way by first devising a theory that concentrated exclusively on *human* consciousness—the one variety about which we will brook no "maybes" or "probablys"—and then *look and see* which features of that account apply to which animals, and why. There is plenty of work to do, which I will illustrate with a few examples—just warm-up exercises for the tasks to come.

In *Moby Dick,* Herman Melville asks some wonderful questions about what it is like to be a sperm whale. The whale's eyes are located on opposite sides of a huge bulk: "the front of the Sperm Whale's head," Melville memorably tells us, "is a dead, blind wall, without a single organ or tender prominence of any sort whatever" (Ch. 76). As Melville notes: "The whale, therefore, must see one distinct picture on this side, and another distinct picture on that side; while all between must be profound darkness and nothingness to him" (Ch. 74).

> Nevertheless, any one's experience will teach him, that though he can take in an indiscriminating sweep of things at one glance, it is quite impossible for him, attentively, and completely, to examine any two things—however large or however small—at one and the same instant of time; never mind if they lie side by side and touch each other. But if you now come to separate these two objects, and surround each by a circle of profound darkness; then, in order to see one of them, in such a manner as to bring your mind to bear on it, the other will be utterly excluded from your contemporary consciousness. How is it, then, with the whale? . . . is his brain so much more comprehensive, combining, and subtle than man's, that he can at the same moment of time attentively examine two distinct prospects, one on one side of him, and the other in an exactly opposite direction?

Melville goes on to suggest that the "extraordinary vacillations of movement" exhibited by sperm whales when they are "beset by three or four boats" may proceed "from the helpless perplexity of volition, in which their divided and diametrically opposite powers of vision must involve them" (Ch 74).

Might these "extraordinary vacillations" rather be the whale's attempt to keep visual track of the wheeling boats? Many birds, who also "suffer" from eyes on opposite sides of their heads, achieve a measure of "binocular" depth perception by bobbing their heads back and forth, giving their brains two slightly different views, and permitting the relative motion of parallax to give them approximately the same depth information we get all at once from our two eyes with their overlapping fields.

Melville assumes that whatever it is like to be a whale, it is similar to human consciousness in one regard: there is a single boss in charge, an "I" or "ego" that either superhumanly distributes its gaze over disparate scenarios, or humanly flicks back and forth between two rivals. But might there be even more radical discoveries in store? Whales are not the only animals whose eyes have visual fields with little or no overlap; rabbits are another. In rabbits there is no interocular transfer of learning! That is, if you train a rabbit that a particular shape is a source of danger by demonstrations carefully restricted to its *left* eye, the rabbit will exhibit no "knowledge" about that shape, no fear or flight behavior, when the menacing shape is presented to its *right* eye. When we ask what it is like to be that rabbit, it appears that at the very least we must put a subscript, *dexter* or *sinister*, on our question in order to make it well-formed.

Now let's leap the huge chasm that separates our cousins, the whale and the rabbit, from a much more distant relative, the snake. In an elegant paper, "Cued and detached representations in animal cognition," Peter Gärdenfors (unpublished) points out "why a snake can't think of a mouse."

> It seems that a snake does not have a central representation of a mouse but relies solely on transduced information. The snake exploits three different sensory systems in relation to prey, like a mouse. To strike the mouse, the snake uses its *visual* system (or thermal sensors). When struck, the mouse normally does not die

immediately, but runs away for some distance. To locate the
mouse, once the prey has been struck, the snake uses its sense of
smell. The search behavior is exclusively wired to this modality.
Even if the mouse happens to die right in front of the eyes of the
snake, it will still follow the smell trace of the mouse in order to
find it. This unimodality is

> particularly evident in snakes like boas and pythons, where
> the prey often is held fast in the coils of the snake's body,
> when it e.g. hangs from a branch. Despite the fact that the
> snake must have ample proprioceptory information about the
> location of the prey it holds, it searches stochastically for it, all
> around, only with the help of the olfactory sense organs
> (Sjölander, 1993, p. 3).

Finally, after the mouse has been located, the snake must find
its head in order to swallow it. This could obviously be done with
the aid of smell or sight, but in snakes this process uses only
tactile information. Thus the snake uses three separate modali-
ties to catch and eat a mouse.

Can we talk about what the snake *itself* "has access" to, or just
about what its various parts have access to? Is any of that
obviously sufficient for consciousness? The underlying pre-
sumption that Nagel's "what is it like" question makes sense at
all, when applied to a snake, is challenged by such possibilities.

I have argued at length, in *Consciousness Explained* (1991),
that the sort of informational unification that is the most
important prerequisite for *our* kind of consciousness is not
anything we are born with, not part of our innate "hard-
wiring," but in surprisingly large measure an artifact of our
immersion in human culture. What that early education
produces in us is a sort of benign "user-illusion"—I call it the
Cartesian Theater: the illusion that there is a place in our
brains where the show goes on, towards which all perceptual
"input" streams, and whence flow all "conscious intentions" to
act and speak. I claim that other species—and human beings
when they are newborn—simply *are not beset* by the illusion of
the Cartesian Theater. Until the organization is formed, there
is simply no user in there to be fooled. This is undoubtedly a
radical suggestion, hard for many thinkers to take seriously,
hard for them even to *entertain*. Let me repeat it, since many

critics have ignored the possibility that I mean it—a misfiring of their generous allegiance to the principle of charity.

In order to be conscious—in order to be the sort of thing it is like something to be—it is necessary to have a certain sort of informational organization that endows that thing with a wide set of cognitive powers (such as the powers of reflection and re-representation). This sort of internal organization does not come automatically with so-called "sentience." It is not the birthright of mammals or warm-blooded creatures or vertebrates; it is not even the birthright of human beings. It is an organization that is swiftly achieved in one species, ours, and in no other. Other species no doubt achieve *somewhat similar* organizations, but the differences are so great that most of the speculative translations of imagination from our case to theirs *make no sense*.

My claim is not that other species lack our kind of *self*-consciousness, as Nagel (1991) and others have supposed. I am claiming that what must be added to mere responsivity, mere discrimination, to count as consciousness *at all* is an organization that is not ubiquitous among sentient organisms. This idea has been dismissed out of hand by most thinkers.[1] Nagel, for instance, finds it to be a "bizarre claim" that "implausibly implies that babies can't have conscious sensations before they learn to form judgments about themselves." Lockwood is equally emphatic: "Forget culture, forget language. The mystery begins with the lowliest organism which, when you stick a pin in it, say, doesn't merely react, but actually *feels* something."

Indeed, that is where the *mystery* begins if you insist on starting *there*, with the assumption that you know what you mean by the contrast between merely reacting and actually feeling. And the mystery will never stop, apparently, if that is where you start.

In an insightful essay on bats (and whether it is like anything to be a bat), Kathleen Akins (1993) pursues the sort of detailed investigation into functional neuroscience that Nagel eschews, and she shows that Nagel is at best ill-advised in simply *assuming* that a bat *must* have a point of view. Akins sketches a few of the many different stories that can be told from the vantage point of the various subsystems that go to making up a bat's nervous system. It is tempting, on learning these details, to ask ourselves "and where in the brain does the bat *itself* reside," but this is an even more dubious question in the case of the bat than it is in our own case. There are many parallel stories that could be told about what goes on in you and me. What gives one of those stories about *us* pride of place at any one time is *just this*: it is the story you or I will tell if asked (to put a complicated matter crudely).

When we consider a creature that isn't a teller—has no language—what happens to the supposition that one of *its* stories is privileged? The hypothesis that there is one such story that would tell us (if we could understand it) what it is actually like to be that creature dangles with no evident foundation or source of motivation—except dubious tradition. Bats, like us, have plenty of relatively peripheral neural machinery devoted to "low level processing" of the sorts that are routinely supposed to be entirely unconscious in us. And bats have no machinery analogous to our machinery for issuing public protocols regarding their current subjective circumstances, of course. Do they then have some *other* "high level" or "central" system that plays a privileged role? Perhaps they do and perhaps they don't. Perhaps there is no role for such a level to play, no room for any system to perform the dimly imagined task of elevating merely unconscious neural processes to consciousness. After all, Peter Singer has no difficulty supposing that an insect might keep its act together without the help of such a central system. It is an open empirical question, or rather, a currently unimagined and complex set of open empirical questions, what sorts of "high levels" are to be found in which species under which conditions.

Here, for instance, is one possibility to consider: the bat

lacks the brain-equipment for *expressing* judgments (in language), but the bat may nevertheless have to *form* judgments (of some inarticulate sort), in order to organize and modulate its language-free activities. Wherever these inarticulate judgment-like things happen is where we should look for the bat's privileged vantage point. But this would involve just the sort of postulation about sophisticated judgments that Nagel found so implausible to attribute to a baby. If the distinction between conscious and unconscious has nothing to do with anything sophisticated like judgment, what else could it involve?

Let us return to our vultures. Consider the hypothesis that for all I could ever know, rotting chicken carcass smells to a turkey vulture exactly the way roast turkey smells to me. Can science shed any light, pro or con, on this hypothesis? Yes, it can almost effortlessly refute it: since *how roast turkey tastes to me* is composed (and exhausted) by the huge set of reactive dispositions, memory effects, and so on, and so forth, that are detectable in principle in my brain and behavior, and since many of these are utterly beyond the machinery of any vulture's brain, it is flat impossible that anything could smell to a vulture the way roast turkey smells to me.

Well, then, what *does* rotting chicken smell like to a turkey vulture? (Exactly?) How patient and inquisitive are you prepared to be? We can uncover the corresponding family of reactive dispositions in the vulture by the same methods that work for me, and as we do, we will learn more and more about the no doubt highly idiosyncratic relations a vulture can form to a set of olfactory stimuli. But we already know a lot that we *won't* learn. We will never find a vulture being provoked by those stimuli to wonder, as a human being might, whether the chicken is not just slightly *off* tonight. And we won't find any amusement or elaborate patterns of association or Proustian reminiscence. Am I out in front of the investigations here? A little bit, but note what kind of investigations they are. It turns out that we end up where we began: analyzing patterns of behavior (external and internal—but not "private"), and attempting to interpret them in the light of evolutionary hypotheses regarding their past or current functions.

The very idea of there being a dividing line between those creatures "it is like something to be" and those that are mere "automata" begins to look like an artifact of our traditional presumptions. I have offered (Dennett, 1991) a variety of reasons for concluding that in the case of adult human consciousness there is no principled way of distinguishing when or if the mythic light bulb of consciousness is turned on (and shone on this or that item). Consciousness, I claim, even in the case we understand best—our own—is not an all-or-nothing, on-or-off phenomenon. If this is right, then consciousness is not the sort of phenomenon it is assumed to be by most of the participants in the debates over animal consciousness. Wondering whether it is "probable" that all mammals have *it* thus begins to look like wondering whether or not any birds are *wise* or reptiles have *gumption*: a case of overworking a term from folk psychology that has losts its utility along with its hard edges.

Some thinkers are unmoved by this prospect. They are still unshakably sure that consciousness—"phenomenal" consciousness, in the terms of Ned Block (1992, 1993, 1995, forthcoming)—*is* a phenomenon that is either present or absent, rather as if some events in the brain glowed in the dark and the rest did not.[2] Of course, if you simply will not contemplate the hypothesis that consciousness might turn out *not* to be a property that thus sunders the universe in twain, you will be sure that I must have overlooked consciousness altogether. But then you should also recognize that you maintain the mystery of consciousness by simply refusing to consider the evidence for one of the most promising theories of it.

Postscript: Pain, Suffering, and Morality

In the discussion following my presentation at the conference at the New School, attention was focused on a question about animal consciousness that is not explicitly addressed above: according to my model, how would one tell which animals were capable of pain or suffering (or both)? Drawing on the presentations and discussions later in the conference, I offer here an oversimplified sketch of the direction my theory recommends for answering this question.[3]

The phenomenon of pain is neither homogeneous across species nor simple. We can see this in ourselves, by noting how unobvious the answers are to some simple questions. Are the "pains" that usefully prevent us from allowing our limbs to assume awkward, joint-damaging positions while we sleep experiences that require a "subject" (McGinn, 1995), or might they be properly called unconscious pains? Do they have moral significance in any case? Such body-protecting states of the nervous system might be called "sentient" states without thereby implying that they were the experiences of any self, any ego, any subject. For such states to matter—whether or not we call them pains or conscious states or experiences—there must be an enduring, *complex* subject *to whom* they matter because they are a source of suffering. Snakes (or parts of snakes!) may feel pain—depending on how we choose to define that term—but the evidence mounts that snakes lack the sort of over-arching, long-term organization that leaves room for significant suffering. That does not mean that we ought to treat snakes the way we treat worn out tires, but just that concern for their suffering should be tempered by an appreciation of how modest their capacities for suffering are.

While the distinction between pain and suffering is, like most everyday, nonscientific distinctions, somewhat blurred at the edges, it is, nevertheless, a valuable and intuitively satisfying mark or measure of moral importance. When I step on your toe, causing a brief but definite (and definitely conscious) pain, I do you scant harm—typically none at all. The pain, though intense, is too brief to matter, and I have done no long-term damage to your foot. The idea that you "suffer" for a second or two is a risible misapplication of that important notion, and even when we grant that my causing you a few seconds pain may irritate you a few more seconds or even minutes—especially if you think I did it deliberately—the pain itself, as a brief, negatively-signed experience, is of vanishing moral significance. (If in stepping on your toe I have interrupted your singing of the aria, thereby ruining your operatic career, that is quite another matter.)

Many discussions seem to assume tacitly: (1) that suffering and pain are the same thing, on a different scale; (2) that all pain is "experienced pain"; and (3) that "amount of suffering" is to be calculated ("in principle") by just adding up all the pains (the awfulness of each of which is determined by duration-times-intensity). These assumptions, looked at dispassionately in the cold light of day—a difficult feat for some partisans—are ludicrous. A little exercise may help: would you exchange the sum total of the suffering you will experience during the next year for one five-minute blast of no doubt excruciating agony that summed up to the "same amount" of total pain-and-suffering? I certainly would. In fact, I would gladly take the bargain even if you "doubled" or "quadrupled" the total annual amount—just so long as it would be all over in five minutes. (We are assuming, of course, that this horrible episode does not kill me or render me insane—after the pain is over—or have other long-term effects that amount to or cause me further suffering; the deal was to pack all the suffering into one jolt.) I expect anybody would be happy to make such a deal. But it doesn't really make sense. It implies that the benefactor who provided such a service gratis to all, *ex hypothesi*, would be doubling or quadrupling the world's suffering—and the world would love him for it.

It seems obvious to me that something is radically wrong with the assumptions that permit us to sum and compare suffering in any such straightforward way. But some people think otherwise; one person's *reductio ad absurdum* is another's counter-intuitive discovery. We ought to be able to sort out these differences, calmly, even if the best resolution we can reasonably hope for is a recognition that some choices of perspective are cognitively impenetrable.

Notes

[1] Two rare—and widely misunderstood—exceptions to this tradition are Julian Jaynes (1976) and Howard Margolis (1987), whose cautious observations survey the field of investigation I am proposing to open:

> A creature with a very large brain, capable of storing large numbers of complex patterns, and capable of carrying through elaborate sequences of internal representations, with this capability refined and elaborated to a very high degree, would be a creature like you and me. Somehow, as I have stressed, consciousness conspicuously enters the scheme at this point of highly elaborate dynamic internal representations. Correctly or not, most of us find it hard to imagine that an insect is conscious, at least conscious in anything approximating the sense in which humans are conscious. But it is hard to imagine that a dog is not conscious in at least something like the way an infant is conscious (Margolis, 1987, p. 55).

[2] John Searle also holds fast to this myth. See, for example, Searle, 1992, and my review, 1993.

[3] For a more detailed discussion, see "Minding and Mattering," pp. 448–454 in Dennett, 1991.

References

Akins, Kathleen, "What is it Like to be Boring and Myopic?" in Bo Dahlbom, ed., *Dennett and his Critics* (Oxford: Blackwells, 1993).
Block, Ned, "Begging the question against phenomenal conscious-

ness" (commentary on Dennett and Kinsbourne), *Behavioral and Brain Sciences*, 15 (1992): 205–6.

Block, Ned, "Review of Daniel Dennett, *Consciousness Explained*," *Journal of Philosophy*, 90 (1993): 181–93.

Block, Ned, "On a Confusion about a Function of Consciousness," *Behavioral and Brain Sciences*, 18 (1995).

Block, Ned, "What is Dennett's Theory a Theory of?" in *Philosophical Topics*, Special issue on the work of Dennett, forthcoming.

Dennett, Daniel, *Consciousness Explained* (Boston: Little Brown, 1991).

Dennett, Daniel, "Review of John Searle, *The Rediscovery of Consciousness*," *Journal of Philosophy*, 90 (1993): 193–205.

Dennett, Daniel, "The practical requirements for making a conscious robot," *Phil. Trans. R. Soc. Lond.* A 349 (1994): 133–46.

Gärdenfors, Peter, "Cued and detached representations in animal cognition," unpublished.

Houston, David C., "Scavenging Efficiency of Turkey Vultures in Tropical Forest," *The Condor*, 88 (1986): 318–23, Cooper Ornithological Society.

Jaynes, Julian, *The Origins of Consciousness in the Breakdown of the Bicameral Mind* (Boston: Houghton Mifflin, 1976).

Leiber, Justin, "'Cartesian Linguistics?'" *Philosophia*, 118 (1988): 309–46.

Lockwood, Michael, "Dennett's Mind," *Inquiry*, 36 (1993): 59–72.

Margolis, Howard, *Patterns, Thinking, and Cognition* (Chicago: University of Chicago Press, 1987).

McGinn, Colin, "Animal Minds, Animal Morality," *Social Research* 62:3 (1995).

Nagel, Thomas, "What we have in mind when we say we're thinking," (Review of *Consciousness Explained*), *Wall Street Journal* (November 7, 1991).

Searle, John, *The Rediscovery of Consciousness* (Cambridge, MA: MIT Press, 1992).

Sjölander, S., "Some cognitive breakthroughs in the evolution of cognition and consciousness, and their impact on the biology of language," *Evolution and Cognition*, 3 (1993): 1–10.

Primate Language
and Cognition:
Common Ground*

Duane Rumbaugh

Research of the past decade has served to underscore the close psychological relationship between humans, chimpanzees, and the other great apes. In his evolutionary theory, Darwin (1860, 1871) posited both psychological and biological continuities between animals and humans. Although the evidence for biological continuity has been strong for decades, the evidence necessary for affirmation of psychological continuity is recent.

Descartes and Animals

The absence of strong evidence for the processes of intelligence and language in animals has permitted the earlier proclamations of Descartes ([1637]1956) in the mid-1600s to spawn the belief of discontinuity. Because animals could not

* Research and preparation of this paper was supported by grant HD-06016 from the National Institute of Child Health and Human Development, by grant NAG2–438 from the National Aeronautics and Space Administration, and by additional support from the College of Art & Sciences, Georgia State University. The author thanks William A. Hillix, David A. Washburn, Rose A. Sevcik, and E. Sue Savage-Rumbaugh for critical comments on this manuscript. He also thanks William D. Hopkins, Steve Suomi, Mary P. Williams, and Leslie Burke for data on capuchin and rhesus monkeys that were pooled with other data for their species and included in Figure 1; Ms. Andrea Clay for the primate figures in Figure 1; John Gulledge for assistance in data; Judith Sizemore and Tawanna Tookes for manuscript preparation; and all of the LRC faculty and staff for their various contributions to the framework herein presented. The views advanced herein are the responsibility of the author.

talk sensibly, it was easy for people to conclude that animals had no sensibilities. They came to be viewed as "subhuman" or "infrahuman"—as not having made the grade to the exalted status of human. They were held to be without reason, thought, affect, intelligence, and language.

Humans, but not animals, were able to think because they had rational souls. Accordingly, they could be held accountable for their actions—and for their sins. When they sinned, God inflicted pain. But God would surely spare all animals from ever experiencing pain. After all, without souls, they were not responsible for their action—and, thus, could not sin. Despite the fact that the rationale for Descartes' beast-machine concept has faded with the centuries, the concept is still very much alive.

Throughout history, humans have sought lines of argument and fact that would make us unique from animals. Different we are, but not so totally different as some would think.

The Null Hypothesis, Darwin, and Continuity

Adherence to the view that our species has a totally unique psychology is to misuse the Null Hypothesis. The Null Hypothesis holds that there is "no difference" between us and animals. Only with evidence sufficient to reject that hypothesis should one argue that "a real difference" exists—one not attributable to chance. By contrast, the general tendency has been to begin with the conclusion that real differences *do* exist between the psychology of humans and animals, and that the onus is upon the researcher to come up with proof to the contrary!

Darwin basically advanced the Null Hypothesis as he argued for continuity from animals to human. He did not go so far, though, as to argue for identity of process. Apes are apes, apes are not humans; but because of the very close genetic similarity which they share (>98% shared DNA; Sarich, 1983; Andrews and Martin, 1987; Sibley and Ahlquist, 1987), there are neurobiological grounds to anticipate some degree of similarity in their psychology.

That said, attention is now directed to behavioral research of recent years, both here and abroad, that has produced an abundance of evidence for rejection of the Cartesian beast-machine concept and for support of Darwin's postulations of continuity. Animals have been found capable of complex communication, symbolizing capacities, rule-learning, number learning, and even language.

Comparative Cognition

The Language Research Center's history rests in our long-standing interests in the parameters of human intelligence that might be traced to our nearest living relatives—the great and lesser apes, the monkeys of the New and Old Worlds, and the prosimians (Napier and Napier, 1994). One tactic for this study posited a relationship between *Transfer of Learning* and *Brain Complexity*.

Transfer of Learning

Transfer of learning is a very important element in generalized competence. Most learning takes place in relatively specific contexts—a student studies biology at his/her desk; a teenager learns how to drive a Chevrolet; a pilot learns to fly a single-engine plane; a student learns sculpting in a studio; and so on.

Transfer of learning can influence subsequent learning (for example, acquiring information about something) and performance (for example, doing a task, taking a test, and so on) in two basic ways. First, the effect can be a facilitating or *positive* one. The learning or performing of one task might facilitate the learning of a second task. Learning to drive a stick-shift Chevrolet entails many principles that can be applied to learning to drive a variety of other cars. One "knows something quite general about driving" after the stick shift car has been mastered. The learning of one subject in biology can, and should, facilitate the learning of another, and so on. On the other hand, the effect can be an impediment or *negative* one, in part if not wholly. Thus, learning to drive in a car with a stick shift can have a negative effect on subsequently driving an automatic shift car. For example, one's left foot may flail around when one attempts to stop a car that has no clutch. Use of the feet is relevant to the stopping of most vehicles, but the specifics of what they are to do can be either positive or negative. All of us have had problems with vehicles where controls are either absent or relocated compared to the car we drive most frequently (that is, controls for turn-signals, headlights, windshield wipers, and so forth). Pilots become generally competent (*positive* transfer) in learning to fly a wide variety of planes, but past expectations about the locations of specific vital controls (that is, the throttle, the flaps control, and so on) can lead to disaster (*negative* transfer) in crisis situations where there is no time to analyze what to do to correct the problem. Finally, it is acknowledged that learning one thing, such as French, might have absolutely no transfer of learning relevance to flying a plane or driving a car.

Generally, persons who are facile at transferring even small amounts of learning have a marked advantage in new situations and are viewed as highly intelligent and clever. They are able to see common principles and logical as well as functional dimensions of relationships among diverse situations and topics. For them, for example, the learning of Latin facilitates the learning of many other languages and academic subjects. For others, Latin has no relevance to anything!

Primate Brains and Transfer of Learning

Interestingly, primate species differ both markedly and rather systematically in the ease with which they can learn *and* transfer their learning. The size and complexity of primate brains can result in comparable amounts of learning being transferred *positively* for some species (notably the larger ones) and *negatively* for other species (notably the smaller ones). The differences in transfer are associated with average brain size per species and are not reliably associated with differences in brain size among individual specimens with a species, except insofar as brain size and complexity increases with growth and development to maturity.

Transfer of learning and skills is basic to the development of competence in both humans and animals. Accordingly, great emphasis has been placed upon the assessment of transfer-of-learning skills in studies of primate intelligence. Both conceptually and procedurally it can be studied in comparable ways across species, for we can assess the efficiency and effect of transfer relative to what the individual is known to have learned. Several of our own studies, designed to relate primate brain evolution to transfer-of-learning competence, have been facilitated by the development and use of the Transfer Index (TI) (Rumbaugh and Pate, 1984).

The TI is a procedure designed to afford equitable assessments of primates' complex learning and transfer skills. Its design attenuates artifactual differences between species' learning and performance that might be produced due to their differences of size, manual dexterity, processes of attention, and so on. Brain complexity was estimated according to Jerison's (personal communication) estimations of the "extra" brain and neurons afforded primates due to the process of encephalization. Encephalization refers to an enlargement of the brain's size/weight relative to the body's weight beyond that which would be afforded by allometric relationships that characterize a variety of proportions and weights of the body's anatomy. In other words, a certain amount of brain is needed just to service the tissue needs, processes, and basic behaviors of the body.

Thus, within limits, allometric relationships imply that large mammals will have larger brains than small mammals. But apart from that, *encephalization* refers to the tendency for the brains, notably those of primate species and in particular those of the great apes and our own, to be extraordinarily large. It is this "extra" brain size and complexity which surely provides the basis for intelligence both in primates and humans, who also are a primate form.

Research with over 121 primate subjects of several species relates their transfer-of-learning skills, as assessed by the TI, to the dimension of brain complexity in a very interesting manner: a qualitative shift in those skills was documented in interaction with the amount of learning established prior to tests of transfer.

The amount of learning prior to test was defined by achievement standards that required the subjects to learn sets of visual discrimination learning problems to two levels of correctness—67% and 84% choices correct—to prepare them for tests of their transfer skills. These achievement standards, in effect, required that the subjects achieve specified numbers of choices correct within specified numbers of training trials prior to test for transfer of learning. As one might expect, the 84% level required that more choices be correct than did the 67% level.

The learning situation in which these performance schedules were applied was a series of two-choice, object-discrimination problems. Each problem consisted of a pair of objects that clearly differed from one another in size, color, and form. One object of each pair, if chosen, resulted in food, whereas choice of the other object resulted in no food. Once the subject achieved the level of mastery required (either 67% or 84%) for a given set of problems, the test of transfer of learning was given.

This transfer test consisted in "switching" the correctness of the objects: the object that was associated with food, if chosen, now became "incorrect," and the object which had not been associated with food now became correct. The TI computes the degree to which transfer of learning was positive or negative—or absent—on the basis of performance on a series of problems.

Figure 1 portrays the change in test performance (percentage responses correct) obtained as the learning standard was increased from the 67% to the 84% level. And it is here that a very important finding must be clarified.

As the amount of learning was increased, the prosimians and

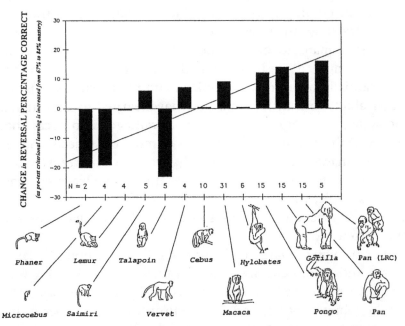

Figure 1. The enhancement of transfer-of-learning in relation to brain complexity of primate (N = 121) is portrayed. The vertical axis quantifies the *change* in the percentage of responses that are correct as a result of the per-transfer test learning criterion being increased from 67% to 84% correct. Each point on the baseline is for a particular species, except for the one (right end) that is for 5 language-competent chimpanzees and bonobos of the Language Research Center. Their enhanced performance is believed to be due to the enrichment afforded by their research participation and their language skills. (See Acknowledgments for contributions of unpublished data from others. See Cooper [1980] and De Lillo & Visalberghi [1994] as sources for data on *Phaner, Microcebus, & Capuchin*. Other unpublished capuchin data was contributed by Hopkins and Suomi.)

smaller monkeys generally tended to do worse (for example, they had fewer choices correct on transfer tests after being trained to the 84% than to the 67% level). For them, increasing the amount learned handicapped their performance on transfer tests and produced *negative* transfer effects with these procedures. By contrast, the great apes and even the larger and more complex-brained monkeys improved in their transfer tests as the level of per-test learning was increased. For them, the more they knew, the better they did on transfer tests—the greater their *positive* transfer of learning.

I again emphasize that the ability to learn and to transfer one's learning to a leveraged advantage is generally held among humans to be evidence of "smartness," not "dumbness." Thus, it is here argued that the great apes are substantially smarter than monkeys and prosimians because of their ability to transfer what they have learned to a leveraged advantage in transfer tests. That advantage is so great that many of the ape and larger species of monkey subjects (that is, Rhesus macaques) do substantially better on the transfer test trials than they had been allowed to during the learning trials (for example, when trained to, say, the 67% level, they would be 78% correct on transfer tests—a value higher than required prior to test).

For primates, a large body means a disproportionately large brain and extra neurons, which, in turn, correlate highly with the values obtained from the y-axis of the figure shown (*extra* brain volume, $r = .82$; and *extra* neurons, $r = .79$, respectively). Average body weights and brain weights per species correlated highly with each other (0.96); body weight correlated highly with transfer-of-learning proficiency (0.88); and brain weight per species also correlated highly with transfer skills (0.84).

Jerison's (1991) Encephalization Coefficient, which relates brain weight to body weight, is only generally correlated with the body weights of the primate species here used. For example, although both the diminutive squirrel monkey and talapoin have higher Encephalization Coefficients than does the massive gorilla, they are substantially below the gorilla in their complex learning and transfer skills. Consequently, that measure did not correlate significantly with transfer-of-training skills significantly.

Simian Intelligence

Before addressing the research which has perhaps most firmly cemented the continuum that relates the psychology of apes and humans—that is, *language*—let us consider the smartness of monkeys. Monkeys have an order of smartness or intelligence that is substantially below that of the apes, but even so at least the larger-sized ones are impressive.

Research at our Center defined their ability to use a joystick in a battery of complex tasks designed to measure their learning, memory, vigilance, eye-hand coordination, planning, relative-value judgments, and so on in relation to physiological changes (Rumbaugh, Richardson, Washburn, Savage-Rumbaugh, and Hopkins, 1989; Washburn, Hopkins, and Rumbaugh, 1989; Washburn and Rumbaugh, 1992a,b). For our purposes, our review of findings is limited to the following:

- Rhesus' performance on the TI and the number of trials which they work each day are the best predictors of training success on our comprehensive battery of tasks referenced above.
- Rhesus prefer to work on tasks of their choice rather than passively to receive incentives for "free."
- Rhesus respond more quickly and accurately when they can choose tasks on which to work than when those same tasks are assigned by experimental procedures.
- Rhesus are "super-learners" in that they readily learned the relative values of the numeral set 0–9 and induced their comprehensive relationships of relative values. They did this in a learning situation in which they did not have to choose the larger number in order to get pellets and then were tested on a series of novel pairs. Even on these novel test trials, they

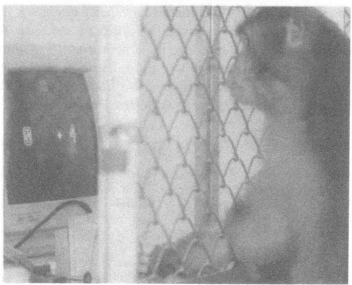

Figure 2a. and 2b. A rhesus monkey (*Macaca mulatta*) works on a numeric task portrayed on the monitor of the Language Research Center's Computerized Test System (LRC-CTS).

reliably picked the number of greater pellet value (Washburn and Rumbaugh, 1991).

- Handedness (for example, reliable use of either the right or left hand rather than both) for joystick manipulation is strongly established in Rhesus, and right-handed monkeys (for example, those who like most humans in writing do best if they use their right hand) tend to be more facile learners than left-handed monkeys.

- Rhesus monkeys manifest characteristics of selective attention similar to those of humans. For instance, they are sensitive to the Stroop-like interference when quantities, between which they are to differentiate, are comprised of Arabic numbers (Washburn, 1994). (The Stroop effect is exemplified when, for example, a human subject is asked to name the *color* of the ink in which various words are printed where the words themselves are *names* of colors—but where there is a lack of congruence between the word and the color in which it is printed. Thus, it is generally difficult to say "green" when it refers to the color of ink used to print the word "red.") When required to choose the array with the *greater* quantity of items, Rhesus monkeys do less well if that array is made up of numerals that stand for relatively small quantities (that is, 2s or 3s) compared to the other array that is made of numerals that are relatively higher (that is, 7s or 8s).

Apes and Language

But what of the apes' language abilities? It was just a few years ago that evidence then available in support of apes' capacity for language was contested by others as attributable to inadvertent cuing by the apes' mentors.[1] Even then, however, we knew that two of our chimpanzee subjects, Sherman and Austin, could look at their word-lexigrams, each being the name of either a specific food they ate or of a specific tool they used to extricate food from puzzle boxes, and categorize them accurately through use of two other word-lexigrams—one standing for "food" and the other for "tool." Their competence in doing thus, in controlled tests, documented their basic capacity for semantics—the meanings of symbols or words (Savage-Rumbaugh, 1986).

But recently our Center has produced findings that even we would not have thought possible 10 years ago. The basic finding (Savage-Rumbaugh, Murphy, Sevcik, Brakke, Williams, and Rumbaugh, 1993) is that the apes can come to understand even the syntax of human speech, at a level that compares favorably with that of a 2–1/2 year old child—if they are reared from shortly after birth in a language-structured environment. By "language-structured" environment I mean one in which the apes' caretakers talk to the infant apes as though they understand all that was being said—announcements of what is about to happen and descriptions of items selected jointly for attention by the ape and its caretaker.

Speech Comprehension by Apes

Reared in this manner, the infant ape's brain develops in a manner that enables it to acquire language—first through its comprehension and then through its expression, a pattern that characterizes the course of language acquisition in the normal child. The ape's comprehension of spoken words is assessed by whether or not it is able to select the appropriate referent for single words that it hears in controlled experimental situations. Their comprehension of syntax is assessed by their capacity to respond logically to novel sentences of request that they hear.

The research program included a child, Alia, 2–1/2 years old. The first ape was Kanzi, a bonobo—a rare species of chimpanzee (Pan paniscus).

Kanzi's comprehension of over 600 novel sentences of request was very comparable to Alia's; both complied with the requests without assistance on approximately 70% of the sentences. If explanation, comment, or encouragement were offered to the subjects when they appeared hesitant about what was being asked of them, their assessed level of correctness was down-graded. (All requests were unrehearsed.

Figure 3. Kanzi listens to words that he hears through headphones during controlled tests of his speech comprehension.

Personnel in the test room listened to loud music so that they could not hear the requests posed to Kanzi.)

How did Kanzi acquire his comprehension of language (Savage-Rumbaugh and Lewin, 1994)? Not through formal instruction, but rather just by being present while his mother, Matata, was receiving her language instruction—instruction with lexigrams, not speech comprehension. Quite possibly it was because Matata was reared by feral bonobos until about age 6-years that she never benefited reliably from her language instruction. It was as though her cognitive structure had been fully committed to topics associated with life in the forest. Though smart by every measure when out in the 55 acres of forest that surround the Language Research Center, she did not learn readily about language. By contrast, the language-learning abilities of her offspring have been impressive.

The discovery afforded by Kanzi has been corroborated with two other bonobos, Mulika (Sevcik, 1989) and Panbanisha, and to a more limited though significant degree with a common chimpanzee, Panzee (Savage-Rumbaugh, Brakke, and Hutchins, 1992). Panbanisha and Panzee were co-reared for their first four years.

Early Environment, Rearing, and Competence

As exciting as it has been to document the apes' language skills, even more important is the principle derived from their research: it is in the logic-structure of the infants' environments that their complex abilities, competencies, and dimensions of intelligence and expression are formed. Their formation is behaviorally "silent" in that their expression might not occur until the age of 2 years or older (Savage-Rumbaugh, 1991). It is during infancy that important basic vectors of competence are formed (Rumbaugh, Hopkins, Washburn, and Savage-Rumbaugh, 1991).

Earlier studies of the 1960s and '70s made clear that chimpanzees subjected to impoverished rearing even during the first two years of infancy only are both socially and cognitively deficient in their capacity for complex learning and the transfer of learning some 12 years later as young adults (Davenport, Rogers, and Rumbaugh, 1973).

It is significant that appropriate language-structured rearing established in the ape what none of the earlier efforts designed to teach apes specific language skills through use of tutorials could—namely, the ability to comprehend speech and its syntax (Savage-Rumbaugh and Lewin, 1994). Such rearing also serves to support the spontaneous appearance (for example, without specific training) of productive language skills that approximate those of a 1–1/2 year old normal child (Greenfield and Savage-Rumbaugh, 1991, 1993).

And given the opportunity to observe a professional flint-knapper, Kanzi learned about stone tools—their use, value, and means of production (Toth, Schick, Savage-

Rumbaugh, Sevcik, and Rumbaugh, 1993). Over time, he has become a reasonably skilled flint knapper; that is, he makes stone tools and does so with good sense. He assesses his flint chips for sharpness and, quite appropriately, makes larger chips to cut thick cables of rope and smaller chips to cut fine ones.

Apes, Language, and the Human Perspective

Elsewhere (Rumbaugh and Savage-Rumbaugh, 1994) we have presented a perspective of apes and their language skills and how both the research and the controversy surrounding it has taken form across the course of the past several decades. Suffice it to say, first, that the history of behavioral science has documented that it is *very* unlikely that theorists will abandon positions for which they have become known. Major changes in perspectives frequently must await the "second" generation of scientists to incorporate the data and the best of diverse perspectives into their science. Nevertheless, the following are held to be factual and without logical refutation: (1) Apes are capable of learning the meanings and representational use of arbitrary symbols that for them have all the functional properties of words. The symbols stand for things that are not necessarily present in time and space, for activities, for the properties of things (that is, temperatures of drinks, ambient noise levels), for the individual's state (that is, hunger, thirst, sleepiness), for other animates (either ape or human or canine), for places to which they would go, for making comments on activities and recent happenings in the laboratory, and so on. These assertions are based on data obtained from controlled scientific tests, replications of studies with different subjects, as well as affirmed by social communication between humans and apes across decades. The processes whereby symbols, signs, and gestures optimally acquire these *semantic* properties are cultivated during early rearing of the ape—just as is the case for the normal child. Formal, discrete trial training of language skills is relatively ineffective and does not establish the ability to comprehend substantial amounts of human speech. (2) Early rearing can establish in the ape the ability to understand the meanings of human speech—even novel sentences of request. Thus, the apes spontaneously acquire a capacity that is normal for the human child—though not for the ape. (3) Early rearing can establish, first, the ability to understand, to comprehend language, and, second, the ability to employ grammar as does the 1– 1-1/2 year-old child.

Critics who emphasize that the apes cannot do "all" that the normal child does and who emphasize speech production as the *sine qua non* of language err in discounting the significance of highly important findings produced by language research with apes. Research with apes has made it very clear that the basis for language is *comprehension*, not speech. Although speech is a highly efficient, unsurpassed medium for linguistic communication, competence in speech rests primarily in comprehension—and only secondarily in the speech and hearing systems that normal humans enjoy. The bedrock of human language processes are traceable to the great apes. And while the great apes are not our ancestors, they are closer to our evolutionary roots than other primate forms.

Apes are not humans—and they probably are quite happy about that fact! Notwithstanding, they are so closely related to us that it is totally reasonable that they have several of the basic elements of human intelligence and language—and indeed they do, though nothing would declare that they must! Common genetics implies not only common ancestry and morphology, it also implies common neuroscientific bases for psychology and behavior. Those who insist on the "total separation" of human and ape psychology, behavior, and competence err in discounting the evolutionary bases which supported the emergence of the primates and notably the great apes and even ourselves!

In sum, the long-held views advanced by some authorities

regarding language as a process unique to humans and the insistence by others that a barrier be "declared," regardless of data, between the basic intellectual processes of animals and humans (as for the ability to plan, reason, symbolize, and so on) have been contrary to the end that relevant research data be given their proper consideration. Research data clearly indicate that nonhuman primates, and notably the great apes, are competent in several, though not all, of the essential dimensions of language and other complex processes.

As stated above, apes are not humans, and humans are not apes. Differences between them are inherent in their genetics. Nevertheless, by the same perspective, they and we are not totally different. Thus, non-trivial similarities between the great apes and us should be expected and happily incorporated into our understanding of life, the natural world, and our "nature."

A New Comparative Perspective

In closing, Descartes and his beast-machine model of animals was wrong. That is now clear. Descartes' beast-machine model of animals has been discredited. His error should no longer be promulgated by the thinking and values of our society.

Apes have vaulted the language barrier. The psychological continuum of humans with them is in place. Apes are not humans—but despite Descartes' proclamation to the contrary, within their own existences they surely reason. Such must be the case because it is certain that their impressive learning and language skills did not evolve for the purposes attendant to those of the research laboratory.

Being competent for reasoning, they, along with us, surely can experience a number of dimensions of *being*. They are surely capable of experiencing pain. (Here it is appropriate to credit veterinary science of recent years with caring for animals that manifest symptoms of pain as probably experiencing pain.)

Apes are great, both in size and intelligence. Their care and conservation are challenges that must be met with renewed energy and commitment. Their appropriate scientific study, conducted sensitively and responsibly, can be of great value to us as we strive to learn more accurately about the nature of our own species and how, through the management of early rearing and early experiences, we can rear generations who will be responsible to themselves, toward others, toward this planet, and toward the natural resources and wildlife that share the planet with us.

Notes

[1] See Rumbaugh and Savage-Rumbaugh, 1994, for a review.

References

Andrews, P. and Martin, L., "Cladistic relationships of extant and fossil hominoids," *Journal of Human Evolution*, 16 (1987): 101–108.

Cooper, H. M., "Ecological correlates of visual learning in nocturnal prosimians," in P. Charles-Dominique, H. M. Cooper, A. Hladik, C. M. Hladik, E. Pages, G. F. Pariente, A. Petter-Rousseaux, J. J. Petter, and A. Schilling, *Nocturnal Malagasy Primates* (New York: Academic Press, 1980), pp. 191–203.

Darwin, C., *Origin of Species* (New York: Hurst & Co., 1860).

Darwin, C., *The descent of man—and selection in relation to sex* (London: Murray, 1861).

Davenport, R. K., Rogers, C. W., and Rumbaugh, D. M., "Long-term cognitive deficits in chimpanzees associated with early impoverished rearing," *Developmental Psychology*, 9 (1973): 343–347.

De Lillo, C., & Visalberghi, E., "Transfer index and mediational learning in tufted capuchins," *International Journal of Primatology*, 15:2 (1994): 275–288.

Descartes, R., *Discourse on method* (New York: Liberal Arts Press, [1637] 1956).

Greenfield, P. and Savage-Rumbaugh, E. S., "Imitation, grammatical development, and the invention of protogrammar by an ape," in N. A. Krasnegor, D. M. Rumbaugh, R. L. Schiefelbusch, and M.

Studdert-Kennedy, eds., *Biological and Behavioral Determinants of Language Development* (Hillsdale, NJ: Lawrence Erlbaum Associates, 1991), pp. 235–58.

Greenfield, P. and Savage-Rumbaugh, E. S., "Comparing communicative competence in child and chimp: the pragmatics of repetition," *Journal of Child Language*, 20 (1993): 1–26.

Jerison, H. J., *Brain size and the evolution of mind*, Fifty-ninth James Arthur Lecure on the Evolution of the Human Brain (New York: American Museum of Natural History, 1991).

Napier, J. R. and Napier, P. H., *The Natural History of Primates* (Cambridge, MA: MIT Press, 1994).

Rumbaugh, D. M., Hopkins, W. D., Washburn, D. A., and Savage-Rumbaugh, E. S., "Comparative perspectives of brain, cognition, and language," in N. A. Krasnegor, D. M. Rumbaugh, R. L. Schiefelbusch, and M. Studdert-Kennedy, eds., *Biological and behavioral determinants of language development* (Hillsdale, NJ: Lawrence Erlbaum Associates, 1991), pp. 145–64.

Rumbaugh, D. M. and Pate, J. L., "The evolution of cognition in primates: A comparative perspective," in H. L. Roitblat, T. G. Bever, and H. S. Terrace, eds., *Animal cognition* (Hillsdale, N J: Lawrence Erlbaum Associates, 1984), pp. 569–85.

Rumbaugh, D. M., Richardson, W. K., Washburn, D. A., Savage-Rumbaugh, E. S., and Hopkins, W. D., "Rhesus monkeys (*Macaca mulatta*), video tasks, and implications for stimulus-response spatial contiguity," *Journal of Comparative Psychology*, 103 (1989): 32–8.

Rumbaugh, D. M. and Savage-Rumbaugh, E. S., "Language in comparative perspective," in N. J. Mackintosh, ed., *Animal learning and cognition* (San Diego: Academic Press, 1994).

Sarich, V. M., "Retrospective on hominoid macromolecular systematics," in R. L. Ciochon and R. S. Corruccini, eds., *New interpretations of ape and human ancestry* (New York: Plenum, 1983), pp. 137–50.

Savage-Rumbaugh, E. S., *Ape language: From conditioned response to symbol* (New York: Columbia University Press, 1986).

Savage-Rumbaugh, E. S., "Language learning in the bonobo: how and why they learn," in N. A. Krasnegor, D. M. Rumbaugh, R. L Schiefelbusch, and M. Studdert-Kennedy, eds., *Biological and behavioral determinants of language development* (Hillsdale, NJ: Lawrence Erlbaum Associates, 1991), pp. 209–33.

Savage-Rumbaugh, E. S., Brakke, K. E., and Hutchins, S. S.,

"Linguistic development: Contrasts between co-reared *Pan troglodytes* and *Pan paniscus*," in T. Nishida, W. C. McGrew, P. Marler, M. Pickford, and F. B. M. de Waal, eds., *Topics in Primatology* (Tokyo: University of Tokyo Press, 1992), pp. 51–66.

Savage-Rumbaugh, E. S. and Lewin, R., *Kanzi: At the Brink of the Human Mind* (New York: John Wiley Publishers, 1994).

Savage-Rumbaugh, E. S, Murphy, J., Sevcik, R. A., Rumbaugh, D., Brakke, K. E., and Williams, S., "Language comprehension in ape and child," *Monographs of the Society for Research in Child Development*, Serial No. 233, Vol. 58, Nos. 3–4 (1993): 1–242.

Sevcik, R. A., "A comprehensive analysis of graphic symbol acquisition and use: Evidence from an infant bonobo (*Pan paniscus*), Doctoral Dissertation, Georgia State University, Atlanta, 1989.

Sibley, C. G. and Ahlquist, J. E., "DNA hybridization evidence of hominoid phylogeny: Results from an expanded data set," *Journal of Molecular Evolution*, 26 (1987): 99–121.

Toth, N., Schick, K. D., Savage-Rumbaugh, E. S., Sevcik, R. A., and Rumbaugh, D. M., "*Pan* the tool-maker: Investigations into the stone tool-making and tool-using capabilities of a bonobo (*Pan paniscus*)," *Journal of Archaeological Science*, 20 (1993): 81–91.

Washburn, D. A., "Stroop-like effects for monkeys and humans: Processing speed or strength of association?" *Psychological Science*, 5: 6 (1994): 375–379.

Washburn, D. A., Hopkins, W. D., and Rumbaugh, D. M., "Automation of learning-set testing: The video-task paradigm," *Behavior Research Methods, Instruments, & Computers*, 21 (1989): 281–84.

Washburn, D. A. and Rumbaugh, D. M., "Ordinal judgments of numerical symbols by macaques (*Macaca mulatta*)," *Psychological Science*, 2:3 (1991): 190–93.

Washburn, D. A. and Rumbaugh, D. M., "Comparative assessment of psychomotor performance: Target prediction by humans and macaques (*Macaca mulatta*)," *Journal of Experimental Psychology: General*, 121:3 (1992a): 305–12.

Washburn, D. A. and Rumbaugh, D. M., "The learning skills of rhesus revisited," *International Journal of Primatology*, 12:4 (1992b): 377–88.

Animal Minds, Animal Morality

Colin McGinn

It would be widely agreed that in order for moral concern to be appropriate for some given entity it is necessary that that entity should satisfy certain psychological conditions. The point of dispute is what precisely these conditions are. What *kind* of psychological being must the entity be if it is to be worthy of moral consideration? What are the necessary and sufficient mental conditions for inclusion in the moral community? We possess a rich system of psychological description—often referred to as "folk psychology"—which we apply to the behavior of human beings, and which is integral to our moral attitudes toward each other. This system runs from the most basic sensations to the most elevated thoughts and emotions. The question is how much of this is strictly necessary for moral notions to apply.

This is not a question about what we can *know* of the mind of some given entity. I am not asking how we come to know that an entity satisfies the minimal mental conditions for moral concern, or whether we can ever know such facts. So I am not about to debate the issue of whether (say) ants have minds and if so what goes on in them. My question is not about the epistemology of other minds. It is about the ontological foundations of moral concern. I am asking what the psychological conditions are that are presupposed in moral judgements and reactions. What has to be the *case*, psychologically, before we can treat something as morally significant—whether or not we can ever know, in concrete instances, that these conditions are satisfied? The question concerns what facts are such that *if* you knew them you would be right to adopt an attitude of moral concern to the entity before you. Not that I think a skeptical attitude toward the existence of other minds, human or animal, is appropriate—I go along with the common sense view that we can know a good deal about the minds of other beings. I am simply distinguishing the two questions, so that problems attached to the epistemological issue are not mistakenly transferred to the ontological issue.

First, I will put forward what seems to me to be the correct answer to my question, bringing out the true import of this answer. Then, I will look at some of the ways in which this answer can be blurred or misunderstood. Next, some mistaken answers will be contrasted with the answer I give. Finally, certain moral consequences of what I take to be the correct answer will be sketched.

I begin with a quotation from Gottlob Frege, the great German mathematician and founder of analytical philosophy. In the course of a discussion of philosophical idealism, he remarks: "It seems absurd to us that a pain, a mood, a wish should rove about the world without a bearer, independently. An experience is impossible without an experient. The inner world presupposes the person whose inner world it is" (Frege, 1967, p. 27).[1] Frege's point here is that the idea of ownerless experience is incoherent: if an experience occurs, it must be an experience *of* some subject or self. This is a necessary truth about experience; it follows from what experience intrinsically is. Experience is necessarily experience for someone. Franz Brentano famously maintained that all conscious states must have an object, that is, have intentional directedness to something other than themselves. Frege is insisting that all conscious states must have a subject, which is built into their very structure. Thus, we cannot make sense of the idea of free-floating experiences—experiences that are not experienced *by* some being. The point is close to the lesson of

Descartes' *Cogito*: the existence of a thinker follows from the existence of thought itself. Wherever there is a thought, an "I" must also be.[2]

Furthermore, as Frege goes on to argue, the self that is implied by experience cannot itself be an experience or anything of that category. For, if the bearer of an experience [e1] were just another experience [e2], then we would be launched on an endless infinite regress, since the [e2] experience would itself need a bearer of its own. Let that bearer be experience [e3]; then still we need a further bearer for that experience, by Frege's principle. Clearly, there will be no end to the bearers we need in order to sustain a single experience, so that experience would be impossible. But experience is possible—we have it all the time—so it cannot be that the subject of experience is itself an instance of experience. The self must be an entity of a completely different ontological type from the experience it bears. If we analyze an occurrent experience into its qualitative type and the subject that bears it, then these two things must exist at different ontological levels, neither being assimilable to the other.[3] A consequence of this, which interested Frege, is that we can deduce the existence of something non-experiential from the existence of mere experience—which puts a big dent in philosophical idealism.

Whenever an experience occurs, therefore, there is something with a complex structure that brings together a particular subject and a qualitative type of experience. Experience can never exist as a simple unanalyzable quality. The experience is always *for* something that is not itself an experience. We have a dyadic structure, consisting of a subject and what that subject experiences. The subject is not represented *in* the content of the experience, of course; it is rather a precondition of there being any experience at all. The self is what *has* the experience, not something that the experience is *about*. No requirement of self-reflection is imposed by Frege's principle, only self-existence.

The principle, however, does not guarantee that the self that must exist is a continuous or persisting self. For all the principle says, each experience might have its own momentary self, ceasing to exist when the experience does. But it is highly plausible to suppose that there is a single self that bears successive experiences, so that many experiences are unified in one subject of consciousness. This is even clearer for *simultaneous* experiences, where again Frege's principle by itself does not strictly imply that my present set of experiences are had by a single subject. Despite that lack of entailment, it is immensely plausible to suppose that my present experiences are unified in this way. The sensations I now have from my different senses are thus borne by a single subject, so that there is no bizarre fragmentation of my self into a multitude of contemporaneous selves. And it is natural, once this point about simultaneous experiences is accepted, to accept also that successive experiences can be unified into a single continuous self—not perhaps for the entire span of one's biological life, but for decently long intervals. At any rate, I shall assume that Frege's principle can be strengthened to allow that different experiences can be had by a single subject, simultaneously and successively. Here, then, is a further reason for distinguishing selves from the experiences they have: the self can persist while its experiences come and go.[4]

What has all this abstract metaphysics got to do with animals and morality? Simply this: the mental structure that is minimally necessary for morality to apply is precisely that the entity in question be an experiencing subject, in the sense just articulated. And the essential point I want to emphasize is that simply by granting experiences to animals we thereby—and necessarily—grant selves to them. This is just a logical consequence of supposing that experiences occur in animals. Animals have, or are, selves—they are subjects *for* which experiences occur. This is an important point morally because I suspect that the denial of selfhood to animals is one of the last hold-outs against including animals in the moral community. People[5] have slowly come to accept that animals have experiences, in just as robust a sense as we do, but they have been reluctant to grant selfhood to animals. Selfhood is the thing that is held to distinguish us from the beasts, to put us on a different moral plane. This matters morally because the primary object of moral respect is precisely the self—that to which experiences happen.[6] But if Frege's principle is correct, as I think it is, then this is an incoherent position. Animal experiences can be no more ownerless than human experiences, since they are *experiences*. The moral community is the community of selves, and animals belong to this just as much as humans. Of course, the selves that belong to different species of animal will differ in certain respects, and some may have more moral weight than others (compare ants and apes), but still each experiencing organism falls into the same fundamental ontological category, that of subject of consciousness. There is no possibility of maintaining that we differ from other species in being *subjects* of experience and not merely *repositories* of experience. It is not that they are just collections of experiences while we are beings that *have* experiences.

Here it is necessary to grasp clearly a point it is easy to overlook, namely, that experiences have moral significance only because they are *for* someone in the full-blooded sense. So if we deny that there is a self that certain experiences are for, then we deprive those experiences of moral weight. To see this, try to imagine experiences of pain that have no bearer. As we have seen, this is really a metaphysical impossibility, but that does not mean that the idea might not have some hold over the imagination. So consider some allegedly ownerless pains—take that concept seriously for a moment. Such pains would not be felt *by* anybody; they would not happen *to* any self. They would simply *be*, ownerless and alone. But if so, why should it matter that they occur? Pain is bad because it happens to someone; but if it happens to *no one*, then no one is suffering, so why should it matter what the quantity of pain in the universe is? We feel compassion for the subject who is experiencing the pain—we think it is awful that this should be happening to *them*. But if there is no subject, then there is no one for whom to feel sorry. The pain is morally neutral. Increasing its quantity does not make life worse for anyone, and so there can be no moral objection to it. But add subjects for this pain, and the situation changes dramatically, from a moral point of view. It is because there are experiencing subjects, who are not themselves experiences, that unpleasant experiences *matter*. It follows, then, that if we are to accord moral weight to the experiences of animals, we need to recognize that animals have selves—that they are more, psychologically, than mere collections of intrinsically ownerless mental states. Frege's principle tells us, in effect, that we cannot exploit the point about the moral inconsequence of ownerless experiences to put animals outside the moral community.

This is the point at which certain metaphysical conceptions of experience and the self assume moral relevance. If we try to think of an animal mind as just a *locus* of intrinsically ownerless experiences, so that experiences occur *in* the animal rather in the way digestion does, then we lose the idea that these experiences are for someone. But then it cannot matter morally what those experiences are. We have to conceive of undergoing experience in that special and peculiar way that selves undergo experience; it is not enough to think of experiences as selfless in their essential nature and bound together by occurring in the same physical body. We must hang onto the common sense idea that for an experience to occur *is* for some subject to undergo it. And, of course, this is exactly the way we do commonsensically think of the experience of each other and of animals. I think of my cat's pain as something that *he* suffers, where this fact is built into the very identity of the pain being suffered. And I include my

cat in the moral community precisely because I recognize that
he is a subject of awareness in exactly the sense that I am
(though we differ in many important respects). Thus, there is a
sense in which the moral community is "ontologically
homogeneous": it consists of a certain special category of
existent, namely, conscious subjects. However different animals may be from each other in other respects, this is one
respect in which we all share a common nature.

Let me emphasize and clarify this point because it is the
heart of my argument. Animal species differ along various
dimensions: they are anatomically and behaviorally different
from species to species, and they are also psychologically
different. Bats, famously, inhabit a different phenomenologi-
cal world from that of humans, so that it is hard—perhaps
impossible—for us to empathize fully with a bat's experience,
since it has a sense modality we lack.[7] Animal species are
phenomenologically heterogeneous. That may make it seem
that nothing unifies the class of sentient beings, so that no
single trait underpins the idea of a moral community including
every member of this heterogeneous group. But this is to miss
the point that however different a bat's experience is from a
human's, both bat and human are alike in being subjects of
experience. We have that trait in common, despite our
physiological and psychological differences. So the moral
community that includes bats and us *is* unified by a certain
metaphysical fact—that we are all experiencing subjects.
Humans may not know what it is like for a bat to experience
the world in the way it does, but we do grasp one highly
important fact about the bat mind, namely, that it is a subject
of those experiences in exactly the same sense that we are the
subject of ours. This is a simple consequence of the fact that
"subject of experience" is a description that applies univocally
to both of us. I know what it is like to be a subject of
experience, and the bat also is a subject of experience, so I
know what it is like for the bat to be one of *those*. What I do not
know is the particular qualitative *mode* of experience that the

bat enjoys. We care (or should care) about what happens to the bat, according to my suggestion, because we care about how things are for experiencing subjects, with the emphasis on the word "subject." We have an ontology of subjects that is a direct consequence of the nature of experience, and this ontology is what underpins and warrants the inclusive character of the moral community that encompasses all experiencing creatures. All across the globe, billions of these subjects are now undergoing an indefinite variety of experiences, some good, others bad, and what matters is that there are subjects having these experiences—not merely that experiences occur with different sorts of qualitative character.

I have been taking it as obvious that there is moral significance to the having of experiences by subjects, notably the having of pain. And I doubt that anyone would seriously dispute this assumption. But there are ways of gliding over it, or distorting it, so that we fail to appreciate its moral import. There are metaphysical views that hinder a clear recognition of the moral significance of a subject experiencing pain. Let me mention some of these, which I suspect operate to dim the force of what is plain to common sense.

The first and most obvious way to avoid taking pain seriously is to be an eliminativist about the mind—either for all minds or just non-human ones. Obviously, if we deny that a creature *has* any experiences, then we shall not need to concern ourselves with moral questions about the occurrence of experiences in that creature. If there is no pain in animals, then we need not worry about causing it. Eliminativism, thus, undermines moral concern in a very straightforward way. But there is a subtler form of the doctrine that I think has more appeal for some people, and which I mentioned earlier—eliminativism about the self. It may be admitted that there are pains but denied that there is any subject that has these pains.

All we can say is that these inherently subjectless pains occur *at* certain locations—that is all we can legitimately mean by the idea of a self. But this undermines moral concern for the reason I gave earlier, since the "for-someone-ness" of pain is not adequately recorded in this way of conceiving things. It is not enough that free-floating experiences should sometimes gather together at some specific location, forming a bundle of mental states, since this does not yet give us the idea that the experiences are *for someone*. It merely gives us the idea that they are *at someplace*. Frege's principle cuts against this eliminativist view of the self by insisting that we cannot make sense of experience conceived as existing prior to the existence of subjects. So there is no conceptual ground from which to claim that animal experience lacks moral weight: if it is really experience, then it must be experience for a subject, and that is what *gives* it moral weight. To put it differently: realism about experience entails realism about the self, and realism about the self is what generates moral concern. So there is no logical gap between ascribing experiences to animals and respecting them as subjects with moral worth.

A subtler error that might be made is this: it might be agreed that Frege's principle forces selves upon us, but then the "for-someone-ness" of experience is misunderstood by modeling it upon ordinary property instantiation. If we think of the undergoing of experience as essentially the same, structurally, as a physical object having a physical property, then we shall not register the distinctive sense in which an experience is *for* its subject—and, hence, we will not see why this relation has moral significance. This is a difficult point to articulate clearly, but I think the intuitive thought should be fairly plain. My pain matters to me because it is pain for me, but my physical properties are not for me in this sense. But now if we insist on assimilating the "for-me" relation to the relation of property instantiation, so that a subject's having an experience is not logically different from a body's having a shape, then we will not be in a position to appreciate properly why pain really matters. In our own case, we are not apt to make this assimilation, because we are directly aware of the "for-me-ness" of the pain; but we are more prone to it for other animals, because we cannot so easily project our own phenomenology into them. This will produce a subtler variant of the ownerless experience view of animal minds, where now we conceive the relation between self and experience in a way that is too external. Only if we keep firmly in mind the *sui generis* character of a subject's having an experience can we hold onto the moral significance of experience. The experience is something the subject *undergoes*; it is not merely something that *attaches* to the subject, in the way one's physical characteristics do. So we not only need to be realists about experience and about the self in order to sustain moral concern; we also need to recognize the unique relation that holds between these two things. The self does not merely instantiate experiences; it is, as we might say, confronted by them.

A less subtle way of blurring the moral status of experience is to adopt a behavioristic view of the mind—a view that has been dominant in animal studies for most of the twentieth century and before. Suppose we identify pain with a disposition to avoidance behavior; then two points will be easy to miss. First, it will be hard to see why Frege's principle should be true: why should the existence of a mere behavioral disposition necessarily imply the existence of a subject of experience? Behavioral dispositions in general do not carry such an implication, since not all of them have an experiential counterpart; so there is nothing in the concept of behavior itself to yield the entailment in question. This is just the wrong sort of concept to invite the application of Frege's principle. Accordingly, behavioral dispositions will not have, or be seen to have, the kind of moral significance that experiences present: they will not give us a subject of consciousness that demands moral respect.

Secondly, it is not clear why having a behavioral disposition should matter anyway. What matters about pain is the way it *feels* to the subject, but this is not captured by talk of dispositions. If we are only allowed to describe an organism in behavioral terms, then we omit the very mode of description that is required for moral concern. A mere assemblage of behavioral dispositions is not the right foundation for moral respect. The experiencing subject is invisible in this way of conceiving an animal. Thus, the behaviorist metaphysics of mind is not morally neutral. I think it is no accident that the most unblushing vivisectors have also been philosophically committed to behaviorism and similarly reductive doctrines. Certainly, the psychologists who have done some of the most questionable kinds of animal experimentation have been explicit behaviorists. For if behavior is all that animal mentality consists in, then there cannot be much wrong with the kinds of "invasive procedures" (as they are euphemistically called) that animal experimenters regularly perform. It is much harder to stiffen oneself to the consequences, for the animal, of one's actions if one views it as a full-blooded subject of experience. Perhaps, indeed, part of the attraction of behaviorism is its moral convenience. The outright denial of animal mentality is the quickest way to justify certain sorts of treatment of animals; but, failing that, one tries to conceive of the animal mind in reductive behaviorist terms, thus evading the moral implications of accepting that animals are subjects of experience too.

I have suggested that the existence of an experiencing subject properly understood is what is minimally required for moral concern. The rationale is that this basic structure is sufficient to ground the idea that the organism has states that matter to it. This is a fairly weak condition, though not so weak as the rejected idea that mere ownerless experience might have moral weight. My point has been that the subject-experience duality is sufficient for it to be rational for us to care what happens to the organism in question. Certain obligations flow from this elementary structure alone—for example, the obligation not to cause gratuitous pain to the organism. But other views have sometimes been adopted, which seem to me to be much too strong and to lack a solid rationale. Let me then mention some of these, indicating why they go beyond what we have seen to be minimally necessary. I will not be able to discuss them fully, but it will be worth seeing how they differ from the proposal I have been making.

Some have held that to be a morally significant entity it is necessary to be a full-fledged moral agent. Only a being who is capable of obligations, of praise and blame, of free action, can count as imposing moral duties on others. This is a very strong condition, excluding not merely all animals but also human infants and many mentally defective human adults. It ignores completely the obvious fact that the capacity for suffering and well-being can exist without having sophisticated moral capacities. One can be a subject of consciousness, with preferences and likes and dislikes, without being a moral agent. And why should moral agency itself have any particular role in determining whether you can be an object of moral concern? Is being tortured (say) worse if you are a being who could himself be blamed for something? The wrongness of torture surely depends on the pain it involves, not on whether the sufferer has such *further* mental characteristics as a capacity for moral action. The same can be said for the suggestion that moral status depends upon a capacity for language—as if only those who can express their feelings verbally deserve to have those feelings respected. Only if we hold—with massive implausibility—that experience is impossible without language could this suggestion have anything to be said for it. But surely the capacity to suffer does not depend upon the capacity to *report* one's suffering. I care about your suffering quite independently of whether you are capable of speech.

Some suppose that self-reflection is a precondition of moral significance. This seems to depend on the idea that one has to be capable of thinking *about* one's experience in order that this experience should count morally. But that is clearly too strong: all that is required is that the experience be experience *for* some subject, not that the subject be aware of itself *as* a subject. Perhaps the self-reflective thesis is a confused version of the thesis I have been defending: it sees the need to bring in the self, but mistakenly supposes that the self has to be part of the *content* of the creature's thoughts. True, morality requires the self; false, morality requires that the self be capable of reflecting on itself.

More initially appealing is a family of views that stress *thinking*. Thus, we have the Kantian idea that moral concern requires that the creature be a rational being. This is not a very clear doctrine until we spell out what rationality is taken to involve. Are creatures which avoid painful stimuli and seek beneficial ones to be counted as rational? If so, then the view will coincide with the view I defend, at least in respect of which creatures get included in the moral community. But if it means a capacity for sophisticated reasoning about abstract matters, then surely it is too strong. Why should my facility with syllogistic forms make my pain matter more? Is rationality not quite *external* to the capacity to feel pain and pleasure? The same can be said of the idea that intelligence is crucial. Again, this notion is vague as it stands; but if it means a flexible adaptability to circumstances, then it is unclear why this should be thought to count for much. If a creature lacks the intelligence to avoid the pain I routinely dish out to it, that is no doubt a sad fact, but it is hard to see why it makes my actions any more excusable. Similarly for the idea that the instinctiveness or reflexiveness of a creature's behavior is reason to exclude it from moral consideration. That in no way undermines the creature's claim to be a self with a potential range of conscious experiences. Even if all my behavior patterns are innate and hard-wired, I may still be capable of sensations, emotions, preferences, and so on. I could, in principle, be an experiencing subject and yet be unable to learn anything new: why should my ineducability be thought to put me beyond the moral pale?

I think it is clear enough that these suggestions have been contrived with the express purpose of excluding animals from the realm of the moral. It is *assumed* that animals should not count in the way humans do, and then the question is how to find a characteristic that we have and they do not. The result is a criterion that does indeed exclude animals, or many of them, but also excludes human infants and adults who are impaired in one way or another. Such theories also tend to lack any cogent rationale for the trait that is selected as crucial. The question is always this: if pain matters morally, as it surely does, how can this be so in virtue of *other* mental characteristics that are not themselves logically necessary for pain to occur? Contrast the account I have proposed, which analyzes the *internal* preconditions for pain to matter morally: that it be felt by a subject; that it not be reduced to something non-experiential; that the subject-experience nexus be correctly understood. I have proceeded from an analysis of experience itself in order to understand what makes experience matter; I have not looked to some distinct set of mental characteristics and then claimed that *these* somehow confer moral importance on experiences.

It is essential to my account that animals be granted selves in addition to mental states of various kinds. My argument for this relied upon Frege's principle, which states a conceptual truth about the nature of experience. This has some nontrivial consequences in the moral sphere. I have already mentioned one: that a moral community consisting of (sentient) animals and humans is ontologically homogeneous, being unified by the fact that all of us are subjects of experience. It is on this basis, and only on this basis, that we can genuinely *empathize* with other animals: we can put ourselves in their position because we recognize that they, like us, are conscious subjects. There is a self in there that is experiencing the world. The animal's consciousness is unified, centered. As we have seen, this implies that no moral dividing-line can be erected around the idea that animals merely experience while we have a self that undergoes experience.

This position also has implications for the morality of killing. Many people maintain that while the death of a human is a great evil, the death of an animal does not matter. No doubt there are many sources for this view, but I suspect that one important source is the belief that animals do not have selves, so that the death of an animal is not the extinction of a self—as the death of a human is. This view seems to depend upon an implicit rejection of Frege's principle. The underlying thought is that the animal's mind is just a collection of mental states with no self to back them. But, as I have argued, this is no more plausible for animals than it is for humans. So there is a self that ceases when an animal's experience ceases. Death is, thus, the same *kind* of thing for animals as it is for humans—it is the rubbing out of a subject of consciousness.[8] The basic wrongness of killing—that it is the destruction of a subject of experience—is common to both animal and human killing. (That is not to say that the death of a human and an animal matter *equally*; it is only to say that the two do not differ *qualitatively* or in principle. There is room for the idea of degrees of wrongness here.)

It would not be morally wrong, so far as I can see, to put an end to experiences that were (*per impossibile*) inherently subjectless, since these would not be experiences *for* anyone; but once we agree that there is a subject that these experiences are for, we must face up to the moral consequences of putting an end to this subject. Killing an animal is snuffing out a self, not simply interrupting a sequence of connected experiences. I suspect that that is something we all know, at least implicitly, but it is sometimes necessary to be reminded of it, especially when the moral stakes are so high.[9]

Notes

[1] See also Strawson, 1994, pp. 129–134, to which I am indebted for a lucid discussion of Frege's principle.

[2] It is important to notice that this point holds in advance of adopting any particular metaphysical conception of reality in general. It is mandatory for materialists as well as idealists, for advocates of substance or of process. The point holds simply at the level of pre-theoretical description of experience.

[3] If experience cannot exist without the self, can the self exist without experience? Frege's principle says nothing to settle this, and the question is difficult. If selves persist during dreamless sleep, then the answer is yes. Given that, we would have an extra reason for denying that the self is an experiential entity. There is no need to take a stand on the issue for my purposes here.

[4] Nothing we have said ventures into the question of what kind of thing the self is—whether, say, it is the same as the body or brain or instead a transcendental ego of some special sort. All we are saying is that the self must exist if experience does, *whatever* its nature turns out to be.

[5] I mean theorists of one kind or another, not ordinary folks. The latter seem not to have doubted animal experience for as far back as memory reaches. I note that recently a number of theorists are catching up with them.

[6] The proper object of moral respect or regard is the person or self, not the mental states he or she possesses. It is a category mistake to accord respect to the experiences of selves. Thus, it is important that animals have selves if they are to be morally respected.

[7] See Nagel, 1979.

[8] I do not mean to deny that humans and animals differ with respect to the significance that death has for them. In the case of humans, long-term projects and hopes for the future enter into the loss that death brings; while this does not seem so for animals. I do believe, however—though I cannot argue it here—that the evil of death does not essentially turn on these differences. What matters fundamentally is simply the cessation of an experiencing subject.

[9] I am thinking particularly of killing animals for food.

References

Frege, Gottlob, "The Thought: A Logical Inquiry," in P.F. Strawson, ed., *Philosophical Logic* (Oxford: Oxford University Press, 1967).

Nagel, Thomas, "What is it Like to be a Bat?" in his *Mortal Questions* (Cambridge: Cambridge University Press, 1979).

Strawson, Galen, *Mental Reality* (Cambridge, MA: MIT Press, 1994).

EVERYDAY LIFE

Introduction

The three excellent papers in this session confirm the underlying truth of this conference—that a discussion of what we humans think about animals is inseparable from what we think about ourselves. From quite different starting points, the papers suggest how humans use the boundary between man and animal to negotiate a number of other boundaries central to the human experience—the boundary between men and women, between the human and the divine, between the tame and the wild, between the engineered and the natural. Just as the boundary between man and animal is a blurry, shifting, and contested one, so also are those other boundaries humans use to make sense of their experience and their history.

Thus, for instance, what is it that distinguishes natural from unnatural sex? We will think differently about that boundary after reading Wendy Doniger's rich presentation on myths and bestiality. Or, in considering Matt Cartmill's masterful overview of hunting, we will be forced to reconsider how humans have frequently changed their understanding of the boundary between technology and wilderness. Andrew Rowan's perceptive account of animal research, where the boundary between human benefit and animal rights is sharply etched, leads us to reflect anew upon human standards of social justice and compassion.

Being "in the company of animals," it is obvious, forces into human consciousness a long list of vexing questions about what it means to keep the company of ourselves.

The three papers perform a second function as well. They are a useful bridge from the wide-ranging and reflective discussion in the past few days to a review of public policy questions (the final session of the conference). For while all three of the papers are philosophical and historical, they each point toward issues which have attracted intense public attention in recent times—laws governing sexual choice and behavior; gun control and who "owns" nature; and, of course, the rights of animals in the context of scientific (or cosmetic) research. The policy questions are not resolved, but they are intelligently situated in a broader context than one gets from the conventional policy conference. In this fashion, then, the authors have well-served the rationale that led to the organization of "In the Company of Animals."

Kenneth Prewitt

The Mythology of Masquerading Animals, or, Bestiality

Wendy Doniger

Cultures throughout the world represent our deceptive relationships with animals as masquerades, which operate in both directions: in our rituals, humans often masquerade as animals, but in our myths we imagine that animals masquerade as humans. The most intense version of this universal theme is the tale of the bestial deception, the masquerade of an animal as a human in the most intimate of all relationships. What do the myths of bestial masquerade tells us about the ways in which humans have fantasized about their relationships with animals?

Waking Up With An Animal

You wake up in the morning and discover that you have been in bed all night with an animal (or a god in the form of an animal): that is the fantasy that underlies both the folktales and the literary retellings of those tales about figures sometimes called "animal lovers." (Unfortunately, this term is often spelt with a hyphen, which produces a potential confusion with animal-lovers, people who are fond of stray cats and dogs. It is easier to distinguish animal husbands—as the Frog Princes are usually called—from those who engage in animal-husbandry. Of course, the partner of an animal lover is, in a most literal

sense, even a bestial sense, an animal-lover.) Freud's Family
Romance (in which the child's parents turn out to be other,
better people than his apparent parents) often involves
animals, for the changeling child may be raised by or among
animals, so that the animal is a maternal surrogate, like a
wet-nurse, impersonating a mother; or the child may be sent
out to be killed, whereupon the compassionate killer relents
and kills an animal instead, taking back its heart (or tongue) as
proof of the murder, so that the animal is a sacrificial
surrogate, impersonating the sacrificial victim. The Family
Romance presents two complementary animal paradigms:
often, lowly animals are assimilated to the lower class people
who adopt a royal child; as animals are below humans, so lower
classes are regarded as naturally below higher classes. But, on
the other hand, animals may be assimilated to gods and
regarded as the high parents of children who appear to be
lower—merely mortal.

Even in folktales which lack an explicit religious agenda, the
union of a human and an animal has theological implications.
Midas Dekkers has suggested that the myth of Leda
impregnated by Zeus as a swan is the source of the myth of
Mary impregnated by God: "Christ was born of a virgin and a
dove; Christianity too is founded on bestiality. . . . Bestiality is
present at the very cradle of Christianity. Bestial tendencies
can be discerned not only in the Christ child himself, but in the
gathering assembled round the crib" (Dekkers, 1994, pp.
9–10). The assembled animals are evidence not so much of the
bestial parentage of the Christ child but of his place in the
mythology of the Family Romance. For Jesus, following the
pattern of the birth of the hero already established by
Oedipus, Romulus and Remus, and many others (and later
continued in Tarzan and Mowgli), is taken from his noble
parents (in this case, God) and nurtured by animals before
being raised by parents of lower birth (Dundes, 1990). Like all
the children of Leda and her swan, Jesus "is at the same time
the product of bestiality (man x animal) and of theogamy (god

x man). . . . Human beings are, so to speak, marrying both beneath and above their station" (Dekkers, 1994, p. 10).

The donkey has special meaning in Christian mythology, as Gerard Kornelis van het Reve argued in 1966/7:

> Whether God is a Lamb with bloodily pierced feet or a one year-old, mousey grey donkey, which allows itself to be possessed by me at length three times in succession in its Secret Opening, what difference does it make as long as He takes away the sins of the world, and has pity on us all? . . . I shall put bandages around His hooves, so that I shall not receive too many grazes when He thrashes about at the moment of orgasm (cited in Dekkers, 1994, pp. 127–28).

Here we may be reminded of Apuleius's famous pornographic novel of the man who turned into a golden ass. The Dutch parliament may well have suspected the lecherous spirit of Apuleius, rather than the devout spirit of Saint Francis, hovering over this argument, for it accused van het Reve of sacrilege.

The Mutilated Equine Foot

Van het Reve's image of the donkey incorporates an important symbol, the wounded foot, in particular the wounded foot of the equine. The horse is one of the most evocative of mythological species, straddling the boundary between the wild and the tame (O'Flaherty, 1981, ch. 6). Mutilated feet are a central theme in European tales of equine masquerades to which the association of witches with horses adds another dimension, for abnormal feet were regarded as "a recurrent sign of contrariness, and, in women, of deviancy" (Warner, 1995, p. 121). Here is an example of such a story:

> [A Czech farmhand went] where the witches were having their feast. . . Now, when he came there, the farmer's wife knew him, and, to hide herself from him, she turned herself into a white horse. But he did not lose sight of the horse. He mounted it and

went to the smith with it, and told him to shoe it. Next day the woman had four horseshoes on, two on her hands and two on her feet. And she had to stay like that always! (Baudis, 1917, pp. 191–92).

Thus, the men in the story (and telling the story) impose culture on the women: if it is a horse, it cannot be a wild horse but must be controlled through its feet, like Cinderella in her impractical glass shoes.

Why does the foot, particularly the mutilated foot, play such an important role in mythologies of the sexual masquerade of animals throughout the world? Feet function as signs that allow a particular individual to be recognized. Moreover, they are signs not merely of individual identity and class identity but of the identity of the species as a whole. In Hindu mythology, one identifying sign of mortals is that their feet touch the ground, while the gods float ever so slightly above it, like hovercraft[1]—just as Jesus walked on the water. Magic animals, on the other hand, cannot always walk on water: A hunter formed an alliance with a beaver woman who requested that he build her a bridge to prevent her feet from touching water. He neglected one spot and she reproached him for his carelessness: I only asked thee to help me dry-footed over the waters. Thou didst cruelly neglect this. Now I must remain for ever with my people (Lang, 1885, pp. 76–80). But why should feet that touch the ground be a sign of mortality? Perhaps, because it is the point of the body where we are earthbound. As Marina Warner has put it, "Feet are ascribed telltale marks of identity and origin, perhaps through the literal-minded wordplay of the imagination, since they are the lowest part of the body and in touch with earth as opposed to the heavens" (Warner, 1995, p. 115). We continue to speak of feet of clay as a metaphor for the weak spot, the mortal spot. The heel of Eve is bruised by the serpent (that sloughs its skin in immortality) as she is banished from Eden for her transgression—a transgression that resulted in her mortality, and in ours. In this context, we may recall the mutilation of the feet of Jesus on the cross—and note that in medieval texts Jesus is sometimes referred to as the hunted stag whose hoof is stained with blood.

The mutilated foot may function as a synecdoche for the mortality of the human body as a whole. We speak of the Achilles' heel and point to our own Achilles' tendons as the sign of our mortality, the place where Achilles was held when he was dipped into the waters that made the rest of him immortal, the place where he remained vulnerable and through which death entered him. (As anyone over fifty will testify, we might more properly refer to our fatal weakness as the Achilles knee: who ever had arthroscopic surgery on a heel?) Like Achilles, the incarnate Hindu god Krsna is killed when a hunter named, surely significantly, "Old-age" mistakes him for an animal and shoots him in the foot (*Mahabharata*, Book 16).

Carlo Ginzburg, following the lead of Claude Lévi-Strauss (1963), offers a magnificent survey of the literature on people distinguished by extraordinarily large or otherwise deformed feet, ending with the "devil's goose foot, equine hoof, or lameness" (Ginzburg, 1991, p. 258). Lameness, here listed as an afterthought among distinguishing marks, provides the key to the meaning of the feet in these myths, as Ginzburg notes with reference to "Achilles—son of a goddess with some equine characteristics, like Thetis, raised by the centaur Chiron." He goes on to spell out some of these meanings:

> Malformations or imbalances in gait differentiate beings (gods, men, spirits) suspended between the realm of the dead and that of the living. . . . The symbolic equivalence of swollen, deformed, scorched, or simply bare feet receives considerable confirmation outside the circle of myths within which we have been moving. In the multiple variants of an apparently marginal detail are enclosed a thousand-year-old history (Ginzburg, 1991, p. 232).

This is a most telling insight, with relevance not only to

distinctions between humans and immortals but between humans and animals.

Bipeds and Quadrupeds

For the feet symbolize our separation not only from the gods, above us, but from the animals, below us (or, by some calculations, also above us). That our feet are indeed the sign of our human condition is confirmed by our basic way (derived from Aristotle) of classifying the animal orders: bi-peds, quadrupeds, and six-legged insects.[2] This assumption was satirized by George Orwell in *Animal Farm*, when the old prize boar, Major, taught the animals, "Whatever goes upon two legs is an enemy. Whatever goes upon four legs, or has wings, is a friend" (p. 21). After the Major's death, the clever, wicked pig Snowball made these sentences into the first two of the Seven Commandments (p. 33), but when he realized how illiterate most of the animals were, Snowball reduced it to the slogan, "Four legs good, two legs bad," which the sheep bleated out to stifle discussion on any point that the pigs did not like (p. 40). When the birds objected that this excluded them (since they had two legs and two wings), Snowball, ignoring the earlier, more subtle Commandment but thinking on his four feet, argued that birds were actually quadrupeds: "A bird's wing, comrades, is an organ of propulsion and not of manipulation. It should therefore be regarded as a leg. The distinguishing mark of man is the *hand*, the instrument with which he does all his mischief" (p. 41). Years later, when the pigs had thoroughly betrayed the revolution, the animals saw a terrible sight: a pig walking on his hind legs, followed by a whole row of them. "Some did it better than others, one or two were even a trifle unsteady and looked as though they would have liked the support of a stick. . ." The animals were about to protest when the sheep bleated out, "Four legs good, two legs *better*! Four legs good, two legs *better*!" (p. 122). And at the end of the book, when the pigs and the humans were celebrating together in the Manor House, the animals noticed something the matter with the pigs' faces: "No question, now, what had happened to the faces of the pigs. The creatures outside looked from pig to man, and from man to pig, and from pig to man again; but already it was impossible to say which was which" (p. 128). To walk upright is the final betrayal, the final denial of the real barrier, the final lie.

Orwell, as usual, merely anticipates and exaggerates "real life," for the line between birds and the quadrupeds called horses was also recently blurred in a Canadian court:

> In *Regina v. Ojibway*, a Canadian court found that a horse carrying a down pillow in place of a saddle had legally become a bird. The Small Birds Act defined a bird as "a two-legged animal covered with feathers," and the court agreed that two legs were merely the minimum requirement. The case report was certainly meant as satire, but textbooks have reprinted *Regina v. Ojibway* without comment, and generations of law students have repeated it (Wachtel, 1995, p. 6).

Of course, if two legs are "merely the minimum requirement," then we, too, are legallly birds when we wear ostrich plumes on our hats.

Mary Douglas, in her analysis of the "abominations of Leviticus," the lists of animals that the Hebrew Bible declares inedible, has taught us the implications of this classifying system based upon not only the number of legs but the means of locomotion:

> In the firmament two-legged fowls fly with wings. In the water scaly fish swim with fins. On the earth four-legged animals hop, jump or walk. Any class of creatures which is not equipped for the right kind of locomotion in its element is contrary to holiness. [One list of forbidden animals] would appear to consist precisely of creatures endowed with hands instead of front feet, which perversely use their hands for walking. . . . Since the main animal categories are defined by their typical movement, "swarming" which is not a mode of propulsion proper to any particular element, cuts across the basic classification. . . . If

penguins lived in the Near East I would expect them to be ruled
unclean as wingless birds (Douglas, 1966, pp. 41–58).

Animals themselves do seem to be structuralists, at least to the
extent that they classify potential predators according to their
gaits: your own horses do not recognize you if you limp badly
enough (which is particularly galling since the limp is often a
result of falling off *them*), and they shy violently at the sight of
lame people and especially at lame horses. The category
"human" or "horse" is defined by the gait, and a creature
without that gait cannot be a human or a horse. "Category
error!" their frightened eyes proclaim (to a structuralist rider),
just as the shuffling gait of the homeless identifies them
instantly to us who are well-heeled (*sic*), and we shy away from
those who are "down at the heel."

The "wrong gait" may be natural (snakes move differently
from birds) or unnatural, the result of a mutilation (a bird with
a broken wing). Animals often suffer mutilations of their feet
when they transgress the boundary between the human and
the animal, a boundary established by their feet. We, by
contrast, suffer injuries to our legs when we move in the
opposite direction, trying to separate ourselves from the
animals. The injury to the foot symbolizes the hobbling of
uprightness and the inauthenticity of our relationship with the
earth. But it also symbolizes the deal that we made with the
devil: the use of our hands for the loss of the power of our
(four) legs. By walking upright, we gain the use of the
opposable thumb ("the organ with which he does all the
mischief," as Orwell's pig put it). We win the privilege of being
artisans by giving up the swift and secure movements of
quadrupeds. The mythological sacrifice leads, in real life, to
chronic lower back pain.

Hephaestus, the artisan of the gods, is lame. Achilles' heel
was also the source of his great gift: Achilles is the man "swift
of foot" *par excellence*. A similar sacrifice, in which the wounded
foot stands as a metaphor for the artist, is implicit in

Philoctetes' festering, stinking foot (wounded from the bite of a serpent, like the heel of Eve), which was the price for his skill as an archer.[3] The metaphor of the foot wounded *by an animal* —here, a serpent—makes clear the source of our problematic humanity. The particular association of horses and wounded legs is an intrinsic part of the myth of Chiron the centaur, who purchased his skill as a physician at the cost of his own constant pain from his wounded foot—and who was the tutor of Achilles with the fatal heel.[4] And, on the other border of the human, we suffer the mutilation of our feet when we cease to be gods (as in Eden) and become real human beings.

Feet distinguish us from animals, so that when we are bestial, the first thing to go are the feet. Hence, as Marina Warner has noted, devils still have animal feet—they are not yet fully transformed—and bestial women already have animal feet— they are beginning to be transformed. The sexist equation of women and animals gives rise to one of the great jokes about standing upright, Samuel Johnson's remark that "a woman preaching is like a dog's walking on his hind legs. It is not done well; but you are surprised to find it done at all" (Boswell, 1791, p. 287).

Carlo Ginzburg has argued persuasively that the human experience underlying our most basic classificatory system is that of the hunter tracking an animal by its footprints. He sees the foot as a sign, the great metonomy: the hunter knows the animal by its footprints, tracks, traces (Ginzburg, 1980). This technique was eventually transferred from the realm of prehistoric hunters and gatherers to that of scientists and remains the basis of many of our taxonomies. (As humankind began to stand upright, we moved from the footprints tracked by Neolithic hunters to the fingerprints tracked by Sherlock Holmes). Noting the importance of the foot in paleontology, for instance, Ginzburg cites a statement made by Cuvier in 1834:

Today, someone who sees the print of a cloven hoof can

conclude that the animal which left the print was a ruminative one, and this conclusion is as certain as any that can be made in physics or moral philosophy. This single track therefore tells the observer about the kind of teeth, the kind of jaws, the haunches, the shoulder, and the pelvis of the animal which has passed (Cuvier, 1834, p. 185).

It may well be that the memory of this ancient and enduring way of knowing the identity of a creature operates, subconsciously, to bring feet and footprints into so many of the myths of masquerade in which identity is in question—the identity of the child, or the identity of the parent.

The riddle of the foot is the riddle of the Family Romance. Oedipus's foot is the the key to the Sphinx's riddle: the creature that goes on four feet, then two feet, then three feet, is the human being who crawls as a child (or an animal), walks upright as a man, walks with a cane as he ages—and then, we might add, dies. Oedipus himself is that man; his name means "Swollen Foot," and his feet are pierced when, at his birth, he is exposed on the hillside—among the animals. His mutilated feet further connect him, especially in Lévi-Strauss's analysis, with other mortals who are paradoxically born from the earth and born from their mothers; they remind us, too, that we were born of the earth, not of the gods (Lévi-Strauss, 1963).

Night and Day

Throughout these stories, we encounter the contrast between day and night. "My night is your day," says the Beast in Jean Cocteau's film of *Beauty and the Beast*. There is an hour of twilight that the French call *"entre chien et loup"* ("between dog and wolf"). That is where these stories take place, for the transitional, marginal, liminal animal transgresses this most basic of all boundaries. Indeed, in Latin (and in many of the Romance languages derived from it), the very words for twilight or dawn (*lux*) and wolf (*luc*) are etymologically related;

the wolf is the twilight animal, or, if you prefer, twilight is the hour of the wolf.

But there is a significant difference between stories in which the primary allegiance of the creature is animal and those in which it is human (or divine). Stories about animal lovers present two variants of a single truth: a human being is really an animal. But the weight of reality is placed differently in different variants, so that when the story ends, and the masquerade is over, either there is a human, or there is an animal. It does matter. Generally speaking, the forms are distributed as in Stith Thompson's motif (B 640.1): "marriage to beast by day, man by night." This is the pattern of *Beauty and the Beast*, the story that is in many ways the paradigm of this entire genre, in which a monstrous groom of unspecified zoological nature is equated with ugliness in contrast with the defining beauty of his bride.[5] She thinks he is a beast, but at night, by the light of the forbidden candle, she discovers that he is a handsome prince. The reverse also occurs, however, when the lover appears to be a human by day but becomes an animal at night. An example of this is Shakespeare's *Midsummer Night's Dream*, in which the magic of the night and the dream and the moonlight gives Bottom the Weaver the head of an ass and bewitches Titania into believing that he is a handsome suitor; in the light of day, Bottom becomes a weaver again, and both he and Titania have only a dim memory of his animal form.

The key seems to be that the true form is the one that appears at night—an interesting assertion of the primacy of what is hidden, the time of dreaming, over what is apparent, the time of the workaday world. Sometimes the true form, the nightly form, is human (when a man or woman has been bewitched into becoming a beast, as in some variants of the Swan Maiden theme, or in the tale of Mélusine), but sometimes the true form is animal (when an animal or a demon masquerades as a human, like the snake lover in India or the fox woman in Japan). This pattern of diurnal and nocturnal images of women is expressed in an Oriya proverb that "captures this ambivalent attitude toward women succinctly: 'Beautiful as a picture by day; a cobra-woman by night' ('Dinore citrini, ratire naguni')" (Marglin, 1985, p. 242).

The two images of animal by day or by night are conflated in the film *Ladyhawke*, said (by the filmmakers) to be "based on a thirteenth-century European legend." A lady and her lover suffer a double curse: she is transformed into a hawk by day and he into a wolf by night: "Only the anguish of a split second at sunrise and sunset when they can almost touch, but not." The curse was to be broken "when there is a day without night, a night without day." This supreme liminality was a solar eclipse; it being day, the knight was already a human, and as the moon obscured the face of the sun, the hawk became a woman and stood beside him. They rode off into the sunset, presumably to live happily ever after. In this story, riding off into the sunset is more meaningful than the usual Hollywood finale; at last the hero and heroine are able to bear the dangerous, liminal moment that separates the human from the animal.

Stories about animals who lose their shells or slough their skins are also about souls shedding bodies, about all of us shuffling off our mortal coil. In many Hindu stories, where one person enters another person's body, people burn the body of the travelling soul, wrongly thinking that it is dead, so that it cannot get back into the body any more (O'Flaherty, 1984). (This is also the premise of the film *Heaven Can Wait* and its variants, such as *Here Comes Mr. Jordan*). The loss of this body is like the terrible moment when the human burns the animal's skin so that it cannot return to its own world. It means you have to keep on going forward. But the Hindu Vedantic model accounts for just one half of the possibilties, assuming that the body is unreal and the soul real. What if the animal—and the body—is the true self, the human—the imagined soul—the masquerade? Then if someone burns the skin, the animal is doomed to be unreal forever, and when we die, and our bodies decay, we die forever. Only our children live on.

Barbara Fass Leavy has useful insights into the way that reality falls on the bestial side more often for the woman than for the man in the Swan Maiden stories that she has studied:

> Animal groom and animal bride stories differ in that the animal groom's disenchantment seems to be based on an assumption that the human form is the true form, the bestial shape some aberration (except in stories where a demon's human form constitutes a deception), whereas a basic assumption about woman is that her beast form defines her essential being. . . . . The woman may prefer the beast to the prince, prefer, that is, debased nature as she resists the restraints of civilization: it is then *she* who is the animal, her beast paramour virtually an extension or projection of herself (Leavy, 1994, pp. 221–22).

The statement made by these stories, that women are animals (most recently translated into the structuralist paradigm: women / men = nature / culture), a bias that has been attributed to the male authorship of most of our texts, has been challenged (by Sherry Ortner, among others[6]), and we might marshal the evidence of a different selection of stories to challenge the sexism of the paradigm by pointing out the equally frequent occurrence of stories in which men are animals (though they are different sorts of animals). Indeed, Marina Warner argues that there are *more* male animals: "The Beast has been primarily identified with the male since the story's earliest forms" (Warner, 1995, p. 279). Or, more subtly, we might argue that the woman who is the animal is the more civilized of the two partners.

Stuart Blackburn, challenging Bruno Bettelheim (1986), comments on the psycho-sexual meanings of the transformation from night to day, viewed from the man's point of view:

> if our tales express not female but male anxieties about adulthood and married life, then we also need a reinterpretation of the 'animal by day, husband by night' pattern in these animal-husband tales. From a female perspective, as Bettelheim

observes, such an alternation makes little sense (animal by night, human by day would more logically express female sexual fears), but he explains away this anomaly by interpreting the animal during the day as a projection of her inability to face reality on the morning after: 'What seemed lovely by night,' he writes, 'looks different by day' [Bettelheim, 1986, p. 297]. Bettelheim does hint at a male perspective when, two pages earlier, he hypothesizes that the distinction between the nocturnal and diurnal identities of the animal-husband represents a man's wish to keep his sex life separate from everyday life. But it remains curious that the non-sexual identity is the animal form. (Is the man, too, overcome by morning-after guilt and denial?) A separation between sexual and ordinary experience exists in our tales, too, but it is maintained neither by a masculine need to isolate sex nor by a female fear of sex. Rather, I believe, the "turtle by day, prince by night" pattern of our tales reflects a male fear of sexuality at home and a male wish fulfillment outside the home. The distinction to be made, then, is not day/night, but internal/external, the domestic versus the public (Blackburn, forthcoming, p. 30).

This is in harmony with A. K. Ramanujan's interpretation of the snake lover story as a tale of two realms, private and public (Ramanujan, 1991). In a French tale, the serpent lover reassured the woman that he could become a man when he chose and asked whether she preferred him in human form by day or by night. "His wife replied that she preferred him to be a man at night, for thus she would be less terrified; by day she would have less fear than by night to have a beast near her" (Delarue, 1956, pp. 178–79). Here, *both* forms are real, but, as usual, the nightly form is the preferred form, the chosen form, as well as the private form. It is, in a sense, the more real form. The same choice is presented, with many other nuances, by the Loathly Lady, who often asks her husband whether he would prefer her beautiful by day (in public) or by night (in private)[7], and is sometimes asked whether she would prefer her husband an animal by day or by night (Leavy, 1994, p. 290). The life of the day and the night may also serve as analogies for the worlds of life and fiction, as Elaine Showalter has remarked of Robert Louis Stevenson's dream of leading "a double life—one of the day, one of the night." She comments: "The double life of the day and the night is also the double life of the writer, the split between reality and the imagination" (Showalter, 1990, pp. 106–7). Or, as I would put it, the twilight of the animal lover is the twilight of the myth.

Animals as People, People as Animals

To cross the boundary from one group to another, from animal to human or from divine to human, is, in a sense, to masquerade, to "pass." Thus, we are punished (or, if you prefer, unmasked) by the mutilation of our feet when we masquerade as human instead of animals or gods, and animals or demons suffer mutilated and/or inhuman feet, too, when they masquerade as humans.

Animals function in the mythology of masquerade in two different ways. On the one hand, since the mythmakers themselves, by and large, lived closely with animals and could see how they behaved, they used animals (snakes, horses) consciously as natural metaphors to express their ideas about human and divine sexuality and masquerade; and by looking at the animals in myths, we can learn something about those ideas. Cuckolders are cuckoos; the word "cuckold" comes from "cuckoo" by the obvious analogy (though with an interesting switch of gender): the bird lays her/his eggs/sperm in another bird's nest. Clearly, we look to the animal kingdom to find the images with which to express our sexual ambivalences.

But in addition to seeing animals through the mythmakers' eyes, we can also look at animals ourselves and see aspects of animal behavior of which the mythmakers may have been unaware; and we can make our own, different mythological judgements about our own animal natures. For myths may express not only human observations of animals but the animal parts of ourselves, the parts that our bodies remember and that our minds—our superegos, formalized by society—may suppress. This realization will help us to understand some of the unconscious levels of symbolism in the myth, some of the ways in which the mythmakers think, without being aware of it, simply because they, too, are animals. As Steven Jay Gould put it, we are to some extent justified in "backreading" our emotions into animals, because we *are* animals, and we understand ourselves best (Gould, 1995).

Let us, therefore, first consider the animals in our myths that express ideas about human nature, and then let us consider what we know about animals that sheds light upon the questions about human nature raised by the myths. What do these animals symbolize? Charles Rycroft offers the best answer I know:

> The aptness of animals in general to provide metaphors must depend on the fact that in some ways they resemble human beings, whereas in other respects they do not, the most obvious way in which they do being that they are born, live and eventually die, the most obvious way in which they do not being that they lack the power of symbolic thought. It must be this fact that animals have drives, passions, motives, a will to live, but cannot speak about them or, so far as we know, reflect upon them, that they have biological destinies but cannot conceive of biological destiny, which makes them such apt and such frequently used symbols for precisely those passions and drives which are hardest to put into words, both intrinsically and because they are liable to repression (Rycroft, 1979, pp. 84–8).

That is, the animals want to do all the things that we want to do, but they lack the language and self-reflexion to tell stories about them. They share our sexuality but not our stories of sexuality.

E. M. Cioran, on the other hand, sees our sexuality not as animal but as an escape from our animality:"Ecstasy replaces sexuality. The mediocrity of the human race is the only plausible explanation for sexuality. As the only mode of coming out of ourselves, sexuality is a temporary salvation from animality. For every being, intercourse surpasses its biological function. It is a *triumph* over animality. Sexuality is the only gate to heaven" (Cioran, 1995, pp. 22–3). In this view, our sexuality is what separates us both from the animals and from the ecstatic saints—or, if you will, from the gods.

For me, contra the prevailing (largely structuralist) argument that anything can symbolize anything, that context alone determines meaning, animals have multiple but not infinite qualities; and particular animals tend to convey particular meanings. A bull is bigger than a rabbit in all cultures (whatever you think "big" means), even though in India it is much smaller than an elephant and in Ireland much bigger than just about anything else they have. A bull does certain unmistakable things. And snakes, both because of their skin-sloughing and because of their uncanny means of locomotion (an important classificatory factor), seem to convey ideas of rebirth (and of death: many snakes are poisonous), of shape-shifting, and of deception; moreover, their shape makes them both phallic and womb-like (when curled up into an ouroboros), hence, a natural symbol of androgyny and a common player in myths of sex-change. Animals as a whole are laden with meanings for us, and, though particular animals have particular meanings for particular cultures, some animals also have particular meanings for all of us.

Deceptive Animals as People

Let us conclude with a consideration of the use of sexual deception by animals, not in myths, but in real life. Telling animals apart is an important part of our taxonomizing, our making sense of the world. Thus, our anthropocentrism drives us to use as the key to our own ability to tell animals apart the animals' ability to tell one another apart in what we now regard (superceding the rule of the foot) as the defining situation of reality (sex): if two animals do not intermate, they are of different species.

It is particularly important for animals to recognize one

another's mating signals, for sex is the one breach in the otherwise nigh impenetrable curtain of their xenophobia, the one moment when they allow a strange animal to get close. Only in sexual matters, then, do they have to tell the difference:

> In nature, red in tooth and claw, every other animal is a potential danger. Even creatures that live in herds, swarms or packs usually keep some distance from each other. At mating time, however, that distance can be reduced to zero. For this there are all kinds of rituals, which in the case of human beings have resulted in the discotheque and in the case of animals in display, sniffing, spawning and chorusing (Dekkers, 1994, p. 29).

Here we may recall the elaborate rituals to which Odysseus and Penelope submit one another before they are willing to recognize one another as true mates.

Yet we know that animals can be fooled. *The Audobon Society Field Guide to North American Birds* tells us: "Mockingbirds are strongly territorial and, like a number of other birds, will attack their reflection in a window, hub cap, or mirror, at times with such vigor that they injure or kill themselves. Thus the boundaries of a bird's territory can be learned by placing a mirror at strategic locations and noting where the attack ceases."[8] We assume that stags' antlers help to conceal them among the branches which they so closely resemble and also to serve them as weapons. But the stags themselves shadow-box with branches when the rutting season begins; so they, too, know that their antlers resemble branches. . . . We also know that animals lie; we know that a mother quail will run out of cover, faking a broken wing, to protect her young lying safe in their nest. This is, in the broadest sense, a sexual lie. But animals also lie in the course of the mating game, a game whose rules were designed to be broken. And they can fool one another with their lies. They are capable of both perpetrating and detecting sexual masquerades.

Certain species have a particularly difficult time of it.

Homophobes often argue that homosexuality, transvestism, and, most of all, transsexuality are unnatural, but these patterns of behavior occur in nature all the time, especially among insects. A particularly dramatic scenario is enacted by certain fish:

> It is noteworthy that in certain conventional male-and-female species, members of one sex may turn such coordination to their advantage by imitating members of the opposite sex. Such activity may be thought of as another nongenetic form of sexual differentiation.
>
> The bluegill sunfish engages in an intriguing form of such gender bending. . . . Male bluegill sunfish exist in three different forms. Large, colorful males court females and defend their territories. A second kind of male—often known as a "sneaker"—becomes sexually mature at a much younger age and smaller size. These small males live on the periphery of a bigger male's territory and clandestinely mate with females while the dominant male is otherwise occupied. Sneaker males mature into a third kind of male, one that assumes the behavior and drab coloration of a female sunfish. These female mimics intervene between a territorial male and the female he is courting. The female mimic, rather than the courting male, usually ends up fertilizing the eggs (Crews, 1994, pp. 113–14).

Apparently the male victim wastes his sperm on the "sneaker" male or mimic, who can then fertilize the female. This is a heightened form of the fakery of the cuckoo, the avine cuckolder. The cuckoo fakes its species—a blow against racism—while the bluegill fakes its gender—a blow against sexism.

Yet, the bluegill is far from unique; nature abounds in such tricksters, and transsexual masquerades are, apparently, more basic than many might suppose. Grouper fish have a kind of mid-life crisis in which, unlike humans, they change their own sex rather than their sexual partners. Here is a different sort of snake in the grass:

> Male red-sided garter snakes enact a similar form of sexual mimicry. At times of peak sexual activity, males congregate around females, forming a so-called mating ball. . . . In 16 percent of the balls, the snake being courted by the males was in

fact a disguised male, what we call a she-male. She-males have testes that produce normal sperm, and they court and mate with females. But in addition to exhibiting male-typical behaviors, she-males produce the same attractiveness pheromone as do adult females. In the mating ball, this second source of the pheromone confuses the more prevalent conventional males, giving the she-male a decided mating advantage (Crews, 1994, pp. 113–14).

Is the mating ball like the great balls held in Europe where young women came to find their suitors? Where Cinderella met her prince? Do red-sided garter snakes lie about, like Alice, waiting for a frog-footman to bring them "an invitation to the mating ball"? The mind boggles.

Sometimes animals mistake us for their mates, often through the process of imprinting, made famous by Konrad Lorenz and his ducklings (Lorenz, 1952). Imprinting works like the magic drug that Oberon has Puck procure in *A Midsummer Night's Dream* and use on Titania: "The juice of it on sleeping eyelids laid Will make or man or woman madly dote Upon the next live creature that it sees" (2.1.170–172). In the case of animals, "the next live creature that it sees" upon emerging from the womb or egg strikes it as a kind of mirror; it thinks it must be like that and, upon sexual maturity, tries to mate with it.

Animals, too, have their sexual illusions. Thus, they provide us with both basic data and basic metaphors with which to formulate our own sexual masquerades. For we, too, are subject to the magic of imprinting, when, like those mocking birds, we use animals as mirrors in the construction of our own self-deceptive self-images.

Notes

[1] As in the story of Nala and Damayanti in the *Mahabharata*, 3.52–54.

[2] The change in length of the thigh is the most dramatic change

from primates to humans, though perhaps it is not the most significant change. The feet grow very little while the femur grows a great deal as we move from primate to human (personal communication from Stephen Jay Gould, April 8, 1995).

[3] See also Edmund Wilson's essay on the price paid for artistic excellence, *The Wound and the Bow.*

[4] See Updike, 1962.

[5] See Warner, 1995, and Leavy, 1994.

[6] See Ortner, 1974.

[7] See Chaucer, "The Wife of Bath's Tale," in *Canterbury Tales.*

[8] Cited in *The New Yorker*, May 4, 1992, p. 77, with the comment: "And where your cat hangs out a lot."

References

Baudis, Josef, *Czech Folk Tales* (London: George Allen and Unwin, 1917).

Bettelheim, Bruno, *The Uses of Enchantment: The Meaning and Importance of Fairy Tales* (New York: Alfred Knopf, 1986).

Blackburn, Stuart, "Coming out of his shell: Animal-husband tales from India," in *Animal-Husband Tales*, Alan Dundes, ed. (New York: Wildman, forthcoming).

Boswell, James, *Life of Dr. Johnson, Vol. I* (Cambridge: Cambridge University Press, [1791] 1992).

Cioran, E.M., *Tears and Saints*, Ilinca Sarifopol-Johnston, trans. (Chicago: University of Chicago Press, 1995).

Crews, David, "Animal Sexuality," *Scientific American* (January 1994): 108–14).

Cuvier, G., *Récherches sur les ossemens fossiles, vol. 1* (Paris: G. Dufour, 1834).

Dekkers, Midas, *Dearest Pet. On Beastiality*, Paul Vincent, trans. (London: Verso, 1994). (Originally published in 1992 in Dutch.)

Delarue, Paul, *The Borzoi Books of French Folk Tales* (New York: Alfred Knopf, 1956).

Douglas, Mary, *Purity and Danger* (Harmondsworth: Penguin, 1966).

Dundes, Alan, "The Hero Pattern and the Life of Jesus," in Otto Rank et al., *In Quest of the Hero* (Princeton, NJ: Princeton University Press, 1990).

Ginzburg, Carlo, "Morelli, Freud and Sherlock Holmes: Clues and Scientific Method," *History Workshop* 9 (1980): 5–36.

Ginzburg, Carlo, *Ecstasies: Deciphering the Witches' Sabbath*, Raymond Rosenthal, trans. (New York: Pantheon Books, 1991).

Gould, Stephen Jay, "Keynote Address," *Social Research* 62:3 (Fall 1995).

Lang, Andrew, *Custom and Myth*, 2/e (London: Longmans, Green, 1885).

Leavy, Barbara Fass, *In Search of the Swan Maiden: A Narrative on Folklore and Gender* (New York: New York University Press,1994).

Lévi-Strauss, Claude, "The Structural Study of Myth," in *Structural Anthropology*, Claire Jacobson and Brooke Grundfest Schoepf, trans. (Harmondsworth: Penguin Books, 1963).

Lorenz, Konrad, *King Solomon's Ring* (New York: Crowell, 1952).

Mahabharata (Poona: Bhandarkar Oriental Research Institute, 1933– 1969).

Marglin, Frédérique Apffel, *Wives of the God-King: The Rituals of the Devadasis of Puri* (Delhi: Oxford University Press, 1985).

O'Flaherty, Wendy Doniger, *Women, Androgynes, and Other Mythical Beasts* (Chicago: University of Chicago Press, 1981).

O'Flaherty, Wendy Doniger, *Dreams, Illusion, and Other Realities* (Chicago: University of Chicago Press, 1984).

Ortner, Sherry, "Is Female to Male as Nature Is to Culture?" in M. Rosaldo and L. Lamphere, eds., *Women, Culture, and Society* (Palo Alto, CA: Stanford University Press, 1974).

Ramanujan, A.K., "Towards a Counter-System: Women's Tales," in *Gender, Discourse, and Power in South Asia*, A. Appadurai et al., eds. (Philadelphia: University of Pennsylvania Press, 1991).

Rycroft, Charles, *The Innocence of Dreams* (New York: Pantheon Books, 1979).

Showalter, Elaine, *Sexual Anarchy: Gender and Culture at the Fin de Siècle* (New York: Penguin Books, 1990).

Updike, John, *The Centaur* (New York: Alfred Knopf, 1962).

Wachtel, Alan, letter to *Scientific American*, February 1995.

Warner, Marina, *From the Beast to the Blonde: On Fairy Tales and Their Tellers* (London: Chattos and Windus, 1995).

Hunting and
Humanity in
Western Thought

Matt Cartmill

Five hundred years ago, when Henry VIII occupied the throne of England, a Portuguese mariner named Raphael Hythloday left the company of Amerigo Vespucci's third expedition and travelled south from India into the imaginary countries of Terra Australis Incognita, one of which (so Thomas More assures us) is an island called Utopia. Here Hythloday found a land full of paradoxical excellences. The Utopians worked very little; and yet they were all rich. They abhorred the death penalty; and yet they had little crime. Most of the few criminals they did have were sentenced to terms of temporary slavery, in which they were made to do the menial jobs nobody else wanted to do. And here was another paradox: one of those menial jobs was *hunting*.

Hunting was an activity that More's fellow Englishmen held to be so delicious—and so expressive of power—that it was reserved for the aristocracy, who spent thousands of man-hours every year riding through the countryside on horseback in search of game. But More wrote,

> The Utopians think that this whole business of hunting is beneath the dignity of free men, and so they have made it a part of the butcher's trade—which, as I said before, they foist off on their slaves. They regard hunting as the lowest and vilest part of butchery, and the other parts of it as more useful and honorable, since they kill animals only to meet human needs, whereas the hunter seeks nothing but pleasure from murdering a poor innocent beast.

Medieval moralists (Rogers, 1881, p. 224; Pike, 1938, p. 23;

Hobusch, 1978, p. 74) had protested against the hunt as oppressive to the peasants, who had to open their lands and furnish their labor to the hunting aristocrats; but nobody in the Middle Ages had ever complained that hunting was oppressive to the *animals*. Erasmus (*Praise of Folly*, 1.18) and others had denounced hunters as idle fools and wastrels; but nobody before More had ever branded them as murderers and perverts.

Yet More did both, and we encounter similar attitudes in the works of later sixteenth-century writers. Since then, anti-hunting sentiment has become ever more common in Western art and literature. It has grown to be more or less the norm in modern America. In recent opinion polls, most Americans express agreement with the proposition that hunting for sport or for trophies, or for any purpose other than putting meat on the table, should be forbidden by law. In what follows, I want to sketch some of the history of this transformation, and to examine some factors that have gone into the growth of our own contradictory and conflicted attitudes toward hunting.

Those contradictions and conflicts grow out of the very concept of the hunt, as we define it. Hunting, for us, is not simply a matter of killing animals. A successful hunt ends in an animal's death, but it has to be a special animal, killed in a special way. It must be free, able to flee or attack the hunter. It must be killed on purpose, and by violence, and in person: no traps, no cages, no poisoned baits, no road kills. Above all, it must be a *wild* animal; and what that means for the hunter is that it must be *hostile*, not friendly to human beings or submissive to their authority. You can kill cows in the dairy barn, but you cannot hunt them. Hunting is thus by definition an armed confrontation between the human world and the untamed wilderness, between culture and nature; and it has been defined and praised and attacked in those terms throughout Western history, from antiquity onward.

Hunting had great mythical and symbolic significance for the Greeks. Three of their major deities—Apollo, Artemis, and

Dionysus—were closely associated with the hunt in myth and ritual. Dionysus and Artemis (or Diana, as the Romans called her) are paired with each other in a symbolic opposition that shows up in various forms again and again in Western thought about hunting. Artemis is a perpetual virgin; Dionysus is the dissolute god of wine. She is a masculinized female (real Greek women did not hunt); he is an effeminate male. Artemis leads a troop of maiden archers in an orderly program of wildlife management; but Dionysus dances at the head of a band of drunken crazies who tear their quarry apart with their bare hands and eat the bloody flesh, like predatory beasts. The followers of Artemis discipline the wilderness, but the followers of Dionysus *participate* in it. The maenad's costume of fawnskins, lynx pelts, and live serpents symbolizes a union with nature just as surely as do the buckskins and the raccoon cap on the mythic American frontiersman.

This tension between the images of the hunter as a park ranger and the hunter as a beast of prey is a corollary of the way we define hunting, as something that occurs on the boundary between the human domain and the wild. Throughout Western history, from the myths of Artemis and Dionysus down to our own stories of mythic hunters like Daniel Boone and Davy Crockett, the hunter has been seen as a liminal and ambiguous figure, now a fighter against wilderness and now a half-animal participant in it, who stands with one foot on either side of the boundary and swears no perpetual allegiance to either side. Perhaps the reason that hunting plays such a large part in Greek myth—and in our own stories about human origins—is that it takes place on that boundary, and thus marks the edge of the human world.

The Greek fascination with hunting was mirrored in the recurrence of mythic hunts and divine huntresses throughout the mythologies of the ancient Middle East, from Egypt all the way over to India. But it had no equivalent in the rest of Europe. Like More's Utopians, the early Romans saw hunting as a farm chore, with no more symbolic or mythical importance than catching rats. Some Roman authors denounced sportive hunting on the Greek model as a silly or vicious affectation, and the first anti-hunting sentiments in Europe appear in Roman history and literature. In 55 B.C., the crowd in the Roman arena rose up to protest the butchering of a score of elephants in a staged hunt. Cicero wrote to a friend about it: "What pleasure can a cultured man get in seeing . . . a noble beast run through with a hunting spear?" (*Ad Familiares*, 7.1.3). This difference between Greek and Roman attitudes, which is partly a rural versus urban difference (provincial Romans sound more like Greeks [Martial, *Epigrams*, 1.49, 12.14; Nemesianus, *Cynegetica*]), recurs in their respective national epics. When the mighty Odysseus kills a stag in Homer's *Odyssey*, the event is described in cheerfully bloodthirsty terms, all full of hock joints and vertebrae (10.161–84). But when one of Aeneas's companions shoots a stag in Virgil's *Aeneid*, it turns out to be somebody's pet, and it runs home to *complain*: "And the animal, / wounded, fled back to his familiar roof; /moaning, he reached his stall, and suppliant / and bleeding, filled the house with his lament" (7.500–502, Mandelbaum, 1961, p. 179).

Literary Roman expressions of distaste for the hunt and pity for the hunter's quarry represent one historical source of modern anti-hunting sentiments, from Thomas More onward. Another ancient line of thought that is woven into those sentiments is the traditional Christian view of human beings as inherently wicked, and of nature as degraded by human sin, which introduced death and predation into the peaceable kingdom of Eden:

> The World did in her Cradle take a fall,
> And turn'd her brains, and took a general maim
> Wronging each joint of th' universal frame.
> The noblest part, man, felt it first; and than
> Both beasts and plants, curs'd in the curse of man.
> (Donne, "The First Anniversarie," lines 196–200)

A third ingredient was the hallowing of the forest in

medieval thought. For the Greeks and Romans, forests were generally threatening and scary places. The Greek and Latin words that mean "woodsy" (*hylaios, sylvaticus*) have the secondary meaning of "savage." In early Christian thought, the wilderness is a sort of natural symbol of hell, and the wild animals that live there in a state of perpetual rebellion against the sons of Adam typify demons and sinners in rebellion against God. But this image was undermined from the very beginning of the Christian era by the counterimage of John the Baptist and other hermit saints in the wilderness, attended by friendly wild animals that the saint's holiness has restored to the docility of Eden. And since the wild animals treated the saints better than most people did, the odor of sanctity soon began to rub off on the forest creatures themselves. In an apocryphal gospel of the eighth century, Jesus, being adored by wild animals, turns to his friends and declares, "How much better than you are the beasts which know me and are tame, while men know me not" (James, 1924).

Other changes in the significance of wild places and creatures during the latter part of the Middle Ages reflect changes in the social significance of hunting. Before the tenth century, small farmers throughout northern Europe had been allowed to hunt more or less freely on their own land. But as new techniques of agriculture produced a surge in medieval crop yields and population, and Europe's forests retreated before the ax and the plow, hunting rights were increasingly taken over by the aristocracy, who put the remaining patches of woods off limits as royal hunting preserves and ruthlessly punished any peasants suspected of taking game (White, 1962, pp. 39ff; Dalby, 1965, p. v; Eckardt, 1976, pp. 27–31; Hobusch, 1978, pp. 117–19).

Hunting thus took on opposite connotations for the peasants and the aristocracy. For the elite, the hunt became an elaborate ritual encrusted with jargon and courtly ceremony, which served to validate the aristocratic credentials of the hunters. The peasantry, on the other hand, associated hunting with freedom, feasting, and rebellion against the authorities in the songs and stories they soon began to tell about Robin Hood and other deer-poaching forest fugitives, from the eleventh century onward (Keen, 1961).

But in both the high and low traditions, there were important late medieval changes in the symbolic meaning of the wilderness. As the forests contracted into exclusive aristocratic playrounds, they became transformed in the European imagination from the gloomy wasteland of earlier tradition into the gay and magical greenwood of late medieval literature. The deer, who are the symbolic inhabitants of the wilderness and give it its English name—etymologically a *wild-deer-ness* —also took on a new symbolic importance. From the eleventh to the fourteenth century, words that had meant "animal" or "wild beast" in several European languages narrowed semantically to mean "deer" or "doe" in particular— English *deer*, French *biche*, German *Wild*, and so on—and words for "deer" and "hunting" became conflated, so that deer became both ideal animals and the ideal objects of the hunt.

In late medieval and Renaissance art and literature, down through the sixteenth century, the deer hunt becomes a recurrent metaphor for erotic love (Thiébaux, 1974); and deer, which had been symbols of cowardice for the Greeks and Romans the way rabbits and chickens are for us, take on symbolic nobility in both folk ballads and high culture. The stag hunt becomes a metaphor for the tragic fall of a noble victim, as in the speech that Shakespeare's Mark Antony makes over the corpse of Caesar:

> Here wast thou bay'd, brave hart;
> Here didst thou fall; and here thy hunters stand,
> Sign'd in thy spoil, and crimson'd in thy lethe.
> O world, thou wast the forest to this hart;
> And this indeed, O world, the heart of thee.
> How like a deer, strucken by many princes,
> Dost thou here lie! (*Julius Caesar*, 3.1.204–210)

In the extreme form of this metaphor, the hunted deer

becomes the crucified Christ, whom one medieval poem describes as the stag "whose hoof is stained/ with blood, for he ransomed us at so great a price," and who appears as a hunted deer (commonly with a cross between his horns) again and again in medieval art and literature (Thiébaux, 1974, pp. 185–228).

Yet in spite of all this symbolic equation of hunting with tragedy and crucifixion, there are no significant medieval sources of anti-hunting sentiment. The hunt did not start to become a symbol of injustice and bloody tyranny until the beginning of the northern Renaissance, when Erasmus condemned the hunt as a "bestial amusement" in 1511 and More denounced it five years later in *Utopia*. A similar revulsion toward hunting is evident in the essays of Montaigne and the plays of Shakespeare, who employs the hunt almost always as a symbol of murder, usurpation, and rape (Shakespeare, *Macbeth*, 4.3; *As You Like It*, 2.1; *Love's Labour's Lost*, 4.1; *Titus Andronicus*, 2.1, 2.2, 3.1). We see it also in some of the artwork of the period—for example, in Albrecht Dürer's disturbing 1504 drawing of the severed head of a stag with a crossbow bolt in its skull and its eye turned backward to look at the viewer.

Perhaps the most surprising place where antihunting sentiment crops up in the sixteenth century is in hunting manuals, which from 1561 on contain rhymed complaints by the game animals denouncing the senseless cruelty of Man the Hunter. The standard English hunting manual of the period, the 1576 *Book of Venerie* by the poet George Turbervile, puts these words into the mouth of the hunted hare:

> Are minds of men become so void of sense,
> That they can joy to hurt a harmless thing?
> A silly beast, which cannot make defense?
> A wretch? a worm that cannot bite, nor sting?
> If that be so, I thank my Maker than,
> For making me a Beast and not a Man.
>
> . . . So that thou show'st thy vaunts to be but vain,

That brag'st of wit, above all other beasts,
And yet by me, thou neither gettest gain
Nor findest food to serve thy glutton's feasts:
Some sport perhaps: yet *Grievous is the glee*
Which ends in blood, that lesson learn of me.
                    (Turberville, [1576] 1908, pp. 176–78)

This is strange stuff to find in a handbook for hunters. One might as well expect to see *Field and Stream* putting out a monthly column on animal rights signed by Bambi and Thumper.

Why, after sportive hunting had been admired and respected all through the Middle Ages, did it start to get attacked in the 1500s? Some of these attacks reflected middle-class antagonism toward the hunting gentry; but that explanation does not hold for Montaigne or Shakespeare or Turbervile. Some of these new negative attitudes toward the hunt were associated with rising doubts about the meaning and reality of other established hierarchies, including the boundary between people and animals (Cartmill, 1993, pp. 87–91). In 1580, Montaigne denied the existence of that boundary, and concluded that "it is [only] by foolish pride and stubbornness that we set ourselves before the other animals and sequester ourselves from their condition and society" (*Essays*, 2.12, "Apology for Raimond Sebeond").

The scientific revolution of the 1600s, and the exponential growth of science that has taken place in the succeeding three centuries, have further eroded the animal-human boundary to the vanishing point. This erosion of human distinctiveness is inherent in the nature of science. Because science tries to find universally applicable explanations for the causes of all things, and justifies its search in practical terms as a means to securing power over nature, it inevitably tends toward a vision of the universe as a collection of lumps of uniform, neutral matter, all obeying the same universal laws and all valuable only as means to human ends. The bargain that science offers us is a Faustian one: in exchange for getting control over these lumps, we must ourselves consent to become lumps of the same uniform stuff.

From the late 1600s on, successive schools of Western thought about the natural order have been looking for loopholes in the contract we have made with science. Descartes and his followers tried to draw an absolute distinction between the world of matter and the world of the spirit, depicting animals as insensate machines made of meat. But almost everybody else has attacked the problem by trying to extend some limited form of citizenship to the animals and drawing them into the moral order—with predictable effects on the symbolic meanings attached to hunting.

The animal condition first began to take on serious moral dimensions in the eighteenth century, when a surprising range of thinkers, from the atheist Julien de la Mettrie at one extreme to John Wesley at the other, agreed in seeing the human mind as only a souped-up version of facilities shared by many other animals (Cartmill, 1993, pp. 98–100). Animal suffering now came to be seen for the first time as a serious natural and moral evil, and the intelligentsia began to try to do something about it by preaching kindness to animals through the mass media of the time: sermons, cheap prints, and the first children's books. In this climate of opinion, hunting came increasingly to be seen as just another cruel entertainment, like bull baiting.

It was a commonplace of eighteenth-century thought that the natives of North America represented the natural state of man, and that they lived chiefly by hunting. If hunting is both morally wrong and the primordial human enterprise, then the hunt naturally begins to be seen as a sign of man's innate depravity. These considerations combined with the early Romantic vision of Nature as a kind of virgin territory, exempt from man's polluting presence, to yield the first foreshadowings of our own familiar image of *Homo sapiens* as a crazed killer ape cutting a bloody swath across the face of sweet green Nature. Alexander Pope, in his *Essay on Man*, identified the advent of human predation with the Fall of Adam (3.147–168); and in 1774, 84 years before the *Origin of Species*, Lord Monboddo came out with the whole killer-ape theory in a pre-Darwinian package. Human beings in a state of nature, Monboddo argued, are not American Indians, but chimpanzees and orangutans. Thousands of years ago, these apes had multiplied to the point where the fruits of the earth could no longer sustain them; and some of them were forced to take up killing and eating other animals, and to invent language and weapons to help them in the chase. "This change of man from a frugivorous to a carnivorous animal must have produced a great change in character," wrote Monboddo. "He grew fierce and bold, delighting in blood and slaughter. War soon succeeded to hunting; and the necessary consequence of war was, the victors eating the vanquished, when they could kill or catch them; for, among such men, war is a kind of hunting" (Burnett, 1774, pp. 270–313, 367, 392–97, 416–20).

Monboddo's misanthropic vision of Man the Hunter was to recur two centuries later in the literature of my own professional discipline, physical anthropology. The anatomist Raymond Dart, who discovered and named our Pliocene ancestor, the so-called man-ape *Australopithecus*, was largely responsible for launching the killer-ape theory of human origins that dominated the anthropological textbooks of the 1960s and '70s. This theory portrayed hunting as a sort of original sin, which had not only started the human lineage off on a different historical course from the apes but had alienated us from nature and turned us into innately vicious and violent creatures. Dart described the fossil man-apes as "confirmed killers: carnivorous creatures, that seized living quarries by violence, battered them to death, tore apart their broken bodies, dismembered them limb from limb, slaking their ravenous thirst with the hot blood of victims and greedily devouring livid writhing flesh." Poring over the cracked and battered man-ape fossils from the South African caves, Dart began to find—or imagine—evidence that *Australopithecus* was not only a killer, but a murderer and cannibal, who "ruthlessly killed fellow australopithecines and fed upon them as he would upon any other beast" (Dart, 1953; Dart and Craig, 1959, p. 201). All the worst traits of human beings were there to be read in—or into—the bones from the South African caves, and they all represented a murderous legacy from our killer-ape ancestry.

The killer-ape image of human beings as sick animals, alienated from the harmony of nature by their destructive technology, has been shared by many influential modern scientists, writers, and and artists (Cartmill, 1993, pp. 11–14, 20–24, 211ff). It is a central part of anti-hunting sentiment in the twentieth century. It seems significant in this regard that some people, who will wax indignant about Southern white males driving their Broncos into the forest and blasting away at the wildlife, are not at all perturbed when Southern Cherokees or Seminoles do the same thing. The reason, I think, is that what disturbs us about the hunt is not the killing of animals as such. If that were what bothered us, we would see more picket lines around slaughterhouses. What we are really disturbed by is the armed confrontation between technology and the wilderness, between sinful human history and the timeless harmonies of nature. And in that confrontation, we continue to see the American natives, as we have always seen them, as being outside of human history: part of the natural landscape of America, along with the buffalo and the prairie grasses and the passenger pigeon. They cannot offend against the natural order, because in our imagination they are part of it.

Oddly enough, this same opposition that we like to pose between the pollution of civilization and the purity of nature provides an important motive for sport hunting as practiced in modern America. It should be stressed here that people hunt for different and sometimes conflicting reasons. Some hunters' rationales contradict those of others, and some hunters' rationales are self-contradictory. For instance, many hunters like to describe their sport as a foraging activity, a thrifty harvest of Nature's gratuitous bounty. The trouble with that description is that (according to the figures I have seen) each pound of venison brought home by the average American deer hunter costs about 20 dollars in cash and five hours of labor. On the whole, the average hunter would be better off harvesting some prime rib in a good restaurant. Another delusory rationale for hunting is the notion that we have a moral responsibility to keep game animals from overpopulating their ranges and starving to death in the snows of winter. The difficulty here is that hunters are among the first to protest when coyotes or feral dogs or wild cats move in and threaten to take over the job of population control. Significantly, this argument is always focussed on those eternal martyrs of the wildwood, the deer. Not even the most conscientious of hunters trudges off into the winter woods to find starving ravens and weasels and put them out of their misery (Cartmill, 1993, pp. 28, 231–32).

But the most literate hunters, the ones who are apt to write books and columns about the joys of hunting, generally agree that the chiefest of these joys is the pleasure of a temporary union with the natural order. "I must know," writes one sporting columnist, "that I am part of, and have common bond with, the wilderness" (Simpson, 1984). Another calls the hunt "a promise with the land" that keeps the huntsman from being "isolated from the natural world" (Holt, 1990). Valerius Geist describes hunting as an "intercourse with nature" (Geist, 1975, p. 153). "The human being," wrote the hunting philosopher Ortega y Gasset, "tries to rest from the enormous discomfort and all-embracing disquiet of history by 'returning' transitorily, artificially to nature in the sport of hunting." Hunting, said Ortega, "permits us the greatest luxury of all, the ability to enjoy a vacation from the human condition" (Ortega y Gasset, 1972, p. 139).

This rationale is a product of the way we define hunting, in terms of a symbolic opposition between the wild kingdom of nature and the polluted domain of human culture and history.

That symbolic opposition is what precludes our hunting in the dairy barn. But at bottom that opposition, and the hunter's motives that are grounded in it, are no less delusory than the notion of the hunt as harvest or the hunt as birth control for Bambi. Our scientific knowledge of the nature of life and the history of this planet impel us to the certain conviction— whether we like it or not—that people are animals and the descendants and cousins of animals, and that the human condition is simply one aspect of the animal condition. We cannot participate in one condition, or enjoy a vacation from the other, by the act of seeking out and killing unfriendly animals of other species.

The facts of evolutionary biology have far-reaching implications, some of which have been traced by other participants in this symposium. They have an obvious bearing on the moral status of recreational hunting. If the human-animal boundary, and the parallel boundary that we like to draw between culture and nature, are as arbitrary a pair of constructs as evolutionary biology leads us to think, then the distinction between wild and domestic animals, between *Wildtiere* and *Haustiere*, is equally arbitrary. If so, it makes eminently good sense to see hunting as More's Utopians did, as just another species of butchery. And if we accept all this, then it seems hard to avoid coming to the same conclusion that Thomas More came to four hundred and eighty years ago: that butchery is not, in the final analysis, an appropriate recreation for a free people.

References

Burnett, J., *Of the Origin and Progress of Language*, 2/e (Edinburgh: J. Balfour, 1774).

Cartmill, M., *A View to a Death in the Morning: Hunting and Nature through History* (Cambridge: Harvard University Press, 1993).

Dalby, D., *Lexicon of the Medieval German Hunt* (Berlin: Walter de Gruyter, 1965).

Dart, R.A., "The Predatory Transition from Ape to Man," *International Anthropological and Linguistic Review* 1 (1953): 201–217.

Dart, R.A. and Craig, D., *Adventures with the Missing Link* (New York: Harper, 1959).

Eckardt, H.W., *Herrschaftliche Jagd, bäuerliche Not und bürgerliche Kritik: Zur Geschichte der fürstlichen und adligen Jagdsprivilegien, vornehmlich im südwestdeutschen Raum* (Göttingen: Vandenhoeck und Ruprecht, 1976).

Geist, V., *Mountain Sheep and Man in the Northern Wilds* (Ithaca: Cornell University Press, 1975).

Holt, C., *Durham (N.C.) Morning Herald*, November 11, 1990, p. B-10.

Hobusch, E., *Von den edlen Kunst des Jägens: Eine Kulturgeschichte der Jagd und der Hege der Tierwelt* (Innsbruck: Pinguin, 1978).

James, M.R., *The Apocryphal New Testament* (Oxford: Oxford University Press, 1924).

Keen, M., *The Outlaws of Medieval Legend* (London: Routledge and Kegan Paul, 1961).

Mandelbaum, A., trans., *The Aeneid of Virgil* (New York: Bantam, 1961).

Ortega y Gasset, J., *Meditations on Hunting* (New York: Scribner's, 1972).

Pike, J.B., *Frivolities of Courtiers and Footprints of Philosophers* (Minneapolis: University of Minnesota Press, 1938).

Rogers, J.E.T., ed., *Loci e Libro Veritatum* (Oxford: Oxford University Press, 1881).

Simpson, B., *Raleigh (N.C.) News and Observer*, October 28, 1984, p. B-14.

Thiébaux, M., *The Stag of Love: The Chase in Medieval Literature* (Ithaca: Cornell University Press, 1974).

Turbervile, G., *The Book of Venerie* (Oxford: Oxford University Press, [1576] 1908).

White, Jr., L., *Medieval Technology and Social Change* (Oxford: Oxford University Press, 1962).

Scientists and Animal Research: Dr. Jekyll or Mr. Hyde?

Andrew N. Rowan

Introduction

In the past two decades, bioscientists have been forced to confront an increasing variety of critics. Nevertheless, at the same time, scientists still belong to one of the most admired professions (Pion and Lipsey, 1981). In the USA, 88 percent of the public believe that the world is better off because of science and scientists are second only to medical doctors in public prestige (NSB, 1989). In the United Kingdom, the three most respected public institutions are medicine, the military, and scientists in that order (Kenward, 1989). Nonetheless, there is still an underlying level of public uneasiness about science and scientists.

One critical group that has grown tremendously in size and influence in the last twenty years is the animal protection movement. As in the nineteenth century, protests over the use of animals in research, testing, and education have touched a responsive chord among the general public. In fact, animal research has long been one of those "hot button" issues that has the capacity to stimulate impassioned opposition. While the level of opposition has waxed and waned over the past one hundred and fifty years, it currently stands at an all-time high. About 15–20 percent of the public would like to see all animal use in research and testing stopped immediately, while another large segment are uneasy about the practice but are prepared to accept it because of its perceived benefits. By contrast, 85 percent of the public agree or strongly agree with the statement that it is acceptable to kill and eat animals. (See Table 1 for public attitudes to different uses of animals.)

Why is the public so sensitive about the use of a few tens of millions of animals in research when they do not object to killing hundreds of millions of pigs and cows and billions of chickens for our meat diet? Why is animal research considered so bad despite the public's high opinion of science (and scientists)? Perhaps it is the image of the scientist as an objective and *cold* individual who *deliberately inflicts harm* (pain, distress, or death) on his (the public image is usually male) *innocent animal victims* that arouses so much horror and concern. This paper does not address the accuracy of this image but rather intends to examine its psychic roots in modern society as well as some of the central themes that appear time and again in the debate. Such themes include cruelty, innocence, suffering, and human benefit.

A Historical Précis

The protest against animal research began in earnest in the second half of the nineteenth century (French, 1975). Some of the more important elements that gave support to the Victorian antivivisection movement were as follows.

First, the Darwinian revolution weakened claims about the uniqueness of human beings and blurred the absolute qualitative differences that had been considered to exist between humans and animals. This narrowing of the gap between humans and animals tended to support Utilitarian arguments that animal suffering was morally important.

Second, philosophical challenges to the dominance of humans over animals began to appear with greater frequency. In the eighteenth century, several clerics argued that animals should be accorded a greater moral status, and then Jeremy Bentham, the Utilitarian, added his influential voice to the debate. He argued that animal suffering should be given significant weight in analyses of what is or is not moral. The ability to employ Reason as opposed to mere Sentiment to challenge the morality of animal research was as empowering to the Victorian antivivisectionists as it is to the animal rights movement today.

Third, the emerging public health movement (the sanitarians) promoted the development of better health and hygiene (for example, cleaning up public water supplies) as a more effective way of improving public health than animal research. They did not specifically oppose animal research, but neither did they support it very strongly.

Fourth, there were some in the medical establishment who were threatened by the new "scientific" medicine based on experimentation. For example, Claude Bernard, the French physician, who is sometimes characterized as the "father" of experimental medicine, was criticized not only by the public for his animal research (including his wife and daughters) but also by leading figures of the French medical establishment. The medical criticism was, however, based more on professional jealousy than on a concern for the animals.

Finally, some of the new Protestant religions tended to undermine claims regarding the uniqueness of human beings or the moral irrelevance of animals by arguing that both animals and humans possessed souls and that God was concerned about all of creation and not simply humans. For example, John Wesley specifically preached that animals had souls (a message ignored in modern Methodism), and many of the early campaigners for animal welfare were clerics in the Church of England (Stevenson, 1956).

The Scientific Image

While the role of biologists, philosophers, clerics, the aristocracy, and others was important in fueling concerns over animal research, the public image of the research scientist was probably also an important factor in fueling public concerns. Towards the end of the nineteenth century, physicians had risen in status to the top rungs of society, having thrown off their earlier association with barbers and butchers. Physicians were no longer to be feared and were perceived to be caring, humanitarian professionals concerned with saving lives and alleviating suffering, often at some cost to themselves. By contrast, the researcher (whether a medical practitioner or not) was perceived by the public to be an unfeeling individual who deliberately and without feeling carried out his experiments.

Henry Salt, a close friend of George Bernard Shaw and an important figure in the Victorian animal "rights" movement, wrote a one act play entitled *A Lover of Animals* in 1895 that not only showed the influence of Shaw but also clearly articulated this dichotomy between physician and researcher. The play concerns the ambitions of Dr. Claud Kersterman, a thirty-five year old hospital surgeon who also does animal experiments. Dr. Kersterman hopes to persuade his wealthy aunt, Miss Moll, to set him up as the attending physician to her proposed Pet Convalescent Home and eventually to inherit her estate. However, he must ensure that his servant, Pate, a deformed half-wit, does not inadvertently reveal his animal experimentation because his aunt, despite enjoying her meat and her furs, is a vehement antivivisectionist. His aunt's companion, Miss Grace Goodhart, learns of the research activities from Pate, which leads to the expected uproar.

Miss Goodhart is not an "animal lover" like Miss Moll, whose concern for animals is limited to the acts of foreigners and scientists but displays much more consistency in her attitudes and behavior. She not only is against animal research but also does not wear furs and is a vegetarian. She is unhappy about her employer's obvious hypocrisy and eventually says as much and is summarily dismissed by the aunt. However, Miss Goodhart is also called upon to express her opinion of vivisection which she does as follows: "I abominate Vivisection as the most horrible of crimes—the more horrible because it is done, as Dr. Kersterman says, deliberately and conscientiously (we must grant him that), and not from mere thoughtlessness, like sport." In other words, it is the premeditated and calculated elements of animal research that aggravate the sensibilities and arouse so much horror.

Another and much better known example of this dichotomy is the Victorian novel of *Dr. Jekyll and Mr. Hyde* by Robert Louis Stevenson, which was published a few years before Salt's play (Stevenson, 1979). In the words of the editor of the Penguin version of the novel, "Jekyll is an apparently respectable man who contains within him a potential for profound wickedness, released in the shape of Mr Hyde. Symonds [Stevenson's friend A.J. Symonds] and many others found this chilling to contemplate" (Calder, 1979). For a significant segment of the public, whether or not they accept the need to use laboratory animals, the Jekyll and Hyde story reflects public perceptions about the dual nature of the animal research scientist. This duality appears time and again in surveys and analyses of public attitudes to science and scientists.

Public Attitudes toward Science and to Scientists

According to Haynes (1994), Western traditions were inimical to science prior to 1600 when the desire for knowledge (except theology) was perceived as dangerous and evil. This attitude is clearly reflected by the Faust legend in which the scientist, Dr. Faust, makes a pact with the devil to gain knowledge and power. Francis Bacon changed public attitudes to the search for knowledge by arguing that scientists were simply developing an understanding of God's laws, but, ever since, the vision of the scientist as a noble seeker after truth has had to vie with a range of more negative stereotypes. For the most part, literature and public attitudes appear to emphasize the baser aspects of scientific character, although

there have been relatively brief periods when public admiration for science and scientists has overcome public concern.
For example, after Newton's death in 1727, he was the subject
of considerable public adulation, and portrayals of medical
researchers in the nineteenth century were often complimen-
tary to the point of eulogy.

In the twentieth century, there have been periods of
widespread public support for scientists, but mad and evil
scientists have never entirely disappeared from view. They
have been a staple of pulp fiction and, according to Haynes,
"with the exception of the superficial characters of much
science fiction, the dominant picture has been of scientists who
recapitulate the unflattering stereotypes of earlier centuries—
the evil scientist, the stupid scientist, the inhuman scientist . . .,
the scientist who has lost control of his discovery . . . " (1994, p.
295).

During the period after the second World War, from the
late 1940s through the 1950s, public support for science in the
United States was very high. It was felt that federally funded
science could surmount any problem the country or world
could throw at it. The development of the polio vaccine was
the clear example. However, beginning in the late 1960s and
lasting throughout the 1970s , science was perceived by more
and more of the public as part of the problem rather than as
part of the solution. Problems arising from chemical pollution,
the destruction of the rain forests, and nuclear power have
tended to undermine the public's confidence in science. More
media attention, that displayed both the human fallibility of
scientists as well as their accomplishments, left the public less
confident in the pronouncements of science.

There is a tendency to view the 1950s, when science and
scientists enjoyed great prestige, as the norm and the current
drop in public approval as an unfortunate trend that must be
reversed. However, Haynes' analysis (1994) indicates that
public attitudes, as reflected in literary figures, were more
usually negative and suspicious than supportive. Allen (1993)

also argues that the positive public attitudes in the 1950s were anomalous, and that the public is usually much more ambivalent about the activities of scientists.

He identifies two main images of scientists in American thought which he categorizes as Reformers (Mechanics) and as Wizards (Megalomaniacs). For example, some of the scientists who appear in works by Hawthorne, Poe, and Melville (for example, Captain Ahab) represent classic examples of the scientist as Wizard (or less flatteringly as Megalomaniac). The Wizard is usually not connected to the community or to his family (if he has one) and is perceived to be elitist. He is very capable but is unconstrained by moral scruples in his search to control or uncover some powerful secret of Nature.

The Reformer/Mechanic (scientist/engineer) is, by contrast, basically a benign character, rooted in the community (that is, democratic and upholding family values). He has some humorous characteristics (for example, absent-mindedness) but is also skilled and well-intentioned. Edison is a classic example of such a Reformer or Mechanic. The Wizard (scientist/theoretician) is anything but benign or humorous.

While Reformers and Wizards appear periodically throughout the development of American literature and the media, Allen notes that the Wizard disappeared for a time during and after the second World War. For example, in science fiction from 1937 to 1950, scientists were portrayed as heroic figures who worked with the military (the Warriors) to preserve civilization. However, in 1951, the Wizard began to reappear as exemplified by Dr. Carrington in the 1951 film, *The Thing—From Another World*.

The perception of scientists' personalities by the public has changed accordingly over the past forty years, but it has always been stereotypical and somewhat distorted. In surveys from the late 1950s, scientists were seen as intellectual and dedicated but difficult to comprehend and erratic in interpersonal relationships. A 1975 survey reported that they were seen as remote, withdrawn, secretive, unpopular, and single-minded souls (Pion and Lipsey, 1981). Other surveys identify qualities such as rationality, objectivity, and coldness with scientists (Gerbner, 1987; Weart, 1988).

In modern times, television is the mirror that reflects society's hopes and fears and, presumably, reinforces public attitudes about whatever they are viewing. Gerbner (1987) has examined the images of scientists portrayed on television and reports that television images of scientists include many ambivalent and troublesome portrayals. Even though there are more positive than negative images of scientists in television (5:1 good to bad), by comparison to physicians (19:1 good to bad) and to law enforcement officials (40:1 good to bad), scientists were more often portrayed negatively.

Gerbner (1987) also reported that exposure to science and technology through television tends to cultivate a less favorable orientation toward science. Heavy television viewers were more negative about science and more likely to want to place restrictions on scientific activity. Among heavy television viewers, a college education had only a small positive effect on attitudes to science. Films also reflect this ambivalence toward both science and scientists. Such popular films as *Project X*, *Greystokes*, and *Splash* reinforce the image of the callous and unfeeling scientist mistreating the charges in his (usually) care. *Jurassic Park* is more of a warning about scientific hubris, while *The Fugitive* has a physician-*healer* winning his mortal combat with a physician-*experimenter*.

Public Attitudes toward Animal Research

Numerous polls of attitudes to animal research and testing have been conducted, and the findings can be summarized as follows: (a) About two-thirds to three-quarters of the American public are prepared to accept the need for animal research. (b) The percentage that actually *supports* animal research is usually about 10 percentage points lower. (c) About 10–15 percent of the public actively oppose animal research. (d) The percentage opposing animal research changes depending on the type of animal used and the type of research (see table 2) (Anon, 1984; DDB, 1983; Gallup Organization, 1982; Gallup and Beckstead, 1988; NABR, 1985; NSB, 1989; 1993). Thus, most people support research that uses rats, but this figure may be halved if dogs are the research animal. Similarly, cancer research is considered very important by the public, but support drops off for alcohol and drug addiction research and product testing, especially of cosmetics and household goods. (e) So-called "basic" research does not receive as much public support as goal-oriented medical research. (f) About half the public is uncertain whether animal researchers treat their animals humanely. (g) It appears as though the public is becoming less tolerant of the use of animals in research. The biennial Science Indicators survey commissioned by the National Science Board in the United States has asked a question on animal research since 1985 (NSB, 1993). Survey participants were asked to express their level of agreement or disagreement with the following statement: "Scientists should be allowed to do research that causes pain and injury to animals like dogs and chimpanzees if it produces new information about human health problems." The level of agreement with this statement has dropped about ten percentage points (from 63 percent agreeing to 53 percent agreeing) from 1985 to 1992. (In the United Kingdom, where a similar question was also asked in 1988, only 35 percent of the public supported the statement.)

Scientific Attitudes toward Animal Research

In the highly polarized debate that characterizes the modern animal research controversy, it is usually assumed that scientists support animal research and animal activists criticize it, with the general public occupying some sort of contested middle ground. However, scientists are also demonstrably ambivalent about what is done to animals in experimental laboratories.

Arluke (1988), Birke and Michael (1992), and Takooshian (1988) have conducted a variety of surveys of scientific attitudes to animal research which reveal more support for the practice than among the general public but still considerable concern. For example, Arluke (1988) reports considerable ambivalence about animal use among research scientists. In one letter he received after his findings were published, the writer notes: "I'm not really the type who usually writes letters to the editor. Nor do I belong to any animal rights groups or "researcher's rights" groups. My only agenda is to share with you the considerable guilt—not "stress" or "uneasiness" but GUILT I've experienced for the past fifteen years since working on rats as a premedical student."

Takooshian's (1988) survey revealed that the researchers were only marginally more supportive of animal use than the general public, while the strongest supporters were people who hunt and the clergy. Overall, the best correlation with animal research scores were the attitudes to animal protection rather than the attitudes toward science. Birke and Michael (1992) conducted a different type of attitude study in which they interviewed a relatively small sample of scientists in depth and reported that some said they could use rats but not cats or dogs, while others objected to the use of animals in household product testing. Their subjects recognized that they were being inconsistent in some cases but, nevertheless, followed their hearts rather than their heads.

These studies indicate that the modern animal researcher has far more in common with the Reformer who is connected to community mores rather than the Wizard who is not. However, the normal process of scientific communication is ideally stripped of any individuality, passion, and feeling. Therefore, written narratives in science are more likely to reinforce the public perception of science as wizardry.

Conclusion

In the modern animal research controversy, "many citizens have begun to judge science according to their own moral standards rather than accepting the measures of professional achievement that scientists apply to themselves" (Ritvo, 1984). The scientific community needs to understand what those moral standards are and recognize the ambivalent perception that the public has of science and scientists. There is a constant tug of war in the public mind between perceptions of the scientist as hero and as villain. For the most part, the scale of public attitudes is tilted toward the scientist as villain, especially when scientific discussion is couched in dispassionate and objective terminology.

This creates problems when scientists attempt to defend practices by arguing that they really do care about animals, patients, or some other compromised group or entity. The stereotypical dispassionate scientist is at a distinct disadvantage when he (or she) tries to convince an already suspicious public that they really do care about the animals they use. Nevertheless, recent research demonstrates that scientists who do animal research are almost as ambivalent about their use of animals as the general public. If scientists were freer in expressing their ambivalence about animal research, it would provide them with a firmer footing in the broader societal mores and make them less likely to be perceived as Wizards and more likely as Reformers. This should lead to a boost in public trust and a more favorable public image for both science and scientists.

TABLE 1. *Attitudes to different uses of animals—United States* (Parents Magazine, 1989). (Based on a randomly selected sample of 1009 American Adults)

Activity	Wrong: Should be illegal	Disapprove: but should *not be illegal*	Acceptable
Killing for Fur	63	22	13
Cosmetics Testing	58	23	13
Killing for Leather	46	23	27
Hunting for Sport	33	27	36
Medical Research	18	18	58
Animal Performances	16	16	63
Capture for Zoos	12	17	66
Killing for Food	5	7	85

TABLE 2. *Attitudes to the Use of Different Species in Medical Research* (NABR, 1985) (Percentage of Sample)

Species	Approve	Disapprove	Do Not Know
Rats	88	9	3
Rabbits	77	19	4
Monkeys	69	25	6
Cows	58	35	7
Dogs	55	40	5

References

Allen, G.S., "Master mechanics and evil wizards: science and the American imagination from Frankenstein to Sputnik," *Massachusetts Review* 33 (1993): 505–558.

Anonynmous, "Gallup poll in Surrey," *FRAME News* (Nov/Dec, 1984): 1–2.

Arluke, A., "Sacrificial symbolism in animal experimentation: object or pet?" *Anthrozoös* 2 (1988): 98–117.

Birke, L. and Michael, M., "Going into the closet with science," *New Scientist* (1992).

Calder, J., "Introduction" to *Dr. Jekyll and Mr. Hyde* by R.L. Stevenson (New York: Penguin Books, 1979).

DDB, *America's Binding Relationship with the Animal Kingdom* (New York: Doyle Dane Bernbach, 1983).

Doble, J. and Johnson, J., *Science and the public: a report in three volumes. Volume 1: Searching for common ground on issues related to science and technology* (New York; Public Agenda Foundation, 1990).

French, R D., *Antivivisection and Medical Science in Victorian Society* (Princeton, NJ: Princeton University Press, 1975).

Gallup Organization, *Attitudes toward the Use of Animals in Laboratory Research and Testing* (Princeton, NJ: The Gallup Organization, 1982).

Gallup, G.G. and Beckstead, J.W., "Attitudes toward animal research," *American Psychologist* 43 (1988): 474–76.

Gerbner, G., "Science on television—how it affects public conceptions," *Issues in Science and Technology* 3:3 (1987):109–15.

Haynes, R.D., *From Faust to Dr Strangelove: Representations of the Scientist in Western Literature* (Baltimore: Johns Hopkins University Press, 1994).

Johnson, R., "The public image of science and its funding," *FASEB Journal* 4 (1990):2431–432.

Kenward, M., "Science stays up the poll," *New Scientist* 16 (September 1989): 57–61.

NABR, *Associated Press/General Media Poll* (Washington, DC: National Association for Biomedical Press, 1985).

NSB, *Science and engineering indicators—1989* (Washington, DC: National Science Board, 1989).

NSB, *Science and Engineering Indicators—1992* (Washington, DC: National Science Board, 1993).

Parents Magazine, "Attitudes to animal use," Poll conducted for Parents Magazine, October, 1989.

Pifer, L., Shimizu, K. and Pifer, R., "Public attitudes toward animal research: some international comparisons," *Society and Animals* 2 (1994): 95–113.

Pion, G.M. and Lipsey, M.W., "Public attitudes toward science and technology: what have the surveys told us?" *Public Opinion Quarterly* 45 (1981): 303–16.

Ritvo, H., "Plus can change: antivivisection then and now," *Science, Technology and Human Values* 9(Spring 1984): 57–66.

Salt, H., *A Lover of Animals* (Rochester, NY: Lion's Den Press, [1895]1984).

Stevenson, L.G., "Religious elements in the background of the British anti-vivisection movement," *Yale Journal of Biology and Medicine* 29 (1956):125–57.

Stevenson, R.L., *Dr. Jekyll and Mr. Hyde* (New York: Penguin Books, 1979).

Takooshian, H., "Opinions on animal research: Scientists versus the public?" *PsyETA Bulletin* 7:2 (1988): 5, 8–9.

Weart, S., "The physicist as mad scientist," *Physics Today* (June 1988): 28–37.

A Consideration of
Policy Implications:
A Panel Discussion

Vicki Croke, Colin McGinn,
Joy Mench, J. Anthony Movshon,
John G. Robinson, James Serpell,
Kenneth J. Shapiro, Nicholas Wade

T HE following questions were posed to the members of the panel in the final session of the conference "In the Company of Animals":

(1) By treating animals as property, do we inevitably or necessarily encourage the treatment of animals as inanimate objects? Should society develop a category that treats animals as something other than property, and, if so, what should that category be?

(2) Animal pain and distress is clearly accorded significant weight in society's attitudes toward animal use. Would you draw a line between animals that can experience pain and distress and those that cannot, and, if so, what are the criteria you would use to determine where that line should be drawn? What about animal suffering—do you perceive it to be identical to animal pain? If not, how would you distinguish between animal suffering and animal pain?

(3) What conflicts arise when society treats animals as groups of organisms (that is, populations) as opposed to individuals? How should society deal with such conflicts? How does society deal with conflicts between human populations and human individuals?

What follows are the responses of the panel members.

VICKI CROKE

The issue of animal suffering versus animal pain is brought into sharp focus at the zoo. Here we find some of the most robust examples of exotic species who are fed well-balanced, nutritious diets and kept in immaculate exhibits. These animals are parasite and predator free. They reach reproductive maturity earlier, and they cycle faster than their counterparts in the wild. Infant mortality rates are low. They have shiny coats and wet noses. But while we keep them in peak physical condition, we plunge many of them into a state of behavioral bankruptcy.

Mothering skills, navigational abilities, and survival techniques may be foreign to these animals. Very few zoo animals could thrive in the wild using only the skills learned in captivity. And it is clear that they suffer for it.

Many zoo animals display hostile, aggressive, withdrawn, and neurotic behavior. The unhappiest ones perform ritualistic acts that are not seen in the wild and appear to indicate a disturbed state—a polar bear pacing the same number of steps out and back over is the most obvious example, but a gorilla regurgitating and reingesting his food or an elephant bobbing her head are signs of the same problem. It is believed that these repetitive displays may allow animals to tap into the body's built-in opiates (the endorphins that produce a "jogger's high").

How many zoo animals suffer from stereotypy, behavior that is not fully understood and is difficult to cure, is not known. The zoo community itself has no estimates, though it vehemently disagrees with the assessment of the British animal rights group, Born Free, which monitors animals in captivity, that mental illness in zoo animals is rampant. Less easily dismissed is the judgement of one distinguished veterinary brain chemistry expert, who feels that as many as 50 percent of the animals at the country's best zoos may be affected.

Do these animals suffer? Clearly, a bear who rubs the pads

of her feet raw with pacing does. A gorilla violently shaking his head to regurgitate food and reingest it appears to suffer also. They are, in the words of one scientist, "compensating for a less than adequate environment."

We do not yet know if this can be characterized as physical pain, but we do know that these animals are suffering. We now know that being humane means caring for more than the physical shell of an animal. But enrichment programs for zoo animals are costly—it is expensive to build adequate habitats, and it is labor intensive to set up activities, such as hiding food throughout the exhibit for chimps or freezing buckets of mackerel for the polar bear pool. As a society, our awareness or our discomfort level has not reached the point at which we demand change.

In 1984, the Minnesota Zoo decided to kill a young and healthy female tiger. For some time, the zoo had tried to sell or even give away the animal to another accredited zoo, but no one was interested. It was decided that the only solution was euthanasia, and the zoo quietly prepared to kill the animal, whose genes were sturdy but overstocked. Survival plans are set up with fairly rigid numbers, allowing for space consider-ations, so extra babies from "too large" litters and nonrepro-ductive older animals are considered surplus—they no longer contribute to the plan. Animals whose family tree is uncertain or ones whose family genes are already well represented in the population are also excess baggage.

Despite precautions, the news hit the press, and the outpouring of condemnation was so great that telephone lines in the community were jammed by protest calls. The tiger lived and was eventually shipped to the Shanghai Zoo.

The issue of the rights of individual animals versus the good of a larger population is not new to the zoo.

There are not always enough spaces in zoos, and we must choose: shall we sacrifice a whole species in order to save a few individuals? Some scientists insist that we must steel ourselves to the hard truths of species preservation and not be so attached to specific animals. This despite the fact that zoos routinely raise money by making the public feel attached to certain cute animals with names.

Some observers say the concerns of individual animals are being voiced increasingly in SSP (Species Survival Plan) meetings, where the course of species preservation within the zoo community is charted.

Many zoo people have asked why they should be condemned when millions of unwanted cats and dogs are killed each year in our country. The question is, why would a zoo compare itself to irresponsible pet owners? We expect our zoo directors to be more humane and ethical and just plain smarter than the kind of people who dump dogs. The comparison is a good one only if zoos want the image of the brute who drowns a basket of kittens.

Terry Maple, director of Zoo Atlanta, says that he would never euthanize an animal for so-called management purposes. He believes that zoo animals who have "served humanity deserve to live."

Perhaps there is a solution. Perhaps we should not accept limitations of space or money so easily. How much space and how much money will it take to save all Siberian tigers—the ones that are valuable to species preservation and the ones that do not quite fit in? Why cannot zoos pool a slush fund and build a tiger sanctuary for the misfits?

In the end, it will not be zoos that decide, but society: what the public cares about, and how willing it is to act on its beliefs.

COLIN McGINN

It is hard not to be a specialist: one looks at animals (including humans) from a particular point of view, in which certain aspects of their nature are selected and focused upon. It is all too easy to fall into thinking that this limited perspective gives the whole truth about animals or at least the most important part of the truth. The molecular biologist is apt to see animals as collections of cells and smaller physical units. The evolutionary biologist sees them as gene survival machines locked in competition for scarce resources. The behavioral psychologist sees them as bodies that jerk and twitch in response to the environmental contingencies. The farmer sees them as food on the hoof, his means of livelihood. The dog fancier sees them as aesthetic exhibits in shows. The pet owner sees them as human companions. The poet sees them as sources of inspiration. The animal welfare activist sees them as victims of human abuse.

Given the pull of specialization, it is necessary to try to obtain a more synoptic view of animals. We need to remind ourselves of how rich and various and many-faceted they are. This is not just a matter of seeing how they differ among themselves— though that too is important; it is more a question of appreciating their full multi-dimensionality. In particular, it is important to see that animals are not defined by their relation to us. Most animals, after all, have lived out their spans in sublime indifference to the habits of those odd chattering bipeds with the removable plumage. Even if we had never existed, they would still be here. We are just as accidental to them as they are to us. Their *esse* is not human *percipi*. And what they are is no simple, single thing. If we are to comprehend them as they really are, we need to recognize the many perspectives from which they can be viewed.

The conference helped in achieving this sort of comprehensive perspective. It offered us animals as viewed in the law, in science, literature, art, ethics, and philosophy. This was eye-opening, both overtly and subliminally. Parents notoriously tend to have a very limited and skewed conception of their children: they see them only from their perspective as parents—as helpless, difficult, dependent. It is often a great surprise to them to find that their children have lives that go quite beyond their status as children. Parents take a specialist view of their offspring, in which their nature is defined in relation to them. Well, we humans tend to take this sort of limited view of animals: they are nothing but what is revealed to one or another specialized perspective. It takes a mental effort, and some input of information, to recognize that animals are not defined by their particular relation to us and the specialized attitudes we adopt towards them. Anthropocentrism is not the problem; the problem is specialty-centrism. I once knew a poet who was so fixated on the idea of animals as objects of rustic beauty and repositories of family traditions that he simply could not see that they also might be viewed as victims of human exploitation or indeed as products of Darwinian natural selection. And the scientists are legion who cannot get over the image of the animal as primarily an object of scientific experimentation, as if that is why they were *put* here. Even welfare activists can become dominated by the idea that animals are defined by their victimhood, as if their primary mode of existence is that of being humanly tormented and done to death.

So I would like to enter a plea, which I think the conference reinforced, that we stay flexible and open and catholic in our thinking about animals. We should take our own particular specialty or interest and suspend it for a while, so as to try out other points of view. I believe the result of this is to create respect for animals, even awe. When a parent comes to understand that her child is not *merely* her child, this can often increase the respect she has for him, for she recognizes that the child has an autonomy and reality that transcends the parental viewpoint. In a sense, the parent's solipsism is breached and the reality of the Other flows in, producing the respect that derives from an appreciation of otherness. In the same way, we need to shed our species solipsism so as to recognize the full reality of other animals; then we shall come to respect them as co-inhabitants of the same world. It sometimes helps here to try to see it from the animal's point of view: the rhino looks at us with the same skewed solipsism we bring to him, and surely we do not want to be as limited in our outlook as he is. We are far more in ourselves than we are for him, and he is more than how he strikes us. I sometimes have the sensation when I gaze at an animal that there is a whole world there that I only dimly glimpse, and that *this* is what the animal really is. I may not be able to fathom this reality fully, but I can at least approximate it by letting my conception of the animal be as complete and various as can be. We all want others to see us in our full reality (well, minus the bad bits); we owe it to animals to extend them the same courtesy.

And perhaps, in the end, that is what it comes down to: we need to improve our *manners* in respect of animals. For too long we have been rude to them, inconsiderate, ignoring their feelings and refusing to respect their autonomous dignity. We must instead show them that we are aware of their existence as beings that are not defined by our practical transactions with them. They too have families, cares, joys, sorrows. They have a *life*. And the first step in respecting their life is to see them from all sides. Then we might start to see them as they really are. Suppose that whales (say) have a God, a whiskery white giant in a celestial sea. This God sees the whole truth about whales, being omniscient (we can only imagine what he says to our God about the behavior of His lookalikes). Well, we should aspire to the outlook of this whale God in our thinking about whales: we should see them in their entirety, from all angles, and with the consideration that goes with such full knowledge.

Seeing Animals As They Are

JOY MENCH

If this conference illustrates one thing clearly, it is this: we need animals, and not just because they are used for food, companionship, or as models for human biology and disease. We need them because they play rich, symbolic roles in our lives. We humanize, demonize, and exalt them, make and re-make them into a reflection of our changing relationships with nature and our conflicting feelings about the wildness and tameness in ourselves.

Animals have long been portrayed as exemplars of human behavior. The medieval Bestiaries extolled the virtues of the female turtle dove, because she "[F]aithfully is true for life to her love. For, once she has a partner, she'll never leave him. *Women would do well to keep her in mind*" (Elliott, 1971, emphasis added). Early twentieth-century American nature writers like William J. Long (dubbed the "nature fakers" by Theodore Roosevelt) spun tales of innate animal generosity and religiosity, and argued that humans could learn much from these "natural" virtues (Lutts, 1990). And in our 1995 version there is "Babe," the cinematic pig who marches to a different drummer and proves that anyone with determination can achieve his aspirations, no matter how far removed from real life.

But can utility and symbol translate into policy that governs our relationships with animals in ways that are beneficial to both animals and humans? I would like to address this issue with respect to two questions that were posed to the panel, the first regarding the treatment of animals as property and the second regarding pain and suffering.

Tannenbaum (1995) argues cogently that treating animals legally as property is neither inappropriate nor bad. As a biologist who studies animal welfare, however, I find that I am not completely persuaded by his argument. Certainly, owners generally take "good" care of their property because they have a personal interest (be it financial, aesthetic, or whatever) in that property. The concept of owner interest, however, has created anomalies in our current treatment of animals, because "good" treatment is usually defined solely or primarily with reference to the particular interest that the owner has in that animal.

"The law is an ass," declared Mr. Bumble in Oliver Twist (illustrating not only our perennial dislike of the legal system but the symbolic value of animals). But under current United States law, an ass is not an ass is not an ass. In fact, the legal status of asses (and other farm animals) is particularly illustrative of the problems that arise when animals are treated as property. Farm animals used in biomedical research are closely regulated both under the Animal Welfare Act and the standards promulgated by the National Institutes of Health (NIH). Farm animals used in agricultural research, on the other hand, are specifically excluded from the Animal Welfare Act and NIH standards and are instead covered under a different, and completely voluntary, set of guidelines (The Guide for the Care and Use of Agricultural Animals in Agricultural Research and Teaching) developed by a non-governmental body. Farm animals kept in zoos are regulated under the Animal Welfare Act, but farm animals kept as pets are not. The latter, however, do fall under the jurisdiction of most state anti-cruelty laws, although apart from prohibiting overt cruelty, these statutes generally do not regulate aspects of day-to-day treatment as federal laws and guidelines do. Animals kept on farms are exempt from the Animal Welfare Act and are also typically exempt from state anti-cruelty statues, providing they are used in "normal" agricultural enterprises. These differences arise principally from the notion that animals are property with a particular function or value, and result in important differences in treatment that require reflection.

Let us take pigs as an example. On swine farms, sows are often housed individually in pens in which they cannot turn. These pens prevent sow-sow aggression and crushing injuries to piglets, allow individual monitoring of sow health and food intake, make handling of sows easier for the farmer, and are economically advantageous because they minimize the space and other resources (for example, energy) necessary for production. Young male pigs destined for market are castrated to prevent "boar taint" in the meat due to male sex hormones, a taste that many consumers find offensive. Anesthesia is not used because it is time-consuming and costly for the farmer, increasing the price of the meat. Long-term close confinement of sows or castration without anesthesia are, therefore, legal (and routine) parts of many agricultural enterprises. These are also acceptable practices for agricultural researchers, who may need to simulate commercial conditions in order for their research to be applicable. Such practices, however, might or might not be considered cruel if they were practiced by a pet pig owner, depending on the state in which the owner resides and the predilections of the local humane societies and judges who enforce the laws. Because they involve long-term confinement and unalleviated (although brief) pain, they would not be acceptable in a biomedical facility or in a zoo, unless there was a strong scientific or educational justification for their use.

All of these standards, then, are legal, but which one (or ones) is appropriate? And who is to make this decision? As society's concern about the treatment of animals grows, conflicts about whose property animals really are, and, thus, who makes the decisions about "good" care, are inevitable. This is illustrated by two opinion polls that I saw recently. The first, which was conducted by an agribusiness group called the Animal Industry Foundation, reported that 69 percent of the public believes that animal production systems should be regulated with respect to animal welfare. The second, a poll of farmers conducted by Farm Futures magazine, found that a similar percentage of farmers (75 percent) thought the public should not have a say in the way they raise animals. Do farm animals "belong" to the consuming public (via public agencies like USDA) or to the farmer? Does a laboratory rat "belong" to the researcher, the research institution, or the federal agency representing the public that funded the research?

In order to resolve these dilemmas, I believe that we need to accord animals a legally and morally appropriate status based on a better understanding of their own characteristics and interests. In order to do this, we will need to come to know them for who *they* are, not for who we need (or want) them to be.

In 1934, the German zoologist Jakob von Uexküll compellingly (if somewhat inaccurately) described the *"Umwelt"* or world-view of the tick. He wrote,

> The 'eyeless tick' is directed to this watchtower [a branch] by a general photosensitivity of her skin. The approaching prey is revealed to the blind and deaf highway woman by her sense of smell. The odor of butryic acid, that emanates from the skin glands of all mammals, acts on the tick as a signal to leave her watchtower and hurl herself downwards. If, in so doing, she lands on something warm—a fine sense of temperature betrays this to her—she has reached her prey, the warm-blooded creature.

Our perceptual world, and their perceptual worlds, can be worlds apart. Many birds are like us in that they have excellent visual acuity and fine color discrimination, but they also can see things in the ultraviolet range that are invisible to us (at least without special technology). Pigs and dogs have comparatively poor vision, but live in a world of odors that we cannot even begin to imagine. Bats and dolphins navigate by sonar, pigeons by magnetic fields (maybe), and salmon by chemical cues (maybe).

The philosopher Thomas Nagel (1979) posed the question: "What is it like to be a bat," and concluded that we can never really know. I suggest that we might be able to come close, however, if we learn to ask the right questions and use a judicious mixture of detached analysis and empathy to interpret the answers we are given.

Animal pain is probably the area that has received the most attention in this regard in recent years. Techniques for evaluating and alleviating animal pain have improved dramatically. However, as Turner (1980) demonstrated in his analysis of Victorian culture, pain and pain alleviation are particularly human obsessions. Perhaps for some other animals pain is less important than it is for us or of little importance at all. I agree with Rowan (1988) that the study of animal suffering and other unpleasant mental experiences distinct from pain deserve far more emphasis than they have been given. But there are also important aspects of animal experience that go beyond pain and suffering, for example, the feelings that we call pleasure and satisfaction (Mench, 1995). Indeed, Rollin (1993) has argued that the "emerging social ethic for animals . . . will demand from scientists data relevant to a much increased concept of welfare. Not only will welfare mean control of pain and suffering, it will also entail nurturing and fulfillment of the animals' natures."

In the last twenty years, scientists have indeed begun to devise techniques to assess suffering and other subjective experiences in animals (Dawkins, 1990). The results of such assessments can be both surprising and illuminating. For example, chickens destined for market are caught by hand and loaded into transport crates, a process that is difficult, time consuming, and labor intensive. Several years ago a mechanized catcher was introduced into Britain, which used rotating rubber "fingers" and conveyor belts to catch and load the birds. The British press, long sensitive to animal issues, decried the mechanization (factory farming, if you will) of animal agriculture, labelled the admittedly gruesome-looking machine "the engine of death," and proclaimed it the worst thing to happen to animal welfare in a decade. However, ethological studies (Duncan, et al., 1986) revealed that the birds found being handled by the machine far less stressful than the standard handling procedures. The novel machine, apparently, meant little to the birds, while humans (predators?) were frightening and aversive. Chicken perspectives and human perspectives, it seems, do not always mesh.

I know that scientists are accused of always saying that they need "more studies," but this time it is really true, at least if we are to create laws and environments for animals that meet their interests, needs, and desires rather than simply reflecting human concepts of what is best for them. Otherwise, I fear that while animals will continue to be in our company, we will never truly be in theirs.

References

Dawkins, M.S., "From an animal's point of view: motivation, fitness, and animal welfare," *Behavioral and Brain Sciences* 13 (1990): 1–61.

Duncan, I.J.H., Gillian, S.S., Kettlewell, P., Berry, P., and Carlisle, A.J., "Comparison of the stressfulness of harvesting broiler chickens by machine and by hand," *British Poultry Science* 27 (1986):109–114.

Elliot, T.J., *A Medieval Bestiary* (Boston: Godine, 1971).

Nagel, T., *Mortal Questions* (Cambridge: Cambridge University Press, 1979).

Lutts, R.H., *The Nature Fakers* (Colorado: Fulcrum Publishing, 1990).

Mench, J.A., "Beyond pain: pleasures, interests and the satisfaction of desires," Animal Behavior Society annual meeting abstracts (Lincoln, NE: University of Nebraska Behavioral Biology Group, 1995), p. 31.

Rollin, B.E., "Animal welfare, science, and value," *Journal of Agricultural and Environmental Ethics* 6, Special Supplement 2 (1993): 44–50.

Rowan, A.J., "Animal anxiety and suffering," *Applied Animal Behaviour Science* 20 (1988):135–142.

Tannenbaum, J., "Animals and the Law: Cruelty, Property, Rights . . . Or How the Law Makes up in Common Sense What It May Lack in Metaphysics," *Social Research* 62:3 (Fall 1995).

Turner, J., *Reckoning with the Beast: Animals, Pain and Humanity in the Victorian Mind* (Baltimore: Johns Hopkins University Press, 1980).

von Uexküll, J., "A stroll through the worlds of animals and men," in C.H. Schiller, ed., *Instinctive Behaviour* (New York: International Universities Press, Inc., 1957), pp. 5–80.

J. Anthony Movshon

I will address my remarks to the two parts of the second question. It is common to attribute sentience and consciousness to animals with whom we are empathetic and about whom it is easy to anthropomorphize (generally mammals, the cuter and furrier the better). Similarly, it is altogether too easy for humans to attribute only basic biological functions to animals with whom we have no particular bond (generally poikilotherms or other creatures of unfortunate aspect). These two views are each too extreme and carry all the vices of extremism. As Daniel Dennett and others have pointed out, the attribution of human-like feelings to some species leads to exaggerated beliefs about their quasi-humanity and often to unwarranted concerns for their welfare based on wildly inaccurate surmises based on fundamentally emotional judgements. Equally, as Andrew Rowan and others have observed, animals with which we have little sympathy can be mistreated without arousing society's ire. For me, the correct point lies in between and must be based on an understanding of each species' natural repertoire of behaviors. Such an understanding should be used to guide and regulate the judgement of levels of pain and distress and the steps taken to attenuate them.

Animal suffering is most certainly *not* identical to animal pain. It is a common misapprehension that pain is itself inherently undesirable and to be avoided. In fact, it is to prevent suffering—or, to use a better word, "distress"—that is most important. Distress can arise from many causes; for example, it is a possible but not inevitable *result* of pain, and it can arise from many situations that are not obviously painful.

Pain is a physiological response to stimuli that have the potential to damage tissue. It has been recognized for some time that pain gives rise to two distinct categories of response, the so-called *sensory/discriminative* and *motivational/affective* aspects of pain sensation. The sensory aspects of pain are not unlike those of other senses. They can be judged and scaled and ranked as other sensory stimuli and are not by themselves distressing. Rather, it is the affective aspects of pain that directly cause distress. There are many natural cases and situations in which these two aspects become dissociated, and analgesic drugs such as opiates appear preferentially to block the affective aspects of pain. Prevention of distress due to pain is thus achieved by controlling the affective aspects of pain. What is equally important, however, is to prevent distress due to other causes.

The most important other source of animal distress arises when animals are prevented from expressing their natural species-typical behavior. It is, of course, much more difficult to prevent all such sources of distress, because this can be done only with a clear understanding of a species' natural behavioral repertoire. However, it safely can be asserted that much unintentional distress is produced in this way by well-meaning people whose understanding of animals is incomplete. I need only offer the example of a pampered and coddled pet as a potential recipient of profound distress to make that point clear.

Finally, it should be noted that *stress*, another word commonly misused, is a natural and often positive aspect of both human and animal lives. Stress can lead to distress, but generally only in circumstances like those described earlier, in which the valuable capacity of stress to promote survival-enhancing behavior cannot be expressed.

In sum: neither pain nor stress is necessarily to be avoided—distress and suffering are.

The Responsibility to Conserve Wild Species

JOHN G. ROBINSON

Until the development of urban society, the lives and deaths of wild animals and people were inextricably intertwined. People killed and consumed animals and vice-versa. This interdependence is reflected in the cultural importance given to wild animals, whose characteristics are symbolically used in a range of traditional cultures. As human beings increasingly buffered themselves, both technologically and culturally, from the actions of wild animals and concomitantly were able to control the lives of these animals, the relationship changed. The evolution of that relationship, as expressed in philosophy, literature, and scientific thought, has been explored in this conference. This essay addresses this same relationship but has a narrower focus: In the modern, increasingly urban society, how should we treat wild animals? Most of us personally experience wild animals only through cultural lenses, such as nature shows on television, or as interesting but vaguely threatening presences during vacation forays into the rural landscape. A more precise question is: What is the ethical justification for people living in the urban society to intervene in the lives of wild animals? This leads into a final question: What kinds of intervention are justifiable?

I will address these questions from the perspective of a conservationist, more precisely one who accepts Aldo Leopold's premise that "A thing's right when it tends to preserve the integrity, stability and beauty of the biotic community. It is wrong when it tends otherwise" (1949, pp. 224–25). This statement can be supported from both the utilitarian position — that to do otherwise would endanger the resource base upon which human society depends — or from a more biocentric position — that wild species, and the natural world in general, have an inherent right to exist. Conservationists have tended to synonymize integrity, stability, and beauty.

A biotic community that has "integrity" has the full diversity of species, which allows the system to function ecologically in an appropriate way. The "stability" of the community, both its resilience to disturbance and its persistence through time, depends on that species diversity. And conservationists consider "beauty," while the term is not in the scientific lexicon, to be defined by that diversity and stability. For Leopold, people were an integral part of this biotic community, and anthropological research has documented the role that traditional cultures play in creating and maintaining biological diversity in many natural communities. Yet it is also clear that in our present world, the actions of both modern and traditional cultures generally tend to degrade natural systems and reduce biological diversity. The present-day rate of species extinction is perhaps higher than at any time in our planet's existence, and the actions of human beings are the single largest contributor to this global degradation.

The impact of humans on the rest of the biota is ubiquitous. Terms like "primeval," "virgin," "primary," "undisturbed," "pristine," even "wilderness," all of which connote biological communities uninfluenced by humans, refer to a certain ideal unattainable in the modern world. From the high deserts of Chang Tang in Tibet to the depths of the tropical forest in central Amazonia, the human presence is everywhere discernible. This is not to state that all nature is a human construct. It is not. Excepting biological communities in urban and agricultural settings, the structure and functioning of biological communities is still predominately determined by species other than humans. But humans do have a pervasive impact on wild species worldwide, and this defines how we must treat wild species. Few truly "wild" species—those uninfluenced by humans—exist today on our planet. And, thus, we cannot abnegate all responsibility for the fates of individual animals or for the continued existence of the species—they cannot be left "to do their own thing." We must take responsibility for our influence on the lives of wild animals.

Our primary responsibility, if we accept Aldo Leopold's premise, is to ensure the survival of species in nature. The least intrusive action is to establish protected areas— parks and reserves for species and the biological communities on which they were a part—and then minimize human impact within these areas. Even here, human impact in and around reserves is significant, and active management is usually necessary to maintain the biological community. Population management, predator restoration, habitat modification, and landscape restoration are necessary tools for protected area managers, and all have an impact on wild species.

More intrusive conservation actions are frequently necessary. If the goal is the preservation of biological diversity, protected areas alone are insufficient. First, it is unlikely that we will be able to protect more than a small fraction of the planet's surface in parks and reserves, and the long-term persistence of many species and communities requires larger areas. Second, governments and regulatory agencies are unable to protect areas if local human inhabitants and other interested parties do not support the park or reserve. Park personnel tend to be inadequately funded, supported, and trained. Through political machinations or illegal actions, local peoples can undermine the best efforts of park managers—as evidenced by the recent difficulties experienced by the United States National Park Service and the Forest Service. Accordingly, conservationists frequently advocate working outside protected areas and enlisting the support of local communities in conservation efforts in and around protected areas. Local community involvement requires that local people value wildlife species, and frequently this means allowing them rights to harvest or otherwise use wild species and wild areas. This approach is considerably more intrusive because it involves treating wild animals as resources. It is also controversial because the consumptive use of wild species is seemingly in conflict with the goal of protecting them. However, it is clear that allowing local people to exploit a species in certain circumstances can vest them in the process of conserving wild species or biological communities.

Another potentially justifiable intrusion is to bring wild animals into captivity. When wild populations are imperiled by habitat conversion, when animals cannot be protected from hunters, or when other species endanger remnant populations, then bringing animals into captivity can be the most responsible action. The removal of the last condors from the wild in California was justified using this argument. Zoos in particular have assumed the responsibility of maintaining populations of endangered species and have become involved in reintroducing animals back into the wild when circumstances are more favorable. Successful reintroductions attest to the utility of this approach—including the efforts of my own institution, then called the New York Zoological Society, in reintroducing bison to the American west at the beginning of this century. Zoos also have brought animals into captivity with the expressed aim of introducing living animals to a generally urban public and educating them on the need for conservation, in effect using individuals as "ambassadors" for their species. And in the United States, some 100 million people annually visit zoos, and some 14 million participate in formal zoo education programs.

If the goal is to preserve the biological community, then the survival of a species takes precedence over the welfare of selected individuals of the species. Human actions which promote the conservation of a species or a population at the expense of individuals are justified. The welfare of the collective as a whole is more important than the welfare of any one individual. However, even from this perspective, there are circumstances in which the individual welfare of an animal attains importance. As animals become rarer, we value individual animals more, and, thus, the mechanism to conserve species increasingly depends on protecting individuals. For instance, consider the proposed establishment of tiger farms in China to provide bones for the traditional medicine trade, or the proposed harvest of black rhinos in southern Africa for the horn trade. In both cases, arguments have been made that these actions would promote conservation of the species. Yet populations of these species are now tiny, and the risks to the population of harvesting are great. Our efforts to conserve these species depend on our success with protecting each individual, and such proposals have received little support within the conservation community.

I have argued that human beings are ethically justified in intervening in the lives of animals if it promotes the conservation of populations or species. Are all kinds of interventions justifiable? From a conservation perspective, the answer is yes. But this answer is incomplete. There are humane considerations that in practice are included. If wild animals are to be harvested, then the humaneness of their killing must be considered. The conservation perspective also does not consider the extent to which a wild species is sentient, yet the actions of conservationists frequently reflect a sensitivity to this issue. For instance, no proposal to bring the mountain gorillas into captivity has been advanced, not even during the recent human tragedy and political unrest in Rwanda. The agonized debates about whether to support harvests of elephants and whales within the conservation community also reflect deep concerns about animal sentience.

The reality that human beings significantly influence the natural world, either directly or indirectly, means that we must take responsibility for the survival of wild species. The inescapable consequence of this is the active management of individual animals, populations, and communities. The more humans intervene, the more responsibility they must assume, and as wild population dwindle, the more responsibility we must take for individual animals. And this creates the paradox. The ultimate goal is to preserve the natural world and the wildness that defines it. Yet the methods we use to conserve species, and care for individual animals, can rob animals of the wildness that we value in them. But to do otherwise is irresponsible.

References

Leopold, Aldo, *A Sand County Almanac* (New York: Oxford University Press, 1949).

Animal Sense (and Non-sense)

JAMES A. SERPELL

The tendency to treat (nonhuman) animals as inanimate or insensate objects is not "inevitably or necessarily" encouraged by treating them legally as property. Under English law, for instance, "owned" animals enjoy greater legal protection from abuse or cruelty than do many "unowned" animals living in the wild. Having said this, however, there can be little doubt that property laws have also been used to justify mistreating animals, and many of the earliest attempts to introduce animal protection legislation were blocked on the principle that people had an absolute legal right to do whatever they liked with their own property (Ryder, 1989). Clearly, the law is an ass when it allows property rights to be used to legitimize the infliction of suffering on others. However, rather than dispensing altogether with rights of ownership—a move likely to be strongly opposed by anyone with a vested interest in continued animal use[1]—it would seem more sensible, as Jerrold Tannenbaum has suggested, to seek ways of providing animals with greater protection within existing property laws. In many countries, the legal owners of historic buildings or endangered species' habitats are also legally prevented from doing whatever they like with their own. In other words, society places a special value on the building or the habitat and, by doing so, restricts the rights of the owner. I can see no obvious reason why society should not also impose comparable restrictions with respect to the treatment of animals.

At some point in the process, a decision would need to be made concerning whether the new restrictions should apply to all animals (unlikely), or just some animals based on their assumed capacity to suffer, and at this point it would probably be necessary to differentiate a special category of animals. The phrase "sentient beings" springs to mind as a possible name for this special category, although the issue of sentience, and how we begin to recognize and measure it in nonhumans, still remains an unresolved question (see below).

Answering the second part of the second question first: in the strict biological sense, "pain" is the subjective sensation arising from the stimulation of nociceptive nerve cells. We may talk about the "*pain* of separation" or the "*pain* of loss" but in these cases we are using the word in a metaphorical sense. Suffering is a much broader term than pain (Dawkins, 1980). I may "suffer" from a pain in the neck, a cold in the head, nausea, or more general feelings of illness or malaise. Feelings of boredom, hopelessness, depression, loneliness, worry, anxiety, or terror may also cause me to suffer. The pain system is phylogenetically ancient, and it is likely, based on reasonably objective anatomical and physiological criteria, that all vertebrates and at least some invertebrates (such as cephalopods) are capable of experiencing painful sensations (Bateson, 1992). The range of stimuli that may cause an individual organism to suffer probably also increases with the complexity of its nervous system, and we need to be alert to the possibility that some animals may suffer in circumstances where we would not. Animals with particularly acute hearing, for example, may be more sensitive then we are to certain wavelengths or amplitudes of sound and may suffer from prolonged exposure. Conversely, we should not assume that all mammals, say, or even all higher primates, necessarily have the same capacity to suffer as humans. Humans may suffer vicariously from the sight, or even the thought, of another's suffering, but there is little evidence that any other animal species is capable of experiencing this degree of empathic suffering (Seyfarth and Cheney, 1992). In the light of these observations, it is probably unwise to draw precise lines of demarcation between those organisms which are "sentient"—that is, capable of suffering[2]—and those which are not. It seems more appropriate at this stage to talk about different degrees and kinds of sentience.

Surprisingly, the capacity of animals to suffer or feel pain has not always been "accorded significant weight" in society's deliberations on the ethics of animal use. From classical antiquity until the early modern period in Europe, philosophical and theological discussions concerning the moral status of animals revolved almost entirely around the issue of "rationality" rather than sentience. Following the teachings of Aristotle, animals were held to be devoid of "reason" and "belief" and, as such, were denied moral consideration (Clutton-Brock, 1995). Many early debaters expressed ethical reservations about the appropriateness of killing and eating animals, but whether animals suffered or not in the process seems to have been largely immaterial (Sorabji, 1993). This curious—from a modern perspective—omission of suffering from early discussions on animal use has never been satisfactorily explained. Perhaps before the development of effective analgesics and anaesthetics in the eighteenth and nineteenth centuries the threshold of public sensitivity to *human* suffering was simply too high to allow for much concern for the suffering of nonhuman animals (Ryder, 1989). Since it is difficult to draw hard and fast distinctions between humans and nonhumans on the basis of sentience (see above), the long-standing appeal of the Aristotelian perspective may also have been the ease with which it could be used to justify the continued exploitation of those deemed to be lower down the scale of rationality (Serpell, 1986).

Once again, answering the last part of the third question first: most human societies have evolved elaborate systems of law for resolving conflicts between individuals and between individuals and the societies to which they belong. Laws may differ markedly between different societies, depending on prevailing ideologies (compare capitalism and socialism), but all of them are essentially concerned with establishing the relative priority of conflicting or competing human interest claims. Since animals, neither collectively nor individually, have legal rights, we have no established rules for assessing the relative strengths of their competing interests. Indeed, the mind fairly boggles at the prospect of trying to devise a system that would enable us to do this.

Treating animals as groups of organisms (populations, species, ecosystems, and so on) creates ethical problems when it encourages people to ignore or devalue the well-being of the individual animals comprising those groups. An extreme example is provided by modern intensive farming. Intensive farming systems are geared towards achieving maximum productivity at minimum cost. In the process, they select for strains of domestic livestock that are able to sustain abnormally high rates of growth and reproduction under unusually suboptimal environmental conditions. Abundant behavioral and physiological evidence now suggests that the individual animals within these kinds of systems experience significant degrees of suffering (Dawkins, 1980). However, because the focus is on net output at the group level, the welfare of individual animals tends to be ignored. Legislation outlawing some of the more extreme farming practices might help to resolve these conflicts, although, ultimately, consumer willingness to spend more on foods produced in less intensive ways is probably the only long-term solution to the dilemma.

In the field of conservation, it is also standard practice to ignore or subordinate the interests of individual animals for the perceived good of their own or other species. Hence, the managerial "culling" of selected individuals as a means of controlling wildlife populations, the preservation of indigenous species through the eradication of introduced competitors and predators, or the capture and confinement of individual members of endangered species for the purposes of captive breeding. In some of these contexts it is unclear whether the population focus necessarily results in a net reduction in individual animal welfare. Culling, for example, may be more humane than allowing animals to starve to death "naturally." In others, however, the conservation ethic is undoubtedly at odds with moral perspectives based on animal rights or welfare. For instance, in national parks throughout the world, park officials will not usually intervene humanely, either with veterinary care or euthanasia, in situations where wild animals are suffering or dying from supposedly "natural" causes. This policy is based on the argument that human intervention in such contexts would represent interference with the very process of natural selection which *created* wild animals in the first place (Rolston, 1992). As long as society places a high priority on the preservation of "wildness" in this sense, it is difficult to see how this sort of conflict of interests could ever be resolved to the satisfaction of all parties.

Reflecting on the conference as a whole, I am repeatedly reminded of David Freedberg's[3] reference to the medieval practice of assessing the character, nature or "essence" of both animals and people from their superficial appearances. It has left me wondering how far we have really progressed in this respect since the Middle Ages, at least with regard to animals. How readily we still dress them in fashionable, symbolic or

metaphorical clothing! How difficult it still seems for us to peel
back these fictitious disguises and see the real animals that lurk
within! When we misrepresent humans, when we say or believe
things about them which are untrue, sooner or later somebody
will put us straight, if necessary by means of a lawsuit. But
animals have no means of challenging our misrepresentations.
Like putty, they are moulded by our perceptions into virtually
any image we choose to impose on them. And nine times out of
ten they are diminished in the process. Admittedly, many of
these distorted images are trivial and harmless in themselves,
but the process which creates them is not. It is indirectly
responsible for perpetuating inappropriate, and frequently
inhumane, attitudes to animals and their treatment.

We need to study animals carefully and objectively. We need
to learn to see them as they really are, not as we imagine them
to be. Only then will we be in a respectable ethical position to
begin to judge how they should be treated.

Notes

[1] These would necessarily include not only livestock farmers,
animal researchers, and accessory industries but also relatively
"benign" users such as companion animal breeders and owners and
the owners/managers of zoological collections.

[2] Strictly, the term "sentience" denotes the capacity to feel anything
at all, not just pain, distress, or discomfort (Ryder, 1992).

[3] David Freedberg, a participant in the "In the Company of
Animals" conference, was unable to revise his talk in time for this
publication.

References

Bateson, P.P.G., "Do animals feel pain?" New Scientist, 134:1818
(April 22, 1992):30–33.

Cheney, D.L. and Seyfarth, R.M., *How Monkeys See the World: Inside the Mind of Another Species* (Chicago: Chicago University Press, 1992).

Clutton-Brock, Juliet, "Aristotle, The Scale of Nature, and Modern Attitudes to Animals," *Social Research* 62:3 (1995).

Dawkins, M. *Animal Suffering: the Science of Animal Welfare* (London: Chapman & Hall, 1980).

Rolston, H., "Ethical responsibilities toward wildlife," *JAVMA*, 200:5 (1992):618–23.

Ryder, R.D., *Animal Revolution* (Oxford: Basil Blackwell, 1989).

Ryder, R.D., "Painism: the Ethics of Animal Rights," in *Animal Welfare and the Environment*, R.D. Ryder, ed. (London: Duckworth, 1992).

Serpell, J.A., *In the Company of Animals* (Oxford: Basil Blackwell, 1986).

Sorabji, R., *Animal Minds and Human Morals* (Ithaca, NY: Cornell University Press, 1993).

An Animal's Take on the "Company Man"

KENNETH J. SHAPIRO

In the symposium paper occasioning this two-part question (1), Tannenbaum offered the thesis that treating animals as property in the law provides them adequate protection. He argued that this is the case since the legal concept of property originally developed in part around cattle ("chattel"); and such property, in his view, can have rights under the law.

In response, I maintain that the concept of nonhuman animals as property does not afford them adequate protection. For it reduces them to the same category of inanimate objects to the subordination of which much law in our system is dedicated, precisely to assure that a property owner's rights prevail with respect to his or her property. Although the terms "inevitably" and "necessarily" are too strong—history is more a matter of contingencies than inexorable consequences—history indicates that we have treated and continue to treat animals as inanimate objects when it suits our needs. A legal system that treats animals as property clearly provides and, there is no reason to doubt, will continue to provide justification for that exploitation.

In regard to the second part of the first question, a category is emerging in the culture, buttressed by the animal rights movement and by certain areas of scientific investigation (for example, cognitive ethology), which suggests an alternative category ontologically and, it follows, in jurisprudence—the animal as subjectivity.

Let me review a few considerations which support these responses to the two questions posed.

United States law, in particular, overwhelmingly supports the rights of a property owner. In the case of animate or movable property, its protection (note I am precluded from using a personal pronoun) is relegated to occasional limitations on otherwise preemptive property owner rights. In effect, an owner is protected in doing whatever he or she sees fit to maintain and enhance the economic value of his property, while that property itself is bereft of any interests and rights.

The fact that animals were those entities around which the concept of property developed offers no argument that that conception provides an adequate framework for their protection, as, no doubt, they were taken merely as things of a certain value from that outset. Yet the concept of property is stretched beyond its carrying power when it refers to an entity as an object that is, more than a thing, the "subject of a life," to use Regan's term in *The Case for Animal Rights* (1983). The thrust of investigations of nonhuman animals clearly establishes what we have always known, that animals are experiencing beings. While they can be taken and treated as objects, as can we, they are subjectivities—beings that intend, anticipate, suffer. . . . In the case of companion animals, they are members of our family; more generally, they are members of their own families.

Historically, we have used the device of ontologically reducing subjectivities to objects to justify institutions in which they are treated as property, as in the case of slavery and the oppression of women. To end those injustices it was not enough to provide limits on their respective property owners.

In the Middle Ages, nonhuman animals occasionally had standing in court.[1] In a recent Supreme Court case, Justice Douglas asserted that trees had standing, borrowing from Stone's argument in *Should Trees have Standing?* (1974). The question of standing is very important because while an entity that is treated as property under the law cannot be brought to trial, nor can that entity seek redress in the courts. Incidentally, standing is also the issue when groups of people seeking to protect those entities are prevented from doing so in the courts. A number of recent cases involving the Animal Welfare Act in which the court ruled for further protection of nonhuman animals under that act were overturned when the animal protection groups which brought the suits were denied standing (*Animal Legal Defense Fund v. Yeutter*; *Animal Legal Defense Fund v. Secretary of Agriculture*).

The medieval examples show that having standing in court does not assure fair treatment in a society as animals in medieval times were often the objects of cruelty—although the intensive institutionalized exploitation of our times has been possible only with modern technology. However, it also shows that the status is possible under the law.

Two contemporary legal scholars have provided extended arguments on the status of animals as property. Francione (1995) presents the negative case that treating animals as property is inadequate protection; while Wise (1995) argues the positive brief that at least chimpanzees are persons and, therefore, have standing under the law. From case material in the common law, Wise deduces the attributes of personhood under the law. He then applies these to current research on chimpanzees, such as that which Rumbaugh presented in this symposium, and concludes that these animals meet the criteria for personhood under the law.

Without rehearsing the litany of contemporary exploitative animal-based institutions, one example will suffice to make the negative case. Animals in the research laboratory are primarily treated as parts of the laboratory. In effect, they are first made into "preparations" through experimental manipulations that isolate and vary a particular part-process of the animal, for example, brain functioning is systematically altered through central stimulation or lesioning. After being connected to recording equipment, such as an EEG and EMG, the prepared part-process (a still living animal) is further reduced as it becomes a conduit for a controlled input and measured output. Grossly, stuff is put into the preparation at one end and comes out the other. Such reductive experimental procedures are supported by the language found in journals and proposals: animals are not named; they are referred to by impersonal pronouns; passive voice denies them the status of autonomous actors; even their suffering and death is laundered into technical talk of, respectively, "aversive stimuli" and "anesthetized and exsanguinated." This reduction of them to parts of the lab, to extensions of lab instrumentation and equipment, smacks of the animal as property. To date, it has largely served to protect the rights of the owner, here the research enterprise, to conduct intensive, invasive research. The "rights" of the animals in the lab retreat to that preciously limited quality of and right to life that remains once that invasion has been given preemptive value.

Ironically, a developing area of social scientific research which investigates the subculture of the laboratory has revealed that at least lab technicians often give its nonhuman inhabitants an ambivalent status. Beneath the official language which gives the prerogatives of ownership to the "lab-lords" and which clearly treats animals in the lab as inanimate objects, an "underlife" evolves in which individual animals in the lab are named, given special care, and grieved at their death.

Regarding the premise of the first part of the second question, the weight given to pain and distress is only limited and relative. The two most prominent ethics in contemporary moral philosophy (rights and utilitarianism) do give these such "significant weight." However, current legislation and its regulations do not adopt a rights position. The utilitarianism they embody gives lip-service to the "costs" to animals while being permissive on the side of the researcher. The primary ethical language they adopt is "humane" talk, where "necessary suffering" is sanctioned and "necessary" is a rubber-fence concept stretched to encompass virtually all extant experimental procedures. While scales measuring suffering have been developed, they are under-utilized. The United States Department of Agriculture's annual report of research facilities consists of three gross levels and is woefully deficient (Orlans, 1993, pp. 125-26). More importantly, on the ground, recognition of and intervention for the amelioration of pain is deficient. In an ethnographic study of animal research laboratories, Phillips (1993) found that researchers rarely used analgesics and denied the presence of pain in their research animals.

Rather than attempting to draw a line between species the members of which have typical capacities with respect to pain (as the question implies is the operative option), assessment of the presence of pain and distress should be done on an individual case by case basis. Assessment requires a combination of behavioral, physiological, and functional (evolutionary) assays. Another helpful criterion is that no procedure or condition is acceptable which would be considered too painful or stressful if employed on the human animal proposing the procedure. Finally, the benefit of the doubt should always be given to the nonhuman animal, that is, pain is assumed present until proven absent.

Regarding the term "suffering," it is a broad term inclusive of pain and distress. If to suffer is, as the dictionary defines it, "to undergo something painful or unpleasant," then there is a spectrum of such unpleasantness—pain, distress, fear, anxiety—all of which must be given "significant weight." But our assessment cannot stop at these, for there are also harm and death which involve, respectively, the loss of some and the loss of all capabilities with or without undergoing the experience of something painful or unpleasant. Incidentally, in these terms, boredom is also an important harm in that in the extreme it involves the loss of interest in capably interacting with the world.

So we can and must distinguish between pain and suffering; and we can and are obligated to add to the already extensive spectrum of the latter two other considerations, harm and death, which also should be accorded "significant weight."

One of the positive developments of the last two decades since the inception of the contemporary animal rights movement is that from an initial concern limited largely to physical pain, we have been moving to consideration of more and more of this extended spectrum of "suffering."

The concept of "well-being" has arisen out of this broadening, which has shifted consideration from the reduction of unpleasantnesses to the provision of positive qualities of life. This concept itself is currently expanding from the notion of enrichment, which is often impoverished in its implementation (a toy in the cage), to the more sophisticated notion of "natural environments" (Gibbons, Wyers, Waters, and Menzel, 1994). In both laboratories and zoos, this means providing a setting which approximates that natural to the members of the species being investigated or displayed. Until we do this, we are harming these animals in a very basic way. For nonhuman animals are radically dependent on their environment; their habits are habitat-dependent and those habits constitute their species identity. Caging such an animal, even in an enriched cage, can be the death of that animal, the loss of that significant part of their identity that is species-specific.

To take the second part of the third question first, American society is well constructed to deal with conflicts between populations and individuals for it was built on the antinomy between individuals and society. The American political system is a complex set of structures consisting of checks and balances. The federal government, for example, has three branches, one of which, Congress, itself has branches, and a second of which, the courts, represents both "the people" and the individual. Complementing these internal checks are those external safeguards consisting of various local governments in each of which these internal checks are reiterated. Clearly, our Constitution and the institutions that followed from it recognized and, indeed, were built on the problem of protecting the rights of individuals and the needs of groups.

An analogy can be drawn between this individual and societal antinomy and the situation with regard to nonhuman animals. We need ethics and political mechanisms that build in ways of dealing with the conflict between species and individual animals so that there are checks and balances that protect both.

However, currently, there are not two but three opposing societal treatments of animals: animals as groups, as individuals, and as things. While Regan's animal rights philosophy would rescue animals from their objectification, their reduction to things, Naess' environmentalist philosophy focuses on the individual/group opposition.

Every moral philosophy has its optimum range of application. In my view, the individual-based rights philosophy is most effective in the laboratory research and factory farm issues, where clearly animals are deanimated as well as deindividuated. However, with respect to free animals, those in the "wild," two values are in play at once—the integrity of an animal as an individual and as a member of a species/ecological system. While the animal as a thing in the laboratory is a self-serving construction to justify exploitation, the animal as a member of a species which in turn is part of a ecological system is not. For animals in this context, we need ethics that respect both integrities. Conceptually, we need only liberate "lab animals" from their infra-individual status; more complexly, the supra-individual status of wild animals must be preserved while protecting their individual rights.

There are at least two current prospects. The feminist ethic of caring emphasizes "beings-in-relation," an entity that cuts across the individual/group tension. After a period emphasizing the incompatibility of animal rights and environmentalist philosophies, a second concept that reconciles obligations to individuals and groups/systems is being articulated. The notion of a mixed community presumes that we are all—human, nonhuman animal, and plant—in this together. Schematically, there is a set of concentric circles that frame this mixed ethic. In the center, there is a community containing yourself and your family, often including a companion animal. But there are other communities to which you also owe ethical consideration—the pigeons in the park, feral animals, factory farm animals, animals in the wild, and so on. All of these communities have some ethical pull. While all consist of individuals, the outermost more than the innermost circles also force consideration as groups/systems.

It is worth noting that even without this emerging philosophical reconciliation, animal rights and environmentalist organizations often have worked together. As a practical matter, the two movements have similar positions on most issues for which there is any possible political purchase.

The conflicts that arise when society treats animals as groups as opposed to individuals, then, must be dealt with by developing policies that recognize the value of both considerations, using as a model those institutional structures built to constructively represent both the individual and society.

Notes

[1] See, for example, E.P. Evans (1906), and the recent movie, *The Advocate*.

References

Evans, E.P., *The Medieval Prosecution and Capital Punishment of Animals* (London: William Heineman, 1906).

Francione, G., *Animals, Property and the Law* (Philadelphia: Temple University, 1995).

Gibbons, E., Wyers, E., Waters, E., and Menzel, E., eds., *Naturalistic Environments in Captivity for Animal Behavior Research* (Albany: State University of New York, 1994).

Orlans, B., *In the Name of Science* (New York: Oxford, 1993).

Phillips, M., "Savages, drunks, and lab animals: The researcher's perception of pain," in *Society and Animals* 1:1 (1993): 61-82.

Regan, T., *The Case for Animal Rights* (Berkeley: University of California Press, 1983).

Stone, C., *Should Trees Have Standing* (Los Altos: William Kaufmann, 1974).

Wise, S., "Why chimpanzees are entitled to legal rights," presented at John Marshall Law School, January 1995.

NICHOLAS WADE

Unlike the animal rights movement, I believe for all the obvious reasons that scientists need to be able to conduct experiments on lower animals. But I realize those same reasons would seem not nearly so obvious to me if cited by advanced extraterrestrials to justify their research on humans. The animal rights movement deserves every credit for confounding the certainty and restraining the behavior of those of us who eat meat, wear leather, and support biomedical research.

The animal rights movement, of course, has been enormously effective at the margin. It has not stopped testing on animals. But campaigns by New York's Henry Spira and others have encouraged many companies to look for alternative testing methods that do not require animals. Patient lobbying by Christine Stevens of the Animal Welfare Institute has secured important changes in the minimum standards for keeping laboratory animals. Though some animal rightists believe these standards do not go far enough, universities complain bitterly of the extra costs.

But I suspect there are limits to what can be achieved by the movement's traditional campaigns. The health establishment, with the public's firm support, is not about to test new drugs on humans without first seeing if the drugs make rats keel over. Though a few scientists conceive of even monkeys as mere biological systems and have grossly abused them, I would like to think that most take good care of their laboratory animals, both for ethical reasons and because poor conditions make poor experimental subjects. Given the victories already won in improving the treatment of laboratory animals and minimizing their use, it may be that further efforts will bring only diminishing returns.

It is certainly an important goal, and a test of our civilized values, to protect monkeys and cats and dogs from maltreatment by researchers. But I think an even higher goal, to be pursued in parallel, is to protect whole species from being driven to extinction by the destruction of their habitat.

It is, of course, for the movement to decide what its goals should be. But many would not be sorry to see some of its formidable energies deployed in the effort to save endangered species and their habitats, both in the United States and abroad. Peter Singer's Great Ape Project—a proposal to accept chimpanzees, gorillas, and orang-utans into the community of moral equals with humans—is an interesting first step in this direction. The Project, as I understand it, is not directly aimed at preserving their habitat, but this goal is implied in the requirement that the great apes not be killed, caged, or tortured without due process.

The core of the animal rights agenda is ethical issues, and it may seem that the preservation of species is more an environmental cause. Singer's initiative points the way to laying an ethical groundwork for species preservation.

Respect for the domestic animals around us is the first step to enlarging our imaginations to those in distant oceans and jungles. Few would wish the animal rights movement to abandon its traditional agenda of seeing that pets are not abused and that farm animals are decently treated. But some extensions of this agenda, such as the absolutist demand to abolish zoos, can be seen as misdirected, since zoos are an effective way of teaching people about the richness of nature.

Primitive peoples respect nature because they consider it populated with supernatural powers. Our respect for nature is based on biologists' understanding of evolution and the intricacy of plant and animal communities. In the wider scheme of things, animal rightists could see scientists as their natural allies, not their opponents. Perhaps the time has come to declare victory in the laboratory and to move on to the wider battlefield of protecting the wild animals in the world's forests and oceans, whose right to existence is under daily challenge.

The questions asked of the panelists are, of course, questions for philosophers, of which I am not one. So instead of jumping into the minefield, I will tiptoe round its edges.

(1) Proudhon said all property is theft, but economists see property as the basis of value. Animals that cannot be owned as property are likely to be animals that have no value and, hence, to be treated as worthless. Slaves, being owned, were generally afforded the necessities of life. Though it is, of course, morally repugnant to treat humans that way, as the basic framework for treating animals, it is not so bad. Farm animals, having value, are given adequate food and shelter because it is in the farmer's own interest to do so.

Ownership rarely conveys absolute rights. I may own my land, but many laws and ordinances constrain how I use it. I cannot keep peacocks, throw wild parties that disturb the peace, raise a satellite dish to more than a certain height, add a building that defies the town's zoning laws, or chop down a wood that contains an endangered species. People's rights over the animals they own are similarly constrained, and the constraints can be loosened or tightened exactly as much as the laws require.

In other words, we have a workable system already in place for protecting animals, as property, to whatever degree society wishes. The idea of emancipating animals from ownership is presumably another way of saying they should enjoy some category of rights as citizens. The time is not ripe for that idea. The proposal to give special rights to the great apes is more likely to fly, and if I wanted to test a trial balloon, that is the one I would launch.

(2) Pain as an avoidance reflex—as when bacteria swim away from a noxious chemical—is different from pain that is consciously felt. The issue here is consciousness. But since we do not yet understand the neural basis of consciousness in humans, we have no way of knowing to what degree, if any, other animals are conscious. I am, therefore, skeptical of schemes to base animal rights on inferred degrees of consciousness.

(3) Our society has many provisions to protect the rights of the individual from the government. It is not clear that there is any broad willingness to extend such individual rights to animals. But society is willing to protect animals as classes or populations, as in the Animal Welfare Act, that protects laboratory animals, and the Endangered Species Act, intended to safeguard the populations of certain animals and their habitats. Better to seize and exploit what society will give—protection to animals as classes—than to hold out for what all but Jains will deny—acceptance of a bug as an individual with rights.

Index

accidents (Aristotle), and classification of animals, 13

Achilles, 347, 350–51

activist views: on human-animal relations, 176–79; on the legal status of animals, 126–27, 129–32, 164, 169–71

Aesop: *Life of Aesop*, 236–39; modern versions of fables, 230–31; Prometheus, 235; versified by Socrates, 233

Africans: attitudes about hyenas, 111–13, 114

Africanus, Leo, 104

agriculture: as exempt from cruelty laws, 155; research in, 401–2. *See also* farm animals

Akins, Kathleen, 293–95

Aldrovandi, Ulisse, 77–78

Allen, G. S., 384–85

American Sign Language, 73

analogies, in evolutionary biology, 199–200

anesthesia: use in agriculture, 401–2

Angell, George, 165–66

Animal Farm (Orwell), 348–49

animal-human boundaries, 63–64, 67–86, 341; and evolutionary biology, 377; and hunting, 372; in myths and fables, 236, 357–59; and science, 372–73; similarities and differences, 275–79

animal-human creatures, 50, 77–79

animal lovers, in mythology, 343–45, 353, 356

animal protection movement, 379. *See also* animal rights

animal research, 379–92; activist prohibition of, 175; and animal experience, 277; animal pain and suffering, 422–24; and justifiable pain, 155, 161, 173; and philosophy of individual rights, 425–26; public attitudes toward, 386–87, 390; views of scientists on, 383–86

animal rights, 21–22; animal pain and suffering, 422–24; animal rights movement, 427–30; cruelty and, 167–71; legal rights, 126, 131–32, 180. *See also* activist views; law

animals, as depicted in fiction, 249–69

animal training, 27–40

anthropomorphism, 249, 253

anti-cruelty laws. *See* cruelty laws

anti-Semitism, 102, 113

antivivisection movement, 380, 382–83

apes: killer-ape theory, 373–75; and language, 311–17; and lower side of human nature, 68; and transfer of learning, 306–8. *See also* chimpanzees; primates

Apuleius, 345

aristocracy, and hunting, 365–66, 369

Aristotle, 5–7; and classification of animals, 9–11, 210–11; Four Causes, 7–9; on hyenas, 94–99, 113; and modern attitudes toward animals, 19–22; on value of beast fables, 234

Arluke, A., 388

arthropods, and body segmentation, 203

Australopithecus, 374–75

baboon heart transplant, 206–7

"backreading" human characteristics into animals, 195, 198, 214, 221

Bacon, Francis, 41–42

Barthes, Roland, 251

bats, consciousness in, 283–84, 293–95

bears: as myth or metaphor, 259–62; in zoos, 394–95

beast-machine concept (Descartes), 302, 317

Beauty and the Beast (Cocteau), 352–53

beavers, and backreading human characteristics, 196

behavioral problems in zoos, 394–95

Beidelman, T. O., 111, 115

Notes on Contributors

Matt Cartmill professor of biological anthropology and anatomy at Duke University Medical Center. He is the author of "Significant Others" (1995) and "Reinventing Anthropology" (1994).

Juliet Clutton-Brock a member of the Department of Zoology at the Natural History Museum in London. She is the editor of the Journal of Zoology and recently published "Origins of the dog: Domestication and early history" in James Serpell, editor, The Domestic Dog (1995).

Vicki Croke the "Animal Beat" columnist for The Boston Globe. She is currently writing The Modern Ark: Zoos Past, Present, and Future (forthcoming).

Daniel C. Dennett is Director of the Center for Cognitive Studies at Tufts University. He recently wrote Darwin's Dangerous Idea (1995).

Cora Diamond is William R. Kenan, Jr., Professor of Philosophy at the University of Virginia. Her most recent work is The Realistic Spirit: Wittgenstein, Philosophy, and the Mind (1991).

Wendy Doniger is Mircea Eliade Professor of the History of Relgions at the University of Chicago. She is the author (under the name of Wendy Doniger O'Flaherty) of Other Peoples' Myths: The Cave of Echoes (1995) and Textual Studies for the Study of Hinduism (1990).

Stephen E. Glickman is professor of psychology at the University of California, Berkeley. He is the author (with L.G. Frank, P. Licht, T. Yalimkaya, P.K. Suteri, and J. Davidson) of "Sexual differentiation of the female spotted hyena" (1992).

Stephen Jay Gould is the Alexander Agassiz Professor of Zoology at Harvard University.

Vicki Hearne is an author, animal trainer, and poet. She is the author of Bandit: Dossier of a Dangerous Dog (1991) and Animal Happiness (1994).

John Hollander is Sterling Professor of English at Yale University. He is the author of The Gazer's Spirit (1995) and the editor of Animal Poems (1995).

Nicholas Howe is the Director of the Center for Medieval and Renaissance Studies at Ohio State University. He is the author of Migration and Mythmaking in Anglo-Saxon England (1989).

Nicholas Humphrey is professor in the Department of Psychology at the New School for Social Research. His most recent publication is Soul Searching (1995).

Colin McGinn is professor of philosophy at Rutgers University. His most recent work is Problems in Philosophy (1993).

Joy Mench was an associate professor in the Department of Poultry Science at the University of Maryland when she prepared her article for this issue. She is currently a professor in the Animal and Avian Sciences Department at the University of California at Davis.

Notes on Contributors

J. Anthony Movshon
is Director of the Center
for Neural Science and
Investigator at the
Howard Hughes Medical
Center. He has written
numerous articles in the
area of neuroscience.

Kenneth Prewitt
is President of the Social
Science Research Council.

Harriet Ritvo
is Arthur J. Conner
Profesor of History at
MIT. She recently
published "Classification
and Continuity in *The
Origin of Species*" in
*Charles Darwin: The Origin
of Species* (David Amigoni
and Jeff Wallace, editors,
1995).

John G. Robinson
is Vice-President of the
International Wildlife
Conservation Society. He
edited (with K.H.
Redford) *Neotropical
Wildlife Use and
Conservation* (1991) and
recently published "The
Wildlife Conservation
Society: 100 years of
rectitude" (1995).

Andrew N. Rowan
is Director of the Tufts
Center for Animals and
Public Policy. He is the
author of *The Animal
Research Controversy*
(1995).

Duane M. Rumbaugh
is Regents' Professor of
Psychology and Biology at
Georgia State University
and Director of the
Language Research
Center in Atlanta. His
most recent publications
include
"Anthropomorphism
revisited" (1994) and
"Language in
comparative perspective"
(with E.S.
Savage-Rumbaugh, 1994).

James Serpell
is associate professor of
human ethics and animal
welfare in the School of
Veterinary Medicine at
the University of
Pennsylvania. He is the
editor of *The Domestic
Dog: Its Evolution,
Behaviour and Interactions
with People* (1995).

Kenneth J. Shapiro
is the Editor of *Society an
Animals*. He is currently
working on *Animal Mode
of Human Psychology:
Science, Ethics, and Policy*
(forthcoming).

Jerrold Tannenbaum
is clinical assistant
professor at the Tufts
University School of
Veterinary Medicine. He
is the author of "Benefit
and Burdens: Legal and
Ethical Issues in
Veterinary
Specialization."

Gerald Vizenor
is professor of english
and Director of the
American Studies
Summer Institute at the
University of California
Berkeley. He recently
published *Shadow
Distance: A Gerald Vizenor
Reader* (1994) and
Manifest Manners (1994).

Nicholas Wade
is a science editor at *The
New York Times*.

CPSIA information can be obtained
at www.ICGtesting.com
Printed in the USA
LVHW112319300721
693981LV00003B/10